P9-EIG-209

BIRDS OF THE GREAT PLAINS

Birds
of the
Great Plains

Breeding Species and Their Distribution

Paul A. Johnsgard

UNIVERSITY OF NEBRASKA PRESS   LINCOLN AND LONDON

Copyright © 1979 by the University of Nebraska Press
Manufactured in the United States of America

Library of Congress Cataloging in Publication Data

Johnsgard, Paul A
  Birds of the Great Plains.

  Bibliography: p. 513
  Includes index.
  1. Birds—Great Plains. 2. Birds—Eggs and nests.
I. Title.
QL683.G68J63   598.2′978   79–1419
ISBN 0–8032–2550–4

For Sarah,
who wants to be an ornithologist

CONTENTS

ILLUSTRATIONS

Figures

Black-and-white photographs

Color Plates   *following page* 368

TABLES

PREFACE

Nebraska and its adjoining states represent a nearly unique ornithological situation. Although they are almost entirely grassland, the plains states are variably dissected by river-bottom forests that provide natural passageways for forest-adapted species to enter the plains and sometimes to cross them. Although eastern and western bird faunas tend to be separated by the plains, they often mingle to some degree, resulting in competition and sometimes hybridization. The central Great Plains thus are exceedingly interesting from ecological, evolutionary, and zoogeographic perspectives, and the plains bird fauna is significant both for what is present and for what has been variably excluded.

I first hoped to consider the entire Great Plains as a comprehensive unit, but a survey of the literature rapidly made it apparent that from either a geological or a botanical standpoint the region was far too large to be dealt with easily. I thus began to consider various compromises between my initial comprehensive vision and the restrictive approach of including only one or two states. I felt that the region covered should essentially saddle the 100th meridian—traditionally considered the dividing line between eastern and western bird faunas—and should include as many of the essentially grassland-dominated states as feasible. An excellent recent book on the breeding birds of North Dakota (Stewart 1975) provided a kind of northern "anchor" for the work, and recent and comprehensive books on the birds of Texas (Oberholser 1974) and Oklahoma (Sutton 1967) did the same for the southern portion of the region. It was thus necessary to establish only eastern and western limits, which was done by choosing those lines of longitude that included the maximum of grassland habitats and the minimum of forest or montane communities.

The books by Stewart on North Dakota birds, by Sutton on Oklahoma birds, and by Oberholser on Texas birds were invaluable to me. I also constantly used the reports by Johnston (1965) and Rising (1974) on the breeding birds of Kansas; the two-volume work by Bailey and Niedrach (1965) on the birds of Colorado; and those by Bailey (1928) and by Ligon (1961) on New Mexican birds. The classic two-volume monograph by Roberts (1932) on Minnesota birds has recently been updated by the smaller but very useful guide by Green and Janssen (1975). Wherever comments are made without citations in the text on distribution, breeding periods, or clutch sizes relative to these states, such sources represent the authorities. The only two states lying entirely within the region under consideration that have lacked any comprehensive ornithological coverage are Nebraska and South

Dakota. Details on Nebraska are based on my own survey of information in the *Nebraska Bird Review* and on other appropriate literature. South Dakota was more difficult, since the report by Pettingill and Whitney on the Black Hills has until recently been the only extended regional publication available. But I was able to obtain page proofs of the new book on South Dakota birds, through the kindness of its editor, Byron Harrell (1978). I was thus able to make some additions and corrections to my original text on South Dakota, which, like that for Nebraska, has necessarily been based largely on diverse published sources.

Because this book is largely directed to a rather broad, nontechnical audience, it has several goals. Primarily, I hope it will serve the layman bird watcher as a predictive guide to the birds that probably breed in any local area within the region considered. Second, the descriptions of breeding habitats, nest locations, and the appearances of nests and eggs should help the reader locate and identify the nests of species that interest him. Third, the accounts of breeding biology supply a nucleus of information on behavior and ecology that should offer some insight into these birds' adaptations and that may stimulate readers to try to fill some of the many gaps in our knowledge. Last, the distributional maps should interest ecologists and students of bird distribution, since the region represents an exciting interface between grassland and nongrassland communities and between eastern and western faunas. The book is not intended as a field guide, except in the very limited sense of helping to identify nests. But it was written in the hope that it might accompany the reader into the field rather than remaining on the shelf at home. Thus its format has been kept moderately small so it can be carried easily. With minor exceptions, the sequence of species and both the vernacular and the scientific names follow those in the fifth edition of the A.O.U. *Check-list of North American Birds*. Frequently encountered alternate scientific and vernacular names have been shown in parentheses; the possessive in vernacular names has been eliminated (e.g., Bell vireo rather than Bell's vireo); and the mottled duck and lesser prairie chicken are considered subspecies rather than separate species, to conform with my earlier books on these groups. Separate accounts are provided for all species except those that are extinct or were unsuccessfully introduced.

As with all such books, a tremendous debt is due to thousands of amateur and professional ornithologists for accumulating and publishing the information on which any regional bird guide must be based. Beyond the books and other references already mentioned, I must thank several persons for individual help with particular problems, for supplying photographs, or for reading portions of the manuscript. These include Dr. Ann Bleed, Mr. Harold Burgess, Dr. Calvin Cink, Dr. Byron Harrell, Mrs. Myra Mergler Niemeier, Jean and Ed Schulenberg, Judy and Phil Sublett, Dr. George Sutton, and Mr. Fred Zeillemaker. I owe a special debt of gratitude to Cristi Nordeen for carefully searching the *Nebraska Bird Review* for breeding records and to the secretaries of the School of Life Sciences, University of Nebraska-Lincoln, for manuscript typing.

INTRODUCTION

Extending from the vicinity of Great Bear Lake in the Mackenzie District of Canada to the Pecos River and Balcones Escarpment of central Texas, the Great Plains of North America stretch across a three-thousand-mile expanse along the eastern slope of the Rocky Mountains. To the northeast, they join the Canadian Shield and feed such rivers as the Saskatchewan, Assiniboine, and Red that drain eastward to mouths in Hudson Bay. To the east, the Great Plains merge with the Central Lowlands to form the drainage system of the Missouri, Mississippi, and Ohio rivers, which ultimately reach the Gulf Coast. Although the northern half of the Great Plains is dominated by Canada's boreal forest of spruces, firs, and other conifers, the southern parts of Alberta, Saskatchewan, and Manitoba and the states extending from North Dakota to Texas are predominantly perennial grassland. The prairies and plains of central North America represent one of the largest and most uniform of the continent's major ecosystems, and the "grassland biome" evokes an image of vast herds of bison amid a sea of grass that once extended across the heartland of North America. Most of these grasslands have now been converted to grainfields or else have been subjected to such grazing pressure as to degrade them almost beyond recognition. Yet major remnants remain in national parks, national grasslands, and wildlife refuges, and tiny fragments are still to be found in rural cemeteries, railroad rights-of-way, and small nature reserves. In such places essentially all the original birdlife of the plains can still be found. This book documents the present breeding distributions of the birds of this region, both as an aid to bird watchers and as a biological analysis of this major component of the North American biota.

TOPOGRAPHY, LANDFORMS, AND CLIMATE

Because of the enormous area of the Great Plains and the severely limited information on breeding bird distribution in such states as Montana, Wyoming, and eastern Colorado, this book covers only from the Canadian border (49° latitude) to the Red River boundary between Oklahoma and Texas, then westward along the 34th parallel to the 104th meridian, which includes the eastern parts of New Mexico and Colorado and closely coincides with the western limits of Nebraska and the Dakotas. The eastern limit was established by using the eastern borders of Oklahoma and Kansas and following the 95th meridian to the Canadian border, encompassing the western portions of Minnesota, Iowa, and northwestern Missouri. The area thus enclosed includes all of five states and parts of six others and represents a maximum north-south distance of slightly more than 1,000 miles and a maximum east-west distance of nearly 550 miles. The total surface area includes 502,000 square miles, or 17 percent of the land area of the United States south of Canada.

Although there are minor variations, the overall topography of this region is an inclined plain, which slopes downward from the west to the east at an average gradient of about 10 feet per mile (see fig. 1). The highest point in the region is Sierra Grande, in northwestern New Mexico, 8,732 feet above sea level, while the lowest point is in southeastern Oklahoma, 323 feet above sea level. To the north, the Black Hills of South Dakota provide a secondary montane influence, with a maximum elevation of 7,242 feet at Harney Peak. Along the eastern limits, the only highlands of significance are the Ouachita Mountains, a part of the Ozark Plateau, which attain a maximum height of more than 2,000 feet. Over nearly the entire area, drainage is to the southeast into the Missouri and Mississippi systems; but in North Dakota the Souris and Red rivers are part of the Hudson Bay drainage system.

Approximately the northern half of this region has been markedly affected by glaciers (fig. 2); the most recent or Wisconsin glaciation has had the strongest effect on present-day landforms. Most of eastern and northern North Dakota, parts of eastern South Dakota, and all of western Minnesota bear such glacial scars as ground moraines, dead ice moraines, and end moraines. The entire Red River Valley represents a lake plain that was once covered by glacial Lake Agassiz, which drained and retreated northward less than 10,000 years ago. Likewise the Souris Lake plain of north-central North Dakota and several other smaller lake plains and areas of deltaic sands associated with glacial lakes show similar evidence of glaciation. West of the Missouri River the uplands consist of many extensively eroded plains, with buttes, bedrock valleys, and highly eroded "badlands" marked by dry valleys, steep, dry slopes, and sharp ridges that provide excellent habitats for a variety of cliff-nesting bird species.

Whereas the glacial till covering the eastern portions of the Dakotas dates from the Wisconsin glaciation, eastern Nebraska, northeastern Kansas, and adjoining parts of western Iowa and Missouri are covered by earlier Kansas glacial till, over which more recent accumulations of loess associated with the Wisconsin glaciation have been deposited. Probably valley systems that were outwash plains during this period of glaciation were major sources of such loess, but the Sandhills region of Nebraska may also have contributed wind-carried materials. This area of sand dunes, the largest in North America and perhaps in the western hemisphere, covers nearly 20,000 square miles and includes what may be the largest area of essentially unmodified grassland left in the entire Great Plains region. Apart from the dunes themselves, a

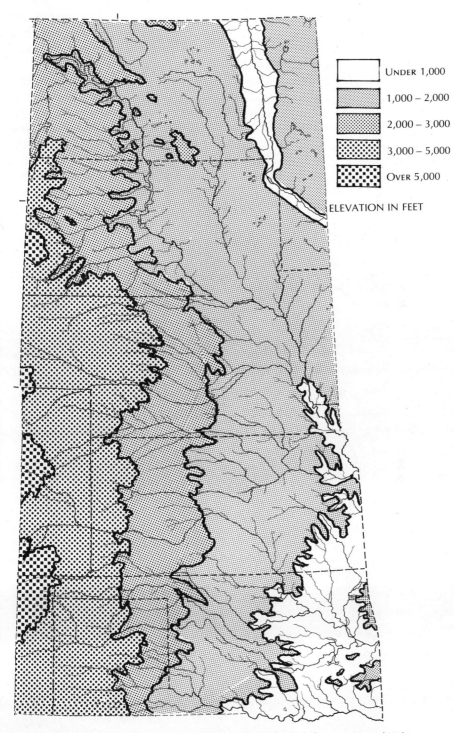

Legend:
- Under 1,000
- 1,000 – 2,000
- 2,000 – 3,000
- 3,000 – 5,000
- Over 5,000

ELEVATION IN FEET

Fig. 1 Topography of the Great Plains states, adapted from a map in the *Oxford Atlas*.

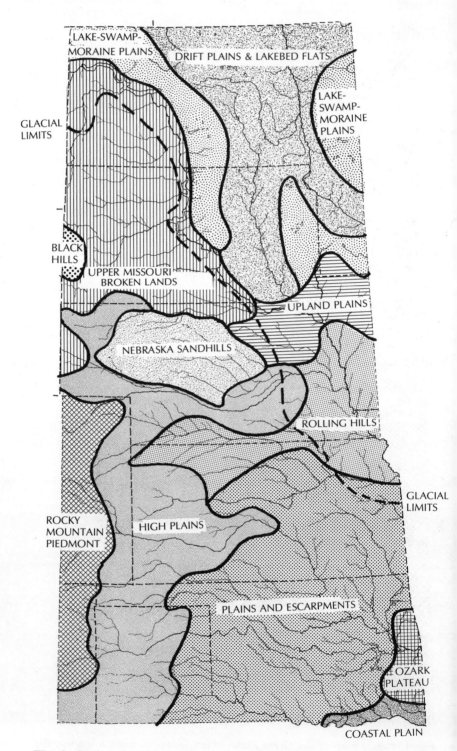

Fig. 2. Surface landforms of the Great Plains states, based on the land surface form map in the *National Atlas*.

high water table has allowed for the development of many low meadows and marshy environments reminiscent of those of the glacial moraine country to the north, supporting many of the same species of aquatic plants and animals.

Nearly all the surface topography of Kansas can be described as undulating plains; as in North Dakota, there is only about a 3,000-foot difference between the highest and lowest points in the state. But unlike North Dakota, very little of Kansas exhibits strong glacial topography or evidence of extensive erosion. Only in a few areas, such as along the Smoky Hill River, are the underlying sedimentary deposits exposed by erosion. There are also some small "badlands" in Kiowa County, south-central Kansas, and areas of sandhills along the Cimarron River, the Arkansas River, and the Kansas River.

Oklahoma is the most topographically varied of any of the states under consideration here. In addition to an overall 4,000-foot gradient from the northwestern to the southeastern corners of the state, four mountainous regions are entirely or partly within the state. The Ozarks, which are primarily in Arkansas, are elevated plains through which many narrow river valleys have been cut, resulting in a series of hills that generally do not rise more than 400 feet from base to top and are either treeless or at most covered with scrubby oaks. The Ouachita Mountains are much more rugged and more heavily timbered; they give rise to many mountain streams fed by the heavy precipitation characteristic of the area. The Arbuckle Mountains are the greatly eroded remains of an ancient mountain range that has been worn down to a high plateau, with deep river-cut valleys. Limestone sinkholes and some caves also occur here, as do artesian wells and mineral springs, such as in the Chickasaw Recreation Area. The Wichita Mountains in southwestern Oklahoma are a series of parallel granitic uplifts that rise 700 to 900 feet above their bases and support only scant vegetation.

The rest of Oklahoma and adjacent Texas is predominantly characterized by rolling hills mantled with forest, woodland, or grasses. There is a small strip of Gulf Plain associated with the lower Red River. An extensive area of gypsum hills extends from the Kansas line to the Red River in west-central Oklahoma and includes many deep river canyons with gypsum ledges, as well as salt springs or salt plains such as Great Salt Plains National Wildlife Refuge. Last, the western tip of the Oklahoma panhandle shows an abrupt transition from the high plains to the Rocky Mountain piedmont, in the form of the Black Mesa. This bed of volcanic lava provides a unique topography and associated bird habitats found nowhere else in the region except in northeastern New Mexico, where an extinct volcanic peak some 8,000 feet high has been designated Capulin Mountain National Monument.

That part of the Texas panhandle lying south of the Canadian River and the adjacent part of eastern New Mexico is the "Staked Plain" (Llano Estacado), a flat, treeless area that was reportedly so named because its absence of landmarks forced early Spanish explorers to drive stakes into the ground to mark their trails. Like the high plains to the north of the Canadian River, it is nearly waterless, and much of the standing water is distinctly alkaline. Both Buffalo Lake and Muleshoe national wildlife refuges lie within the region of the Staked Plain.

The pattern of rainfall throughout this region is relatively simple (fig. 3). In general, it increases from northwest to southeast, at the approximate rate of about one inch per 40 miles at the northern edge of the region, to one inch per 10 miles at the southern edge. The wettest locality in the region is the Ouachita Mountains of southeastern Oklahoma, with more than 50 inches of precipitation annually, while there is less than 13 inches of precipitation each year near Clayton, Union County, New Mexico. About three-fourths of the rainfall occurs during the growing season, which ranges from about 100 days in northern North Dakota to 240 days in extreme southeastern Oklahoma. However, evaporation increases correspondingly as one proceeds south. The highest rates of annual evaporation (as measured from lake surfaces) occur in the Staked Plain region, characterized by evaporation rates more than four times greater than precipitation rates. Much lower evaporation rates are typical of the cooler, more northerly states, and parts of eastern North Dakota can thus support a lush tall-grass prairie vegetation with less than 20 inches of precipitation a year, whereas the same amount of precipitation on the Staked Plain allows only for the barest stands of buffalo grass and other xerophytic plants.

NATURAL VEGETATION

Although enormous changes have occurred in the vegetation of the region, numerous historical records and sufficient relict communities still exist to provide a reasonable basis for mapping the original distribution of vegetation types through the region. Largely on the basis of the vegetation map assembled by Küchler (1964), it is possible to estimate the relative abundance of major plant communities that once covered the land surface of the region. Using such criteria, it seems likely that the 502,000-square-mile area was once 81 percent grasslands, 13 percent hardwood deciduous forest or forest-grassland mosaic, 3 percent sagebrush grasslands, and 2 percent coniferous forest or coniferous woodland. The remaining 1 percent is now covered by surface water, predominantly of recent origin resulting from river impoundments.

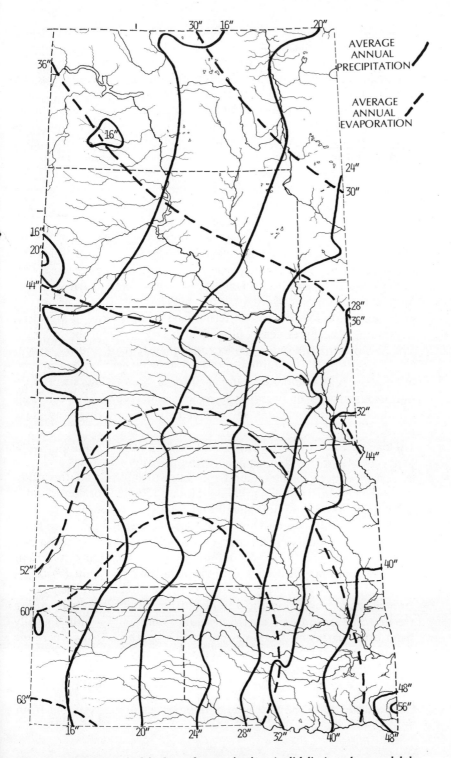

Fig. 3. Average annual inches of precipitation (*solid line*) and annual lake evaporation isopleths (*broken line*) in the Great Plains states, based on U.S. Weather Bureau data.

The grassland-dominated communities in the area consist of several associations, ranging from tall-grass prairies to short-grass plains or steppe vegetation (fig. 4). The tallest and most species-rich of the American grasslands are the bluestem (*Andropogon*) prairies of the eastern Dakotas, western Minnesota and Iowa, and portions of eastern Nebraska and Kansas, terminating in northern Oklahoma. Stewart (1975) listed four "primary intraneous" species of breeding birds (upland plover, bobolink, western meadowlark, and savannah sparrow) abundantly associated with such prairies in North Dakota, as well as thirteen secondary intraneous species with fewer numbers. To the west of the bluestem prairies in the Dakotas lies the eastern mixed-grass prairie (Stewart 1975) or the wheatgrass-bluestem-needlegrass prairie (Küchler 1964). The dominant plants are shorter than those of bluestem prairie, but a large number of flowering forbs are also characteristic. Stewart lists eleven primary intraneous species and twelve secondary intraneous species of breeding birds associated with this vegetation type in North Dakota, including Baird sparrow, chestnut-collared longspur, Sprague pipit, and lark bunting. The wheatgrass-needlegrass prairie (or western mixed-grass prairie, in Stewart's terminology) occupies nearly all of North Dakota from the Missouri Valley westward and extends over more than half of South Dakota. The vegetation is predominantly short-grass species and scattered midgrasses, plus a moderate number of forbs. Stewart lists six primary intraneous species of breeding birds and nine secondary species, all of which he also indicates as characteristic of eastern mixed-grass prairies.

This short-grass prairie, or grama–buffalo grass association, occurs on localized slopes and dry exposures in the western Dakotas and over extensive portions of the region from western Nebraska southward to the Staked Plain. The "high plains" biota is adapted for considerable aridity, and its array of both plants and animals is somewhat restricted. Stewart lists only two bird species (horned lark and chestnut-collared longspur) as primary intraneous forms, and ten more as secondary intraneous species. In Texas and southwestern Oklahoma a variant of this vegetation type occurs, with the inclusion of mesquite (*Prosopis juliflora*). Sutton (1967) reports that only the presence of the golden-fronted woodpecker and the absence of the family Parulidae during the breeding season separate the mesquite community ornithologically from other parts of Oklahoma, but farther south in Texas where the mesquite component is better developed a fairly distinctive avifauna exists (Hamilton 1962).

In several areas, extensive regions of sandy soil or sand dunes have greatly affected the vegetation. The largest of these is the Nebraska Sandhills, where the vegetation is mainly widely spaced bunchgrasses,

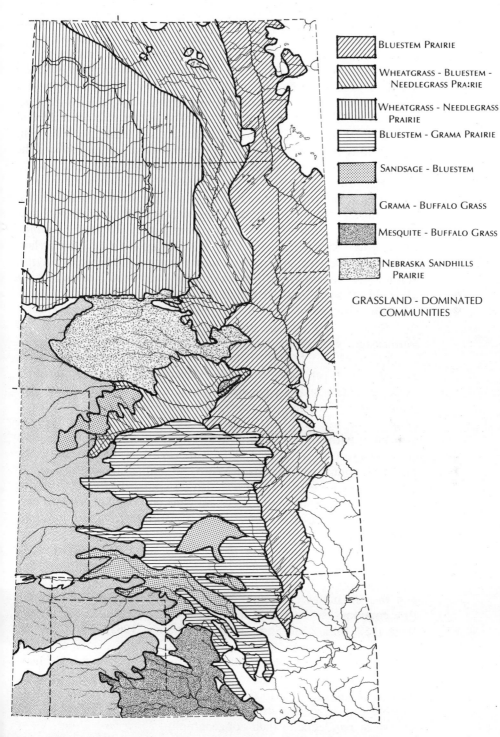

Fig. 4. Distribution of natural plant communities in the Great Plains states, showing grassland-dominated vegetation types. Based on Küchler 1964.

Legend:

- Bluestem Prairie
- Wheatgrass - Bluestem - Needlegrass Prairie
- Wheatgrass - Needlegrass Prairie
- Bluestem - Grama Prairie
- Sandsage - Bluestem
- Grama - Buffalo Grass
- Mesquite - Buffalo Grass
- Nebraska Sandhills Prairie

GRASSLAND - DOMINATED COMMUNITIES

with the intervening areas either unvegetated or sparsely vegetated. Wet meadows and marshes at the bases of these hills allow for a birdlife essentially the same as that of the midgrass prairies to the north. In southwestern Kansas and Oklahoma the large areas of sandy soil associated with the Cimarron and other river systems support a vegetation composed of sand-adapted grasses and sand sagebrush (*Artemisia filifolia*). Rising (1974) lists three species of birds as typical of grassland and xeric scrub in southwestern Kansas (scaled quail, lesser prairie chicken, white-necked raven), but only the quail and the prairie chicken are independent of trees, telephone poles, or similar elevated objects for nesting. A somewhat similar sage-dominated community type occurs on clay soils in southwestern North Dakota, where big sagebrush (*Artemisia tridentata*) and silver sage (*A. cana*) grow in conjunction with short-grass vegetation and cactus (*Opuntia polycantha*). Stewart lists only three bird species (sage grouse, lark bunting, and Brewer sparrow) as primary intraneous species for this community type, along with ten secondary intraneous forms that are virtually the same as those found in other grasslands.

Communities dominated by deciduous or hardwood tree species (fig. 5) are diverse and are particularly abundant in the eastern and southeastern parts of the region. The northern deciduous forest communities of western Minnesota and eastern North Dakota are mapped as a composite of types recognized separately by Küchler (1964) as "oak savanna" and "maple-basswood forest." Additionally, a substantial area of aspen grovelands, such as those in the Turtle Mountains of north-central North Dakota, are included. The bird species associated with all these types are essentially those of the eastern deciduous forest, as outlined by Stewart (1975) for the forests of the Turtle Mountains and the northeastern and southeastern upland deciduous forests. Kellehur (1967) has also described the ecology of this prairie-forest ecotone, and Green and Janssen (1975) have listed seventeen breeding birds as typical of the southeastern Minnesota hardwood forests.

Along the river systems of the Dakotas, Nebraska, and Kansas, a distinctive gallery forest, called by Küchler the northern floodplain forest, provides an extremely important forest corridor linking eastern and western biotas. The significance of these river systems as geneflow corridors has been established by a variety of studies (Sibley and West 1959; Sibley and Short 1959, 1964; Short 1965; West 1962) on hybridization between species or well-marked subspecies of birds, such as the lazuli and indigo buntings, the rose-breasted and black-headed grosbeaks, and the red-shafted and yellow-shafted flickers. Stewart (1975) has described the associated birds of several of these riverine forest systems in North Dakota, nearly all of which are eastern or pandemic in their zoogeography.

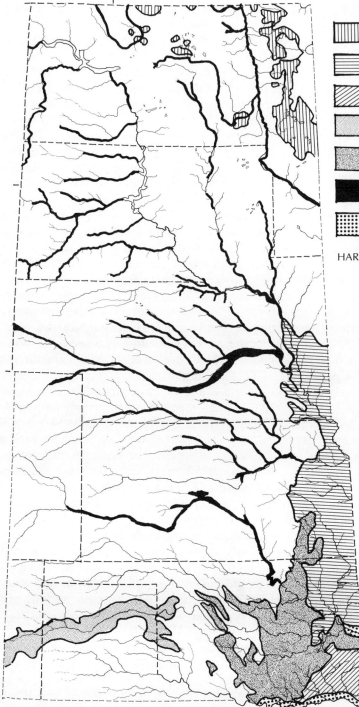

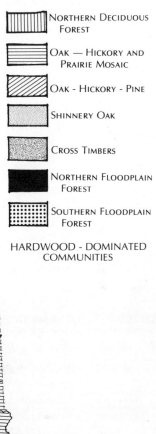

Northern Deciduous Forest

Oak — Hickory and Prairie Mosaic

Oak - Hickory - Pine

Shinnery Oak

Cross Timbers

Northern Floodplain Forest

Southern Floodplain Forest

HARDWOOD - DOMINATED COMMUNITIES

Fig. 5. Distribution of natural plant communities in the Great Plains states, showing hardwood-dominated vegetation types. Adapted with modifications from Küchler 1964.

From the Missouri Valley of the Nebraska-Iowa border southward, a forest type dominated by oaks and hickories (*Carya* spp.) tends to replace the northern floodplain forests along major river systems and also extends to the uplands in wetter sites. Over much of eastern Kansas the oak-hickory forest occurs as a mosaic community with bluestem prairies, with dominance of one or the other dependent upon local conditions of soil, slope, and exposure. In eastern Oklahoma this mosaic pattern is replaced by the "cross timbers" community of large oaks (*Quercus marilandica* and *Q. stellata*) in extensive groves or growing singly, interspersed with medium-tall grasses such as bluestems and other prairie grass species. In the wetter portions of southeastern Oklahoma the forest becomes denser, and the oaks are supplemented with hickories (*Carya* spp.) and pines (*Pinus echinata* and *P. taeda*), resulting in an extensive oak-hickory-pine community in the southeastern United States and the Atlantic piedmont. Only here do such species as the brown-headed nuthatch and red-cockaded woodpecker breed. Along the floodplains of the lower Arkansas and Red rivers a distinctive local southern floodplain forest also occurs, with oaks, tupelo (*Nyssa aquatica*), and bald cypress (*Taxodium distichum*) sharing dominance. In such forests a number of distinctly southern and southeastern species may be found, including several warblers, such as the Swainson, prairie, pine, and yellow-throated.

Last, along the drainage of the Canadian River of Oklahoma and across the panhandle of Texas, a scrubby oak community dominated by shin oaks (*Quercus mohriana* and *Q. havardi*) and little bluestem (*Andropogon scoparius*) occurs, together with various deciduous shrubs and occasional low evergreens such as junipers. To a still undetermined extent, it may also serve as a distribution corridor between eastern deciduous forest species and those of the piñons, junipers, and other coniferous species to the west. The scrub jay, plain titmouse, and other scrub-adapted forms also reach their easternmost limits in this habitat type.

The coniferous-dominated communities (fig. 6) of the area are relatively few and distinctive. In northeastern New Mexico and the adjoining Black Mesa country of Oklahoma, a woodland community type dominated by low junipers (*Juniperus* spp.) and arid-adapted pines (*Pinus edulis* and *P. monophylla*) occurs on uplands and along dry river channels. Here the plain titmouse, common bushtit, and pinyon jay all commonly occur, as does the green-tailed towhee at higher elevations.

The Black Hills coniferous forest, together with the other forests of ponderosa pine (*Pinus ponderosa*) of southwestern North Dakota and western Nebraska, provides the most typically Rocky Mountain biota to be found in the entire region. The analysis of Pettingill and Whit-

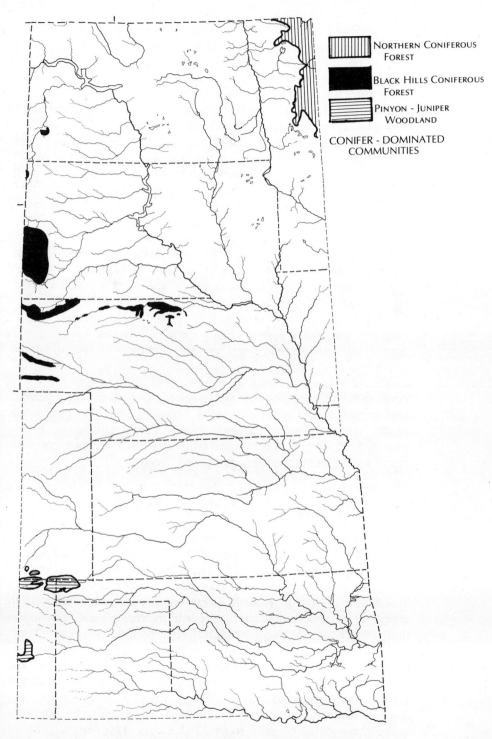

Northern Coniferous Forest

Black Hills Coniferous Forest

Pinyon - Juniper Woodland

Conifer - Dominated Communities

Fig. 6. Distribution of natural plant communities in the Great Plains states, showing conifer-dominated vegetation types. Adapted from Küchler 1964.

ney (1965) indicates that not only are the largest percentage (30 percent) of the Black Hills plant species typical of the Rocky Mountain region, but of the ninety-three resident species of birds, at least twenty-two have distinct affinities with the Rocky Mountains. Many of these are restricted to the Black Hills so far as this book's coverage is concerned. Those associated with the Rocky Mountains include the Lewis woodpecker, dusky and western flycatchers, dipper, and Mac-Gillivray warbler. Although the brown creeper, ruby-crowned kinglet, and northern three-toed woodpecker are also typical of the region, they have their primary affinities with high montane or boreal forest.

The "northern coniferous forest" mapped in figure 6 includes three vegetation types recognized separately by Küchler (1964), including coniferous bogs (dominated by larch, black spruce, and white cedar), the "Great Lakes spruce-fir" forest (dominated by balsam fir and white spruce), and the "Great Lakes pine forest" (dominated by red, white, and jack pines). With few exceptions, the breeding birds of all three community types are the same and are essentially those associated with the Canadian boreal forest. This small intrusion of northern floral elements into the region is of considerable interest, since it supports many uniquely boreal bird species. Green and Janssen (1975) have identified the boreal avian component in their analysis of Minnesota's breeding birds and have listed sixty species as typical of this community type. Species that do not occur elsewhere in the region covered by this book and that are associated with Minnesota's coniferous or mixed coniferous-deciduous forests include the bald eagle, osprey, spruce grouse, common raven, boreal chickadee, hermit thrush, palm warbler, Tennessee warbler, Nashville warbler, magnolia warbler, Cape May warbler, and purple finch.

AVIAN ZOOGEOGRAPHY AND ECOLOGY

As was mentioned earlier, the region encompassed by this book includes 17 percent of the land mass of United States south of Canada, an area about twice the size of Texas. It has been estimated (Peterson 1963) that at least 435 species of birds breed in the United States south of Canada. Based on records assembled here, a minimum of 330 species of birds have bred at least once in the region under consideration, and about 260 might be considered "regular" breeders. Thus, well over 50 percent of America's continental-breeding bird fauna is included within the limits established for this book, even though the region constitutes less than a fifth of the total area of the continental United States excluding Alaska. This is a rather surprising statistic, since the endemic Great Plains bird fauna is known to be relatively meager, consisting of only some 32 (Udvardy 1958) to 38 (Mengel

1970) species, or about 5 percent of North America's total avifauna. Rather than regarding the Great Plains as a center of evolutionary diversity and speciation in birds, it is better to see it as a natural barrier or isolating agent in avian speciation. Thus the Great Plains might be thought of as like an ocean, coming into contact with the bird faunas associated with Canada's boreal forest, the deciduous forests of the eastern and southeastern United States, the Rocky Mountain coniferous forests, and the aridlands of the American southwest, but also separating them to some degree. Mengel (1970) extensively investigated the influence of the plains during speciation in several environments peripheral to them and suggested that the central and southern plains may have been periodically covered by savanna, pine parklands, and even boreal forests during periods of glaciation, being relatively treeless only during the maximally xeric interglacial periods, which may have seriously interfered with speciation at those times. Hubbard (1974) has also examined Pleistocene history as a possible basis for present-day distribution patterns and speciation characteristics of birds in the southwestern aridlands.

As a basis for analyzing the zoogeographic and ecological affinities of the birds included in this book, each species was classified into two categories, representing its general geographic affinities and its broad ecological associations. The zoogeographic categories are as follows:

Endemic: Largely limited to the grasslands or marshes of the Great Plains.

Pandemic: Having a large continuous or disruptive distribution pattern not clearly associated with specific major vegetational types.

Introduced: Added to the fauna by man, purposely or accidentally.

Eastern: Having a breeding distribution generally associated with boreal forest areas to the east or southeast of the region in question.

Northern: Having a breeding distribution generally associated with deciduous forest areas to the north or northeast of the region in question.

Southern: Having a breeding distribution generally associated with deserts or scrublands to the south or southwest of the region in question.

Western: Having a breeding distribution generally associated with montane forests to the west or northwest of the region in question.

The ecological distributional categories are as follows:

Grassland species: Having a breeding distribution generally associated with grasslands.

Woodland and forest species: Having a breeding distribution associated with deciduous or coniferous forests or woodlands, or their successional stages.

Xeric scrub species: Having a breeding distribution associated with sage or other arid-adapted and shrub-dominated vegetational types.

Limnic species: Having a breeding distribution associated with marshes, rivers, lakes, or other surface-water habitats.

Miscellaneous: Having a breeding distribution not specifically associated with any of the categories listed above.

Three other categories, relating to abundance, are also included:

Accidental: One or two recent breeding records, but the normal breeding range is well beyond the limits of the area under consideration.

Extirpated: No longer breeds in the area under consideration but still exists elsewhere.

Extinct: The species evidently no longer exists anywhere.

A listing of all the species included in this book, together with their geographic and ecologic affinities, is given in tables 1 to 5, and a numerical summary of these listings appears in table 6. From these tables it is clear that the greatest single component of the breeding bird biota is woodland and forest species, which represent more than half the total bird fauna, though woodlands and forests occupy only 15 percent of the region's surface area. In contrast, although grasslands cover 81 percent of the area, grassland species make up only 11 percent of the total bird fauna. The small percentage of xeric scrub species (4 percent) corresponds closely with the approximate percentage (3 percent) of sagebrush grasslands relative to the total area. An estimate of the value of marshes and other surface-water areas to breeding birds is provided by the fact that, although such habitats occupy only about 1 percent of the region's total area, limnic species make up 21 percent of the breeding bird fauna.

The zoogeographic influences of regions contiguous to the Great Plains become evident when one considers the geographic affinities of the bird fauna, as summarized in table 6. The largest single component (27%) is eastern, and as might be expected these are nearly all deciduous forest species. The next largest component (23%) is of species having pandemic distributions, particularly forest and limnic species. Species with western affinities make up nearly a fifth of the total, and these are in large measure birds of the Rocky Mountain coniferous forests. Species with northern affinities compose 16 percent, and most of these are birds of the coniferous boreal forests of

Canada and the Great Lakes states. The number of southern species is relatively low, contributing 7 percent of the total, and many of these are xeric scrub or desert-adapted forms of the Chihuahuan aridlands. Interestingly, in spite of the vast area of the Great Plains grasslands, only 5 percent of its bird fauna may be considered endemic, half of which comprises grassland-adapted sparrows, while several others are shorebirds associated with low meadows or prairie marshes. A more detailed analysis of these data has been published elsewhere (Johnsgard, 1978).

Although this book is not specifically intended as a bird-finding guide, the list of bird-watching localities at the back of the book (Appendix A) is inclusive enough that virtually all of the 260 "regular" breeders may be seen at one or more of the localities listed, as is indicated by the associated checklist of species recorded at these localities (Appendix B).

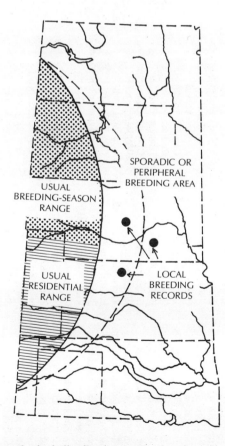

Fig.7. Sample hypothetical distribution map, showing symbols used in the actual maps.

Table 1
Species Associated with Woodlands and Forests

Eastern	Northern	Pandemic	Western	Southern
Mississippi kite	Bald eagle	Cooper hawk	Broad-tailed hummingbird	Golden-fronted woodpecker
Red-shouldered hawk	Goshawk	Sharp-shinned hawk	Northern three-toed woodpecker	Ladder-backed woodpecker
Broad-winged hawk	Merlin	Red-tailed hawk	Lewis woodpecker	Scissor-tailed flycatcher
American kestrel	Spruce grouse	Wild turkey	Western kingbird	Black-crested titmouse
Bobwhite	Woodcock	Ruffed grouse	Cassin kingbird	Black-capped vireo
Black-billed cuckoo	Saw-whet owl	Mourning dove	Ash-throated flycatcher	Great-tailed grackle
Barred owl	Black-backed three-toed woodpecker	Yellow-billed cuckoo	Say phoebe	Blue grosbeak
Chuck-will's-widow	Olive-sided flycatcher	Screech owl	Western flycatcher	Painted bunting
Whip-poor-will	Yellow-bellied sapsucker	Long-eared owl	Dusky flycatcher	
Chimney swift	Gray jay	Great Horned owl	Western wood pewee	
Ruby-throated hummingbird	Boreal chickadee	Common flicker	Scrub jay	
Pileated woodpecker	Brown creeper	Hairy woodpecker	Pinyon jay	
Red-bellied woodpecker	Winter wren	Downy woodpecker	Black-billed magpie	
Red-headed woodpecker	Veery	Willow flycatcher	Plain titmouse	
Red-cockaded woodpecker	Hermit thrush	Common raven	Common bushtit	
Eastern kingbird	Swainson thrush	Common crow	Red-breasted nuthatch	
Great crested flycatcher	Ruby-crowned kinglet	Black-capped chickadee	Pygmy nuthatch	
Eastern phoebe	Golden-crowned kinglet	White-breasted nuthatch	Townsend solitaire	
Acadian flycatcher	Solitary vireo	House wren	Mountain bluebird	
Least flycatcher	Philadelphia vireo	American robin	MacGillivray warbler	
Eastern wood pewee	Northern waterthrush	Cedar waxwing	Western tanager	
Blue jay	Palm warbler	Loggerhead shrike	Black-headed grosbeak	
Fish crow	Nashville warbler	Warbling vireo	House finch	
Carolina chickadee	Tennessee warbler	Yellow warbler	Lazuli bunting	
Tufted titmouse	Magnolia warbler	Yellow-breasted chat	Lesser goldfinch	
Brown-headed nuthatch	Cape May warbler	Brown-headed cowbird		
Bewick wren	Blackburnian warbler	Northern oriole		
Carolina wren	Bay-breasted warbler	American goldfinch		
Mockingbird	Yellow-rumped warbler	Rufous-sided towhee		
Gray catbird	Mourning warbler	Chipping sparrow		
Brown thrasher		Song sparrow		

Table 1 (cont'd)

Species Associated with Woodlands and Forests

Eastern	Northern	Pandemic	Western	Southern
Wood thrush	Canada warbler			
Eastern bluebird	Evening grosbeak			
Blue-gray gnatcatcher	Purple finch			
White-eyed vireo	Red crossbill			
Bell vireo	Pine siskin			
Yellow-throated vireo	Dark-eyed junco			
Red-eyed vireo	White-throated sparrow			
Black-and-white warbler				
Prothonotary warbler				
Swainson warbler				
Golden-winged warbler				
Blue-winged warbler				
Northern parula				
Black-throated blue warbler				
Black-throated green warbler				
Cerulean warbler				
Yellow-throated warbler				
Chestnut-sided warbler				
Pine warbler				
Prairie warbler				
Ovenbird				
Louisiana waterthrush				
Kentucky warbler				
Hooded warbler				
American redstart				
Common grackle				
Orchard oriole				
Summer tanager				
Scarlet tanager				
Cardinal				
Rose-breasted grosbeak				
Indigo bunting				

Table 2
Species Associated with Limnic Environments

Pandemic	Western	Eastern	Northern	Endemic
Pied-billed grebe	Western grebe	Black duck	Common loon	Wilson phalarope
Double-crested cormorant	Eared grebe	Wood duck	Red-necked grebe	Franklin gull
Canada goose	White pelican	Hooded merganser	Horned grebe	
Green-winged teal	Trumpeter swan	Green heron	Ring-necked duck	
Mallard	American wigeon	Little blue heron	Bufflehead	
Pintail	Gadwall	Yellow-crowned night heron	White-winged scoter	
Blue-winged teal	Cinnamon teal	Cattle egret	Common goldeneye	
Northern shoveler	Canvasback	American bittern	Common merganser	
Great egret	Redhead	Least bittern	Greater sandhill crane	
Snowy egret	Lesser scaup	King rail	Yellow rail	
Great blue heron	Ruddy duck	Black rail	Common snipe	
Black-crowned night heron	Snowy plover	Common gallinule	Swamp sparrow	
Virginia rail	American avocet	Purple gallinule		
Sora	Black-necked stilt	Short-billed marsh wren		
American coot	California gull			
Piping plover	Ring-billed gull			
Killdeer	Yellow-headed blackbird			
Spotted sandpiper				
Willet				
Common tern				
Forster tern				
Caspian tern				
Black tern				
Least tern				
Long-billed marsh wren				
Red-winged blackbird				
Common yellowthroat				

Table 3

Species Associated with Grasslands

Endemic	Western	Pandemic	Eastern	Northern
Pinnated grouse	Ferruginous hawk	Marsh hawk	Eastern meadowlark	Sharp-tailed grouse
Mountain plover	Swainson hawk	Short-eared owl	Bachman sparrow	Sharp-tailed sparrow
Long-billed curlew	Prairie falcon	Horned lark	Field sparrow	
Marbled godwit	Burrowing owl	Bobolink	Henslow sparrow	
Upland sandpiper	Poor-will	Savannah sparrow		
White-necked raven	Western meadowlark	Grasshopper sparrow		
Dickcissel	Brewer blackbird	Vesper sparrow		
Baird sparrow	Lark sparrow			
Le Conte sparrow				
Lark bunting				
Clay-colored sparrow				
Cassin sparrow				
McCown longspur				
Chestnut-collared longspur				
Sprague pipit				

Table 4

Species Associated with Xeric Scrub

Western	Southern
Sage grouse	Scaled quail
Rock wren	Roadrunner
Canyon wren	Verdin
Sage thrasher	Curve-billed thrasher
Green-tailed towhee	Gray vireo
Brown towhee	Brown towhee
Brewer sparrow	Rufous-crowned sparrow
	Black-throated sparrow

Table 5

Miscellaneous, Accidental, Introduced, and Extirpated or Extinct Species

Miscellaneous	Accidental	Introduced	Extirpated	Extinct
Turkey vulture (P)	Anhinga (E)	*Successfully*	Blue grouse (W)	Passenger pigeon (E)
Black vulture (S)	White-faced ibis (W)	Ring-necked pheasant	White-tailed kite (S)	Carolina parakeet (E)
Golden eagle (W)	Fulvous whistling duck (S)	Gray partridge	Swallow-tailed kite (S)	Ivory-billed woodpecker (E)
Osprey (P)	Mottled mallard (S)	Rock dove	Whooping crane (N)	
Peregrine (P)	Tricolored heron (E)	House sparrow		
Barn owl (P)	Harris hawk (S)	Starling		
Common nighthawk (P)	Great gray owl (N)	*Uncertain Status*		
White-throated swift (W)	Vermilion flycatcher (S)	Chukar partridge		
Belted kingfisher (P)	Worm-eating warbler (E)			
Violet-green swallow (W)	White-winged crossbill (N)			
Tree swallow (N)				
Bank swallow (P)				
Rough-winged swallow (P)				
Barn swallow (P)				
Cliff swallow (P)				
Purple martin (P)				
Dipper (W)				

NOTE: Letters in parentheses stand for areas of distribution; e.g., P = pandemic.

Table 6

Geographic Affinities of Species: Summary of Listings of Species in Tables 1 to 5

Ecologic Affinities of Species	Eastern	Northern	Western	Southern	Pandemic	Endemic	Introduced	Totals
Woodland and forest spp.	63	37	25	8	31	—	—	164 (51%)
Limnic spp.	14	12	17	—	27	2	—	72 (22%)
Grassland spp.	4	2	8	—	7	15	—	36 (11%)
Xeric scrub spp.	—	—	7	8	—	—	—	15 (4%)
Other spp.	6	4	5	7	11	—	5	38 (12%)
Totals	87	55	62	23	76	17	5	325
%	27%	17%	19%	7%	23%	5%	2%	100%

xlv

Tallgrass prairie in the proposed Prairie National Park, Pottawatomie County, Kansas

Pothole lake in the Sandhills of Nebraska

Niobrara River, Nebraska, a corridor for eastern and western flora and fauna

Pine Ridge country, Massacre Canyon, northwestern Nebraska

Badlands, Toadstool Park, northwestern Nebraska

FAMILY GAVIIDAE
(LOONS)

Common Loon

Common Loon
Gavia immer

Breeding Status: Restricted during the breeding season to north-western and west-central Minnesota, and the Turtle Mountains area of north-central North Dakota.

Breeding Habitat: Nesting is mostly limited to larger and deeper lakes having an abundant supply of fish. In Alberta, studies have shown that loons prefer to nest on lakes with many islands and with a minimum of human disturbance. Of a survey of nineteen lakes, the smallest that supported a pair of nesting loons was 916 acres in area, and a maximum of five pairs were found on a lake of 3,038 acres.

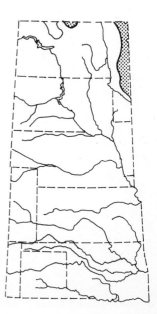

Nest Location: Studies in Minnesota and Alberta indicate that loons have a strong preference for nesting on islands, presumably as an antipredator adaptation, and for nesting in sheltered situations, probably to avoid waves. Virtually all nests are within 4 feet of the water, and most are directly on the water's edge. When loons nest in water, nests may be situated in emergent vegetation or, at times, on the side of a muskrat house. Most nests are situated so as to allow the adult an underwater escape, and when islands are used they are usually less than 2 acres in area.

Clutch Size and Incubation Period: Two eggs, olive brown marked with black. Incubation is by both sexes and lasts 29 days.

Time of Breeding: Minnesota egg dates range from May 8 to July 8. Unfledged young have been reported as late as November 3. North Dakota egg dates are all for mid-June, but young have been seen as early as June 9.

Breeding Biology: Loons are highly territorial, and shortly after arriving on their breeding grounds they establish a territory that may be up to 25 hectares (about 60 acres) in area, which they advertise by the familiar "yodeling" call. Most of the elaborate displays include bill-dipping, raising the head and breast, a "circle dance" between territorial opponents, "splash-diving," rearing upright in the water with folded or spread wings, and a low flying rush over the water. Copulation occurs on shore and is not marked by elaborate display behavior. Both parents care for the young, which often ride on their backs during their first few weeks of life. The fledging period is about 10–11 weeks.

Suggested Reading: Olson and Marshall 1952; Palmer 1962.

3

FAMILY PODICIPEDIDAE
(GREBES)

Pied-billed Grebe

Red-necked Grebe
Podiceps grisegena

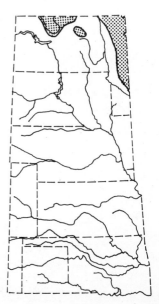

Breeding Status: Restricted during the breeding season to north-western and west-central Minnesota and north-central and north-eastern North Dakota.

Breeding Habitat: In North Dakota, this grebe nests on fresh-water or slightly brackish permanent water areas, usually at least 10 acres in area. It also occurs on shallow river impoundments. Submerged plants, such as pondweeds, are usually present.

Nest Location: Red-necked grebes are in general solitary nesters, but "loose colonies" have been reported at a few Minnesota lakes. Pairs are usually well scattered on larger ponds and lakes, and their nests are either on open water or along the edges of emergent vegetation near open water. The water in such locations may be a foot or two in depth, and the nest is typically floating but anchored to vegetation. The nest is often constructed of submerged aquatic plants, and the location is probably chosen by the male. Both sexes participate in building the nest, with the male performing most of the work, and typically several nests may be constructed by a single pair before one is completed and used.

Clutch Size and Incubation Period: Usually 4 to 5 eggs, sometimes 3–6 (7 North Dakota clutches averaged 3.1 eggs and ranged from 2 to 4). Eggs are initially white or pale blueish, but gradually become stained with brown. Incubation is by both sexes and lasts 22–23 days.

Time of Breeding: North Dakota egg dates range from May 19 to July 16, and dates of dependent young range from June 13 to August 22.

Breeding Biology: Birds apparently arrive on their breeding grounds already paired. Territories are established that may include from 75 to 125 yards of shoreline, within which all breeding activities occur. Territorial and pair-forming displays are not yet well understood, but several mutual displays are performed by the pair. These include simultaneous calling while swimming side by side and erecting the crest, rising breast to breast in the water, and emerging together from a dive, followed by rising in the water and facing each other, sometimes with vegetation in the bill. Copulation occurs on the nest platform, following an invitation posture by the female. After hatching, both sexes brood and feed the young. The young remain on the nest for the first day, then ride on the backs of the parents for the first 3 weeks of life. They are fed for at least 7 weeks, and family bonds break up after 8 to 10 weeks. The fledging period is still unknown.

Suggested Reading: Palmer 1962; Chamberlain 1977.

7

Horned Grebe
Podiceps auritus

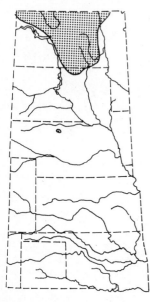

Breeding Status: Breeding occurs over most of North Dakota east of the Missouri River and extreme northwestern Minnesota and occurs rarely or only locally in north-central South Dakota. The bird has also bred in the Nebraska Sandhills.

Breeding Habitat: Nesting in North Dakota occurs on fresh to slightly brackish water areas that range from seasonal to permanent and vary in size from ⅓ acre to several hundred acres. Typically, abundant growths of submerged aquatic plants are present in breeding areas, but emergent vegetation is usually relatively sparse.

Nest Location: These birds are typically distributed as single pairs on ponds or widely scattered pairs on larger lakes or marshes, but as many as five nests have been located on a pond of 43 acres. The nest is usually built over dense beds of submerged vegetation, either in open water or in emergent vegetation near open water, with water depths varying from 6 to 48 inches.

Clutch Size and Incubation Period: From 3 to 6 eggs, usually 4 or 5 (13 clutches in North Dakota averaged 4.5). Eggs are white initially but become stained with brown. Incubation lasts 24–25 days and is probably by both parents. Typically the bird not incubating remains near the nest.

Breeding Biology: In North Dakota, horned grebes tend to select ponds that are fairly small (less than 2.5 acres) and contain mostly open water, which is evidently related to the importance of visual cues in territorial behavior. Displays are mostly mutual and include head-shaking, bill-touching ceremonies, weed-presentation ceremonies, standing vertically in the water facing the mate, and rushing over the water, often carrying vegetation in the bill. The nest is built by both sexes, and copulation occurs on the nest platform. The young are initially tended and fed by both parents and often ride on their backs. After a few weeks one of the parents may leave the area while the other remains with the chicks. The fledging period is not known.

Suggested Reading: Fjeldså 1973; Faaborg 1976.

Eared Grebe
Podiceps nigricollis

Breeding Status: Breeds over nearly all of North and South Dakota, northwestern and southwestern Minnesota, northwestern Iowa, the northwestern part of Nebraska, and probably adjacent Colorado.

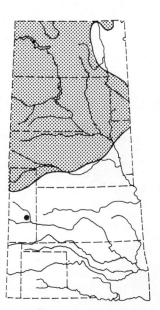

Breeding Habitat: In North Dakota, eared grebes breed on water areas that vary from slightly brackish to subsaline, and from seasonal to permanent. They also use shallow river impoundments, and most nesting areas have extensive beds of submerged aquatic plants. Compared with horned and pied-billed grebes, eared grebes prefer larger, more open ponds that provide abundant feeding areas but also offer a sheltered location where a colony of nests can be placed.

Nest Location: Nests are generally less than 100 yards from the nearest shore and may be in open water or in emergent cover ranging from sparse to dense. Often nests in colonies may be separated by as little as 10 feet, or in extreme cases may even touch each other. In seven colonies, the water depth ranged from as little as 4 inches to as much as 48 inches.

Clutch Size and Incubation Period: From 2 to 8 eggs are typical (101 nests in North Dakota averaged 3.8). Eggs vary from whitish to greenish or buffy. The incubation period is 20½ to 21½ days, and incubation is by both sexes. Apparently only one brood per season.

Time of Breeding: Egg dates in North Dakota range from May 21 to August 9, and extreme dates for dependent young are from June 8 to September 2.

Breeding Biology: Pair-forming displays occur during spring migration while the birds are in flocks but continue after arrival on the breeding grounds. Courting occurs in the center of the breeding areas, and no territorial behavior is evident. Displays are mutual and include an advertising call by unpaired or separated birds, "habit-preening," head-shaking and a "penguin-dance" by both members of a pair standing upright in the water facing each other, and a "cat-attitude" with withdrawn head and fluffed body feathers. The female builds the nest, and copulation occurs on the nest platform, without elaborate associated displays. The nest is abandoned when the last egg hatches, and thereafter the young are tended by both parents, often riding on their backs. Young are relatively independent by their third week, but the fledging period is still unknown.

Suggested Reading: Palmer 1962; McAllister 1958.

Western Grebe
Aechmophorus occidentalis

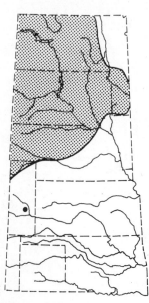

Breeding Status: Breeds in North Dakota east to the Red River Valley, in southwestern Minnesota, throughout most of South Dakota except the extreme southeast, in the northern and western parts of Nebraska, and in adjacent Colorado.

Breeding Habitat: In North Dakota, breeding occurs on permanent ponds and lakes that vary from slightly brackish to brackish and that contain large expanses of open water. Breeding also occurs on semipermanent water areas but usually is restricted to ponds of at least 50 acres.

Nest Location: Nesting is colonial, with nests at times numbering in the hundreds and birds in the thousands. In North Dakota, colonies have been found in areas where the water is from 2 to 4 feet deep in semiopen growth of emergent vegetation, usually hardstem bulrush or phragmites. In other areas cattails are sometimes used. Nests are often very close to one another; in a colony on Sweetwater Lake in North Dakota the average distance between nests was only about 2 yards. Sites offering protection from wave action and deep enough to allow underwater access to the nest are preferentially used.

Clutch Size and Incubation Period: Clutches range from 3 to 7 eggs (12 North Dakota nests averaged 4.2). Eggs are very pale bluish green or buff initially, but soon become stained with brown. Incubation lasts 22–23 days, probably starting with the first egg laid, and is performed by both sexes.

Time of Breeding: Egg dates in North Dakota range from May 15 to June 10. Dates of dependent young range from June 13 to October 5.

Breeding Biology: Territorial activity by pairs is maintained only in the immediate vicinity of the nest, and most display activity occurs before the start of nesting, apparently serving primarily for pair-bond formation and maintenance. Most or all displays are performed by both sexes, often mutually. They include crest-raising while the birds swim together, with associated whistling notes and occasional withdrawal of the head and neck to the back, a "high arch" posture with neck stretched and bill pointed downward and the tail raised, a similar but less extreme "low-arch" posture, ritualized "habit preening", and the "race." In this last-named display, two birds (sometimes one and sometimes as many as six) call, then rise in the water and race side by side over the water surface with arched necks, bills pointed diagonally upward, and wings partially raised. Behavior leading to the race display usually includes threat-pointing with the bill and mutual bill-dipping as the birds approach; diving often terminates the display. When more than two birds perform the race the additional birds are always males. Copulation normally occurs on the nest site, but it has been observed on the edge of a beach where nests were on dry land.

Suggested Reading: Palmer 1962; Nuechterlain 1975.

Pied-billed Grebe
Podilymbus podiceps

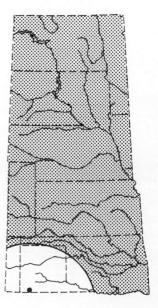

Breeding Status: Virtually pandemic throughout the region, but commoner in the Dakotas, Minnesota, and northwestern Iowa. Uncommon to occasional in Nebraska, eastern Colorado, and Kansas. Breeds locally in Oklahoma and has bred in the Texas panhandle (Muleshoe National Wildlife Refuge).

Breeding Habitat: In North Dakota, pied-billed grebes breed on seasonal or permanent ponds ranging from fresh to moderately brackish, on river impoundments and lakes, particularly those having extensive stands of emergent vegetation with adjacent areas of open water. Compared with eared and horned grebes, this species occupies a wider variety of pond types but is always associated with heavy emergent vegetation, and its distinctive vocalizations adapt it well to establishing territories in low-visibility habitats.

Nest Location: Nests float and are usually in semiopen to dense emergent vegetation, frequently bulrushes. Grasses (whitetop, mannagrass), sedges, cattails, and burreeds are also sometimes used in North Dakota, and water depths at more than 80 sites averaged 25 inches, ranging from 11 to 37 inches. Two or more nest platforms may be constructed by a pair, from 4 to 10 yards apart, often at the edge of vegetation to allow an underwater approach to the nest. (However, in Iowa, 138 nests averaged about 26 feet from nest to open water.)

Clutch Size and Incubation Period: Usually from 4 to 7 eggs, bluish white to greenish white initially, but gradually becoming stained with brown. A sample of 74 nests in North Dakota averaged 6.7 eggs per clutch; 97 successful nests in Iowa averaged 6.18 eggs. The incubation period is 23 days, probably starting with the first egg, since hatching is usually spread over several days. Most incubation is by the female, but both sexes participate. In some areas two broods may be raised each year.

Time of Breeding: Extreme egg dates in North Dakota range from May 7 to August 20. In northwestern Iowa, egg dates range from May 2 to August 8. In Kansas, 19 breeding records are from May 1 to June 30, with a modal egg-laying date of May 15. Oklahoma egg dates are from May 15 to June 12. Thus the

11

species is probably single-brooded in southern areas, with second or replacement clutches regular in more northerly areas.

Breeding Biology: Pairs are very territorial; in Iowa the territory consisted of an arc of about 150 feet around the nest, with the male defending the area, but the pair shared water areas outside territorial limits with other birds. The birds are highly vocal, and most displays are evidently acoustic rather than postural. Copulation has been reported on open water but normally occurs on the nest platform. Both parents care for the young, which regularly ride on the backs of their parents. The fledging period is not known, but by 3 weeks of age the young are fairly independent, which may allow adults to begin a second clutch while the young are still flightless.

Suggested Reading: Glover 1953; Palmer 1962.

FAMILY PELECANIDAE
(PELICANS)

White Pelican

White Pelican (American White Pelican)
Pelecanus erythrorhynchos

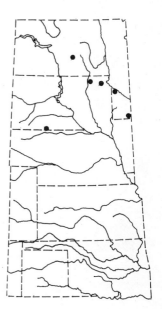

Breeding Status: Breeds in colonies in North Dakota (Chase Lake) and South Dakota (Waubay, Sand Lake, and Lacreek N.W.R.). At least periodically through 1972, pelicans also have bred in Minnesota (Marsh Lake, Lac qui Parle County, and Heron Lake, Jackson County). The only breeding colony in Colorado (Riverside Reservoir, Weld County) is slightly to the west of the region under consideration here. Nonbreeding birds are frequent during summer throughout most of the region, especially on large lakes or reservoirs.

Breeding Habitat: Breeding typically occurs on isolated and sparsely vegetated islands in lakes or reservoirs. For nesting, birds prefer islands that are nearly flat or have only gentle slopes, that lack obstructions that might interfere with taking flight, and that have loose earth easily worked into nest mounds.

Nest Location: Unlike some other pelican species, the white pelican nests only on the ground. A variety of nest materials have reportedly been used in nest construction, including shells, vegetation, dirt, sand, and stones. At Chase Lake the nests are simply depressions in the ground lined with weeds and are often less than 10 feet apart. On the Molly Islands of Yellowstone Lake the birds build mounds in sandy areas that have shallow depressions by simply reaching out from the nest site and pulling in materials with the side or tip of the bill.

Clutch Size and Incubation Period: From 1 to 4 eggs (212 nests on the Molly Islands averaged 1.67); usually 2. Eggs are dull white, with a coarse texture. Incubation period is about a month (up to 36 days reported); incubation is by both sexes.

Time of Breeding: Extreme egg dates for North Dakota are June 1 to June 30; dates of dependent young range from June 1 to July 28. Since the fledging period is about 70 days, it is apparent that many birds must remain dependent through August.

Breeding Biology: Pelicans are at least seasonally monogamous, and little display activity occurs on the nesting areas. Territorial defense is limited to the area immediately around the nest site, and most described displays occur at or near the nest. These include a "head-up" display with inflated or expanded gular pouch, which may serve as a greeting display to the mate and threat toward others, a "bow," with the bill pointed toward the feet and waved from side to side, and a "strutting walk" with the male following the female. Copulation occurs on land and is preceded by wing-quivering and squatting by the female. Both sexes incubate and share in feeding the young, but they feed only their own chicks. At the age of 50–60 days the young of the colony form a large "pod," and they fledge at about 10–11 weeks.

Suggested Reading: Palmer 1962; Schaller 1964.

FAMILY PHALACRO-
CORACIDAE
(CORMORANTS)

Double-crested Cormorant

Double-crested Cormorant
Phalacrocorax auritus

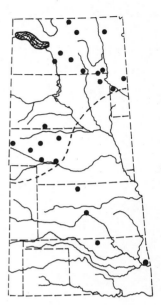

Breeding Status: Breeds in colonies scattered throughout much of the Dakotas, western Minnesota, and the western half of Nebraska. In northeastern Colorado, a colony at Riverside Reservoir, Weld County, is just outside the limits of the region covered by this book. In Kansas, this species has bred (1951) at Cheyenne Bottoms, Barton County, and since 1959 has nested at least periodically at Kirwin N.W.R., Phillips County. There is also a nesting record for Wilbarger County, Texas. In Oklahoma, nest records have been established for Great Salt Plains (1945–50) and Sequoyah National Wildlife Refuge (1972–74).

Breeding Habitat: This species nests on rocky islands or cliffs adjoining water, or in trees in or near water. A supply of fish must be present within 5 to 10 miles of the nesting site. When trees are used, they may be either deciduous or coniferous, but they eventually die from the accumulation of excrement.

Nest Location: In North Dakota, nests in the Missouri River area are situated in the tops of dead trees, primarily cottonwoods, but colonies on islands of natural lakes are usually on the ground. One colony was also reported in willows that were in water 3½ feet deep. Nests on solid substrates usually have a foundation of sticks, herbaceous vegetation, and rubbish, with finer materials added for lining. The male brings such material to the female, who incorporates it into the nest, and additional materials are added through the season. Garbage, including excreta and the remains of dead animals, also accumulates in the nest.

Clutch Size and Incubation Period: Normally from 4 to 7 eggs (51 North Dakota nests averaged 4.6), but up to 9 have been reported. Eggs are pale bluish with a chalky surface. Incubation period is 25–29 days. One brood per season.

Time of Breeding: Extreme egg dates for North Dakota are May 13 to July 18, and extreme dates of dependent young are May 31 to July 25.

Breeding Biology: Cormorants are at least seasonally monogamous, usually breeding initially when 3 years old. Courtship occurs on water and includes much chasing and diving. Males choose the territory, which includes the nest and adjacent perching spot. Copulation occurs on the nest, mainly during the nest-building period. Both sexes assist in incubation, which begins before the clutch is complete; thus hatching is staggered over several days. The young leave the nest by about 6 weeks but continue to be fed by their parents until 9 weeks of age, when family bonds disintegrate.

Suggested Reading: Palmer 1962; Mitchell 1977.

FAMILY ANHINGIDAE
(ANHINGAS)

Anhinga

Anhinga
Anhinga anhinga

Breeding Status: Accidental; has bred in McCurtain County and at Sequoyah N.W.R., Oklahoma, the former in 1937 and the latter in 1971 and 1972.

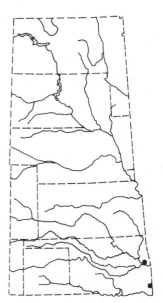

Breeding Habitat: Breeds on sheltered, quiet waters, often in cypress swamps, mangrove-lined bays, and other subtropical areas, primarily freshwater.

Nest Location: Anhingas are colonial, with nests usually clustered in groups of 8–12 pairs, in shrubs or trees from 5 to 20 feet above water. Large twigs and branches provide the foundation, with smaller twigs and leaves forming the lining. As in cormorants, the male selects the nest site and gathers materials, which the female incorporates. Sometimes the same site is used in subsequent years.

Clutch Size and Incubation Period: From 2 to 5 eggs (29 Arkansas nests averaged 3.9). Eggs are chalky and pale bluish green. Incubation period is probably 25–28 days, starting before the clutch is complete. One brood per season.

Time of Breeding: Texas egg records extend from April 14 to July 27. On the Gulf Coast, breeding is concentrated in late winter and spring, with reduced summer nesting, but in southern Florida breeding occurs throughout the year. The recent Oklahoma nesting was in late April and May, with dependent young seen as late as September.

Breeding Biology: Anhingas are monogamous and probably do not breed until their second year. Studies in Florida indicate that males establish territories by taking over old nests or building new "preliminary" nests, on which they display by wing-waving, feather-ruffling, and bowing. Females are attracted by such behavior and approach the nest while performing the same displays as the male. Copulation occurs almost immediately, and soon thereafter the male begins to gather nesting material, which is incorporated by the female. The permanent nest is finished within a few days, and egg-laying then begins. Both sexes incubate and both help rear the young, which are fed by regurgitation. The young are normally raised in the nest, but after a few weeks they tend to leave the nest when disturbed, sometimes climbing back to it later. The fledging period has not been established.

Suggested Reading: Palmer 1962; Allen 1961.

FAMILY ARDEIDAE
(HERONS AND BITTERNS)

Black-crowned Night Heron

Great Blue Heron
Ardea herodias

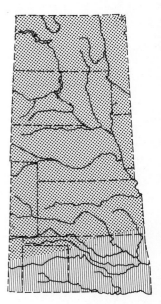

Breeding Status: Pandemic and common throughout the region, breeding locally along many rivers, lakes, and reservoirs.

Breeding Habitat: The species breeds colonially in a variety of aquatic habitats, usually where there are trees, but birds have also been found nesting on the ground, on rock ledges, among bulrushes, and in other elevated situations. Within the Great Plains region, herons often nest in association with cormorants, especially where reservoirs have flooded tall trees.

Nest Location: Nests are usually placed in a crotch or on a large limb of a tall tree, sometimes more than 100 feet above the ground. Usually more than one nest occurs per tree in large colonies, and old nests are frequently reused. Nests that have been used for several years tend to be massive; newly made ones are often flimsy. Adults continue to add materials to the nest until the young are well grown.

Clutch Size and Incubation Period: From 3 to 6 eggs (36 Kansas clutches averaged 4.4), pale bluish green, smooth to slightly rough in texture. The incubation period is 25–29 days. One brood per season.

Time of Breeding: Egg dates in North Dakota range from April 27 to May 15. Minnesota records are from April 25 to May 20. Kansas egg records are from March 1 to April 30, and Oklahoma records are from March 15 to April 26.

Breeding Biology: Great blue herons are seasonally monogamous, and both sexes arrive at the nesting ground about the same time. Birds probably breed initially when 2 years old, but some variation is likely. The male selects the breeding territory, which usually centers on an old nest. Several obviously hostile displays are associated with territorial defense. Additionally, numerous highly ritualized territorial advertising displays occur, including the "stretch," "snap," and others. These are predominantly male displays, given at the nest site, and serve to attract females and aid pair-formation. Mutual behavior between members of a pair includes twig-passing, feather-nibbling, bill-stroking, and similar activities. Copulation is sometimes preceded by displays, such as feather-nibbling, but it may also occur without obvious display. When building or improving the nest, the male gathers materials and the female works them into the nest. Both sexes incubate, and nest-relief ceremonies are performed. The eggs typically hatch over an interval of 5–8 days, and adults feed the young by regurgitating food into the bottom of the nest. Although the young can make short flights in the nest vicinity when 7 weeks old, they usually continue to use the nest and are fed by the adults until they are about 10-11 weeks old.

Suggested Reading: Pratt 1970; Mock 1976.

Northern Green Heron
Butorides striatus virescens

Breeding Status: Breeds over most of the region, except for the northern half of Minnesota and the western portions of the Dakotas, Nebraska, Kansas, and Oklahoma, and the included portions of Colorado, New Mexico, and Texas. Rare in North Dakota, but common farther south in range.

Breeding Habitat: This heron occupies a broad range of habitats and water types, usually near trees, but also sometimes breeds in marshlands well away from tree cover. One of the most adaptable of North American herons, usually breeding as solitary pairs or in loose colonies.

Nest Location: The nest is usually between 10 and 15 feet above the ground, depending on the habitat, but may be directly on the ground or up to 30 feet above it. It varies in form from very flimsy to very bulky, the latter usually when it has been used many times. At times the old nests of other herons are also used, and the birds thus sometimes nest among other species of herons or egrets.

Clutch Size and Incubation Period: From 3 to 6 eggs, with 4 or 5 most common in northern part of range and fewer toward the south (17 Kansas clutches averaged 3.1). Eggs are pale greenish or bluish green, with smooth surface. Incubation period is 19-21 days, but incubation begins before the last egg is laid; thus there is a staggered period of hatching. At least in some areas, two broods are produced per season.

Time of Breeding: Minnesota egg dates range from May 12 to June 11. In Kansas, egg records span the period April 21 to June 20. Oklahoma egg dates are May 1 to June 20.

Breeding Biology: Males select and defend territories on their return to the breeding grounds; separate feeding territories may also be defended. Initially quite large, the male's territory soon shrinks to the area around the nest or nest site. The territory is advertised by a "flying-around" display over the breeding site and by an advertising call from a conspicuous perch. Males also perform "stretch" and "snap" displays similar to those of the larger herons, and after a female has been attracted to the territory both sexes perform "circle-flight," "crooked-neck-flight,"

28

and "flap-flight" displays. After pair bonds have formed, the female completes the nest; the male helps in gathering materials. Copulation occurs on the nest platform or an adjacent branch and continues through egg-laying. Both sexes share in incubation and perform nest-relief ceremonies. The young hatch at intervals and are fed by regurgitation. They remain in the nest for about 16–17 days but do not actually fledge until they are about 21–23 days, with adults continuing to feed them until that time. In areas where two broods are raised, the second clutch may be begun only 9 days after the first brood has fledged.

Suggested Reading: Palmer 1962; Meyeriecks 1960.

Little Blue Heron
Florida caerulea

Breeding Status: Breeding is restricted to the eastern half of Oklahoma, with 1952 (Finney County) and 1974 (Barton County) breeding records for Kansas. Summer visitors may be seen throughout most of the region, and nesting recently occurred at Salyer National Wildlife Refuge in North Dakota (*American Birds* 30:969).

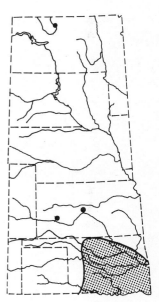

Breeding Habitat: Although found both in freshwater and saline environments, this species is mostly limited to inland habitats such as woodland ponds.

Nest Location: The species is colonial in nesting; nests are situated from a few feet above the ground or water to as much as 40 feet. In a Florida study, the birds usually nested on horizontal limbs with the nest wedged against the main trunk, at an average of about 7 feet above the substrate. They were thus less exposed than snowy egret nests in the same area and tended to be slightly higher. In Oklahoma a variety of broad-leaved trees have been used for nesting, often shared with snowy and great egrets.

Clutch Size and Incubation Period: From 3 to 6 eggs (58 Florida clutches averaged 3.7). Eggs are pale greenish blue with a blue gloss. The incubation period is 22–25 days, averaging 22.8 days, with staggered hatching of the young. One brood per season.

Time of Breeding: Oklahoma egg records range from April 19 to July 7, and observations of dependent young extend from May 20 to July 7.

Breeding Biology: Males begin to establish territories a few weeks before egg-laying by defending an area about 25 feet in diameter around an old nest or nest site. Besides various threat displays, the "stretch" display is perhaps the most important sexual display. Unmated females are attracted to such males but are initially repulsed. Besides the stretch display, the "snap" display, with mandible clicking, is common. Early stages of pair-formation including mutual billing, neck-crossing and intertwining, but virtually no aerial displays as in the green heron and snowy egret. A strong pair bond is formed, but some promiscuous copulatory behavior has been observed. Copulation occurs on the nest platform or close to it. The female completes the nest started by the male, and the male passes twigs to her in an elaborate ceremony. Little nest-building occurs after incubation gets under way; both sexes participate equally in incubation. The young are fed by both parents, who regurgitate food into their mouths or into the nest. The young probably fledge in about a month.

Suggested Reading: Meanley 1955; Palmer 1962.

Cattle Egret
Bubulcus ibis

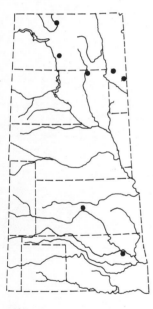

Breeding Status: A few scattered breeding records exist from several states for this self-introduced species, including Oklahoma (Tulsa County, 1962–64), Kansas (Barton County, 1973–74), South Dakota (Brown County, 1977), North Dakota (McHenry County 1976–77), Minnesota (Grant County, 1972–73, and Pope County, 1959 and 1971). No doubt additional breeding records will accrue in these and other states as the species continues to spread.

Breeding Habitat: Cattle egrets occur in a wide variety of freshwater to saline habitats and are more terrestrial than any native North American herons. They are highly social and normally nest among other herons. They are usually found near cattle in North America and forage largely on grasshoppers and other insects rather than on fish like most herons.

Nest Location: Compared with other small herons, cattle egrets tend to nest in relatively dense vegetation, at heights that are variable but usually under 20 feet, averaging about 7 feet. At least in Florida, cattle egrets nest somewhat later than other herons, and their nests are more complete at the time of egg-laying.

Clutch Size and Incubation Period: From 1 to 6 eggs (85 Florida clutches averaged 3.5), very pale blue or bluish white with a smooth surface. The incubation period is 22–23 days, usually 23 days. The chicks hatch at intervals, usually 2 days apart. One brood per season.

Time of Breeding: Dependent young have been observed in Oklahoma between June 30 and August 21. In Minnesota they have been seen between June 19 and August 4.

Breeding Biology: The cattle egret maintains a smaller breeding territory than other heron species, which is related to its high degree of coloniality. Males establish territories that initially cover only a few square yards and are soon reduced to the immediate area around the nest. The male performs several threat displays within this territory, and he also performs several visual courtship displays ("stretch," "twig-shaking," "wing-touching," "forward-snap," "flap-flight," and "forward") which are similar to those of other herons. Females are attracted to a displaying male and form a pair bond by flying to him, landing on his back, and subduing his aggressive tendencies by repeated blows on the head. These blows gradually change to nibbling after the male has ceased to fight back. Mutual back-biting is used thereafter by the pair as a greeting display, and it often precedes copulation. Some instances of polygamous pair bonds have been seen, but monogamy is the general pattern. The female completes the nest started by the male, which may require up to 6 days. Both sexes assist in incubation, with the female apparently sitting most of the daylight hours and the male at night. Compared with other herons, cattle egrets are very attentive to their young, and nestling mortality is low, compensating for their relatively small clutch size.

Suggested Reading: Jenni 1969; Lancaster 1970.

Great Egret (Common Egret)
Casmerodius albus

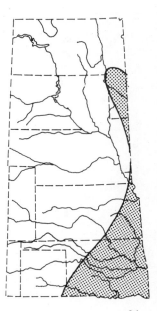

Breeding Status: Breeds in scattered colonies along the Red River in Texas and the eastern half of Oklahoma. Breeds sparingly and locally in Kansas (Cowley County) and the southern half of Minnesota (Grant, Pope, and Lac qui Parle counties). Becoming more common and possibly expanding its range northward in Minnesota. No breeding records for Nebraska or the Dakotas, but summer visitors may be expected there.

Breeding Habitat: The species occurs in freshwater, brackish, and occaionally saltwater habitats but forages in fairly open situations. It is found on streams, swamps, and lake borders, usually close to trees during the nesting season.

Nest Location: Nests are either solitary or in colonies, usually with other species of herons, such as great blue herons. The nests are generally between 10 and 30 feet above the ground in trees, frequently beeches or maples in the northern states and cypress in the south. They tend to be very high up and to be less bulky than those of great blue heron, but larger than those of the smaller heron species.

Clutch Size and Incubation Period: From 2 to 4 eggs, but 3 eggs apparently are most common, at least in Louisiana. The clutch

31

may average larger in more northerly areas. The eggs are blue to greenish blue, with a smooth surface. The incubation period is 23–24 days. One brood per season.

Time of Breeding: In Oklahoma, nest-building has been reported from April 4 to May 20, eggs noted May 23, and dependent young seen July 2 to August 17.

Breeding Biology: In the first phase of breeding males establish territories that center on nest sites or old nests, preferably the latter, since these allow for earlier advertisement displays. When a nest platform is available, the males perform several courtship displays, including a ritualized preening movement or "wing-stroking," the "stretch" display (a vertical neck-stretch followed by a bobbing movement), the "bow" (a repeated twig-shoving movement followed by a bob), the "snap" (a downward extension of the head and neck, accompanied by a mandible snap and a bob), and a circular flight. Males thus attract females to the nest site, where copulation occurs. Within a few days a pair bond is formed, and shortly thereafter egg-laying begins. Both sexes incubate, and they perform greeting ceremonies when exchanging places on the nest. Likewise, both sexes feed the young, which require approximately 6 weeks to attain flight.

Suggested Reading: Wiese 1976; Tomlinson 1976.

Snowy Egret
Egretta thula

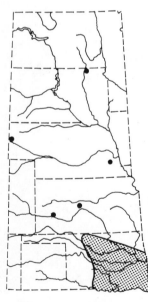

Breeding Status: Largely limited as a breeding bird to the eastern half of Oklahoma. There is a breeding record (two nests) for Kansas (Finney County) in 1952 and another (eight nests) for Barton County in 1974. Postbreeding dispersal is frequent as far north as Nebraska and Iowa; breeding occurred in South Dakota in 1977 (*South Dakota Bird Notes* 29:72), and has occurred twice in Nebraska.

Breeding Habitat: Snowy egrets occupy habitats ranging from freshwater to saline but prefer relatively sheltered locations. Ponds with low willows, buttonrush, and similar shrubs are favored, as are thick stands of mangroves. In Oklahoma the birds are usually found in heronries of little blue herons, great egrets, and black-crowned night herons.

Nest Location: Nests are usually in shrubs or low trees, from 2 to 10 feet above the ground, but up to 30 feet has been recorded. The birds are typically colonial but may nest singly at the edge of their range. The nests are rather flat and elliptical rather than round and are loosely constructed. They are often built of slender twigs a foot or two long, gathered close to the nest site.

Clutch Size and Incubation Period: From 2 to 5 eggs (102 Florida clutches averaged 3.9), with 4 probably the most typical number. The eggs are pale greenish blue with smooth shells. The incubation period is 22½ days. One brood per season.

Time of Breeding: Oklahoma records of nesting span from April 4 (birds carrying nest material) to July 6 (young being fed by parents).

Breeding Biology: After returning to their breeding grounds, males establish a territory that centers on a potential nest site but need not include an old nest. Besides hostile displays, the male performs several sexual displays that include both a stationary and an aerial "stretch" as major advertisement displays. A single "circle flight" around the potential mate is also common, and a more spectacular flight is a towering circular flight from 50 to 150 yards above the female, followed by a spectacular tumbling downward to land beside her. A mutual display called the "jumping over" display, in which one bird makes a short jump flight over the back of the other, is a probable indication that a pair bond has been formed. The male gathers material and the female constructs the nest. Copulation occurs on the nest site or on a limb close to it. The first egg may be laid before the nest is completed, and eggs are laid about 2 days apart. Since incubation (by both sexes) begins before the clutch is complete, the first young hatches about 18 days after the last egg is laid. After 20–25 days the young are ready to leave the nest.

Suggested Reading: Jenni 1969; Meyeriecks 1960.

Tricolored Heron (Louisiana Heron)
Hydranassa tricolor

Breeding Status: Accidental; a 1974 breeding record for the region, in Barton County, Kansas (*American Birds* 28:919), and a 1978 nesting attempt at Long Lake N.W.R., Burleigh County, North Dakota.

Nesting Habitat: During the breeding season this species is primarily found near salt water, inhabiting mangroves, tidal marshes, and similar habitats.

Nest Location: This species builds nests closer to the substrate than most other herons, usually less than 7 feet up, and rarely above 10 feet. The nests also tend to be in more sheltered and sturdier locations than those of snowy egrets. The species is highly social, at least in most areas.

Clutch Size and Incubation Period: From 3 to 6 eggs (38 Florida nests averaged 4.1). Eggs are pale greenish blue with a smooth

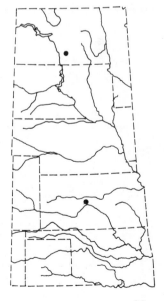

surface. The incubation period is 23–25 days, averaging about 24 days. One brood per season.

Time of Breeding: No information for region concerned; in Florida this species breeds at the same time as snowy egrets and little blue herons.

Breeding Biology: As in the other herons, the male establishes a territory that includes a nesting site and displays within it, threatening other males and attracting unpaired females. Several threat displays are present, as well as various sexual displays. The most elaborate of these is a combined stretch and snap display, which includes sudden extension of the head and neck, seizing a twig and dropping it, and a series of strong pumping movements. Females are initially evicted from the territory but are gradually accepted, and soon the pair begins mutual nibbling and billing. The male builds the foundation of the nest before pair-formation, but the female completes it while the male gathers material. Copulation occurs on the nest or beside it, before and probably during the egg-laying period. Both sexes incubate and care for the young, which hatch at intervals and remain in the nest about two weeks. As in other herons, many of the nestling losses result from starvation of the youngest chick. By the time they are 24 days old the young are fed away from the nest, and feathering is complete at about four weeks.

Suggested Reading: Rogers 1977; Jenni 1969.

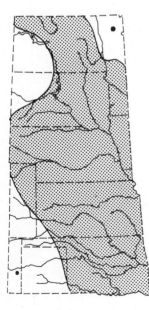

Black-crowned Night Heron
Nycticorax nycticorax

Breeding Status: Breeds locally over most of the region except for northern Minnesota and the drier portions of the Dakotas, Colorado, New Mexico, and the Texas panhandle.

Breeding Habitat: Habitats are extremely varied; both freshwater and saline environments are used, and the surrounding terrestrial vegetation varies from swamps to marshes and even includes orchards and city parks.

Nest Location: The species nests colonially, on dry ground, in bulrush or cattail marshes, or in trees up to 160 feet above the substrate. The nests are closely placed and often conspicuously situated. Newly made nests are flimsy, but they gain size and substance with repeated use. Nests are often situated in heronries that include other species. In our region, nesting is most frequent in bulrush or phragmites marshes or in groves of trees near rivers, often cottonwoods.

Clutch Size and Incubation Period: From 2 to 6 eggs, perhaps with larger clutches in the north (13 North Dakota nests averaged 4.7 eggs; Kansas clutches are typically about 4 eggs). Eggs are pale blue or greenish blue with a smooth surface. Incubation period is 24–26 days; possibly double-brooded in some areas.

Time of Breeding: Minnesota egg dates range from May 6 to July 11; dependent young reported from May 26 to July 19, Kansas egg dates range from May 1 to August 1.

Breeding Biology: As in other species, the male begins the breeding cycle by establishing a territory around a nest or nest site, which gradually shrinks to include only the nest itself and the immediate surroundings. Besides various threat postures, males also perform snap displays and a modified stretch display, called the "snap-hiss," accompanied by a raising of the ornamental crest plumes. These attract other birds, and eventually a female is allowed to enter the nest or approach the display site, after which the incipient pair begins mutual behavior such as nibbling and billing. Later the snap-hiss display serves as a greeting ceremony between the pair. The female completes the nest begun by the male, which may require up to a week. The first eggs are laid about 3–4 days after copulation, which may begin a day or two after the pair bond is formed. Incubation is by both sexes and begins with the first egg, so that hatching is staggered over several days. Until they fledge at about 6 weeks, and for a time afterward, the young continue to beg for food from their parents, following them to their foraging areas. Much of the foraging is done at night, which is the basis for the common name.

Suggested Reading: Palmer 1962; Noble, Wurm, and Schmidt 1938.

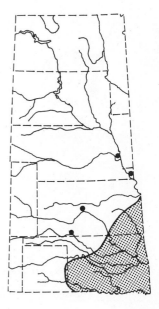

Yellow-crowned Night Heron
Nyctanassa violacea

Breeding Status: Primarily limited as a breeding species to Oklahoma, excluding the panhandle and northwestern areas, and also a local resident in southeastern Kansas, breeding at Cheyenne Bottoms (Barton County) in 1974. It occasionally breeds in northwestern Missouri (Squaw Creek N.W.R.), and there is one breeding record for Nebraska (1963, Sarpy County).

Breeding Habitat: Like the black-crowned night heron, this species is found in diverse habitats ranging from saline to freshwater, and even breeds on rocky, nearly waterless islands. In our region it is usually associated with tree-lined river habitats.

35

Nest Location: Nests in Oklahoma are usually in small, loose colonies separate from other heron species, in trees such as elms, ashes, oaks, box elders, and pecans. Nests there are 30–40 feet high; in other areas the spread has been reported from no more than a foot above the ground to more than 50 feet. Old nests of the previous season are often used, and nests tend to be thick and well built, with materials added through the period of hatching.

Clutch Size and Incubation Period: In Kansas the clutch is reported to be about 4 eggs, and 3–5 is the general range for the species, but little specific information on clutch size variation is available. The eggs are pale bluish green with a smooth surface. The incubation period is unknown but presumably is similar to that of the black-crowned night heron. Sometimes double-brooded.

Time of Breeding: In Kansas, eggs are laid in May and June. Eggs have been reported in Oklahoma as early as March 25, and broods have been seen as late as August 8. Two broods were raised by one pair in Norman in 1927, which fledged their first brood June 7 and the second 2 months later.

Breeding Biology: This species has been studied surprisingly little, but what is known suggests that it is very similar to the black-crowned night heron. The male evidently establishes a territory around a nest or nest site and advertises it with displays that probably include the stretch, accompanied by a loud whooping call. After pairs are formed, both sexes help complete the nest. Both sexes also incubate, and nest-relief ceremonies include billing, feather-nibbling, and plume erection or the stretch display. Both parents also feed the young, but there is no specific information available on their fledging period or rate of growth. However, fledging must require no more than 4-5 weeks, based on the timing of the second brood mentioned above.

Suggested Reading: Palmer 1962; Nice 1939.

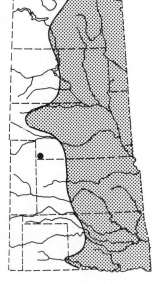

Least Bittern
Ixobrychus exilis

Breeding Status: Probably breeds over the eastern half of the entire region under consideration, but apparently fairly uncommon to rare throughout. It is rare and local in North Dakota and relatively rare in Nebraska, judging from the few nesting records.

Breeding Habitat: In this region the least bittern is associated with freshwater or slightly brackish marshes and lakes that have extensive stands of cattails, bulrushes, and other rank vegetation. It is thus not usually found around large impoundments or rivers, where water levels may fluctuate. Marshes with scattered bushes or similar woody growth are favored.

Nest Location: Nests are built above shallow water, in living or partly living stands of bulrushes or cattails, often close to open water. The nest is made of both dead and living materials and is usually about 6–8 inches across, round or oval. It has a foundation of dried leafy material and twigs that are arranged in a spokelike manner a foot or two above the water, with arched-over vegetation above.

Clutch Size and Incubation Period: From 3 to 6 eggs; the clutch is often about 5 in northern areas and somewhat smaller farther south. In Iowa, 59 clutches ranged from 2 to 6 eggs and averaged 4.4. Eggs are very pale bluish or greenish with a smooth surface. Incubation period is 17–18 days. Regularly double-brooded, at least in some areas or in favorable years.

Time of Breeding: In Iowa, nests are initiated from the beginning of June until the middle of July. In North Dakota, nests with eggs have been seen between June 15 and June 28. Kansas egg dates range from May 21 to July 20, and egg records for Oklahoma range from June 5 to July 21.

Breeding Biology: In Iowa, birds arrive on their breeding marshes about 2 weeks before the start of nesting. Little information is available on territoriality, but since nests are often fairly close to one another, territories must be rather small. The male's advertising call is a series of soft cooing notes, and presumably visual displays are also performed. Males evidently choose a nest site and do the early nest-building, as in other herons. Pair-forming displays still are unknown but probably involve mutual preening and crest-raising, since these occur during nest-relief ceremonies. Both sexes incubate, sharing incubation time about equally. Likewise, both sexes feed the young, but the male assumes the major role in this. The young usually remain in the nest for about 10–14 days but may leave it for short periods when only 6 days old. The adults continue to feed the young after they have left the nest, but at least at times they soon begin a second clutch.

Suggested Reading: Weller 1961; Palmer 1962.

American Bittern
Botaurus lentiginosus

Breeding Status: Nearly pandemic through the region, except for the Texas panhandle and northeastern New Mexico (but breeds in

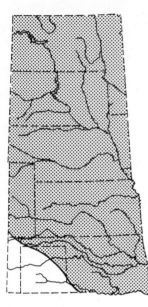

southeastern New Mexico at Bitter Lake N.W.R.). Throughout the area its breeding distribution is local, associated with marshes or tall grasslands.

Breeding Habitat: The species is generally associated with marshes, swamps, and bogs, where there is an abundance of bulrushes, cattails, and similar emergent vegetation. Also breeds in wet swales and in tall fields of grass.

Nest Location: The American bittern normally nests in a solitary manner in tall vegetation, usually cattails, bulrushes, or dense grasses, either on dry ground or on a mound several inches above the water. The nest platform is relatively scanty but usually is very well hidden from above and from the side, by the arching over of vegetation above it.

Clutch Size and Incubation Period: From 2 to 5 eggs (19 North Dakota nests averaged 3.9); eggs are buffy brown to olive buff, the surface smooth and slightly glossy. Incubation period is 24–28 days. Single-brooded.

Time of Breeding: Egg dates in North Dakota range from May 31 to August 2. Minnesota egg dates are from May 20 to June 20. Oklahoma breeding dates range from May 5 (eggs) to August 5 (well-developed young).

Breeding Biology: Relatively little is known of the social behavior of this elusive bird, but males evidently establish and advertise territories with their distinctive "pumping" call, especially at dawn and dusk. However, the male starts no nest during this period. Females are attracted to such territories and form apparently monogamous (possibly polygamous) pair bonds. Copulation has been observed on open ground, after the male displayed his white "shoulder" plumes and persistently advanced toward the female while repeatedly lowering and swaying his head from side to side, as if he were regurgitating food. After overtaking the retreating female he simply climbed on her back, grasped her nape, and copulated. No specific postcopulatory behavior was noted (personal observations). The female evidently chooses the nest location (about 50 yards from the area of copulation in a case personally noted) and apparently does all the nest-building and incubation. The male takes no part in defending the nest, but the female defends it fiercely. The young remain in the nest for about 2 weeks, but the fledging period is still unrecorded.

Suggested Reading: Palmer 1962; Mousley 1939.

FAMILY THRESKIORNITH-IDAE (IBISES AND SPOON-BILLS)

White-faced Ibis

White-faced Ibis
Plegadis chihi

Breeding Status: A rare and erratic nester. Bred in Barton County, Kansas, in 1951 and 1962 and near Tucumcari, New Mexico, in 1973. There is a 1904 breeding record for Clay County, Nebraska, and breeding records for Jackson County, Minnesota, in 1894 and 1895. In 1978 nesting occurred at Long Lake N.W.R., North Dakota, and near Edgemont, Fall River County, South Dakota.

Breeding Habitat: Freshwater marshes or brackish marshes in the western and southern states are the typical breeding habitat of this subspecies; cattails, bulrushes, or phragmites are the usual vegetation.

Nest Location: The species breeds colonially. Nests are on the ground in dense vegetation and are constructed of dry emergent plants. When built in bushes or in trees surrounded by water, the nest may have a substantial platform of twigs, but more leafy materials are present than in heron nests.

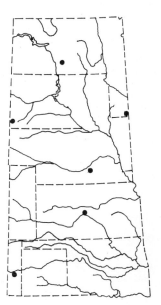

Clutch Size and Incubation Period: From 3 to 4 eggs, dull blue with a smooth or finely pitted surface. Incubation period is 21 days. One brood per season.

Time of Breeding: In Texas, egg records extend from April 20 to July 2. At least 13 nests were present in Kansas in 1962; nesting records there are for June and early July, as are those for Minnesota.

Breeding Biology: Remarkably little is known of the biology of this species. Monogamous pair bonds are formed, and both sexes help construct the nest, which takes about 2 days. Incubation begins with the last egg. Both sexes also incubate, and during nest relief they do mutual billing and preening and utter gutteral cooing notes. The adults continue to add material to the nest during incubation and the fledging period, about 6 weeks. The adults feed the young by regurgitation, with the young inserting their bills into that of the parent, or at times disgorge food into the nest to be picked up by the young. By the time the young are about 7 weeks old they fly with their parents to the foraging grounds, returning with them at night for roosting.

Suggested Reading: Palmer 1962; Burger and Miller 1977.

FAMILY ANATIDAE
(SWANS, GEESE, AND
DUCKS)

Trumpeter Swan

Trumpeter Swan
Cygnus buccinator

Breeding Status: Reintroduced. Originally extirpated, but reintroduced in the 1960s to Lacreek National Wildlife Refuge, Bennett County, South Dakota, and since 1963 has bred there and in adjacent parts of South Dakota (Shannon, Pennington, Meade, Zieback, Haakon, Jackson, Washabaugh, Mellette, and Tripp counties), as well as in Nebraska (Cherry, Sheridan, and Morrill counties).

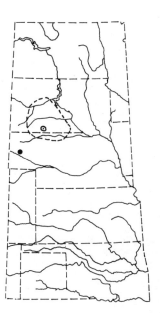

Breeding Habitat: Typical breeding habitat consists of large, shallow marshes to shallow lakes, with an abundance of submerged plants and emergent vegetation, and stable water levels. The emergent plants provide important nesting cover, and the submerged vegetation is the major food source.

Nest Location: Nests are greatly scattered, owing to extreme territorial behavior of adults, and nest sites are usually used for several years. Island locations are preferred over shoreline sites, and when nests are built in emergent vegetation the water is usually between 12 and 36 inches deep. Sometimes muskrat houses or beaver lodges serve as nest sites.

Clutch Size and Incubation Period: From 3 to 9 eggs, averaging about 5. Eggs are creamy white and somewhat granular. Incubation period is 32–37 days, usually about 34 days. Single-brooded.

Time of Breeding: In Montana, egg records extend from April 26 until June 26, and newly hatched cygnets have been seen from June 10 to July 3. In South Dakota, the Lacreek Refuge records indicate that nest-building occurs from April 3 to about May 20 and hatching from May 20 to July 1. Fledging is from September 20 to October 16.

Breeding Biology: Trumpeter swans pair for life, and each pair returns to its nesting area in spring as soon as the weather allows. Territories are established that average more than 30 acres, sometimes more than 100 acres, and are vigorously defended; the adults even exclude their own offspring of previous years. The male performs such territorial defense, but the female participates in mutual "triumph ceremonies" after territorial disputes and also helps defend the nest site. Both sexes help construct the rather bulky nest, which may require a week or more. The eggs are laid at 2-day intervals, and no incubation is performed until the clutch is complete. Thereafter the female performs all the incubation (a few records of males incubating exist but must be regarded as abnormal), while the male defends the nest. Most of the cygnets hatch within a few hours of each other and are led from the nest within 24 hours of hatching. The nest may later be used for resting or brooding, but often the brood is led some distance from the nest for rearing on quiet and secluded ponds. The fledging period is approximately 100 days in the Montana region, which occupies the entire summer and makes it impossible for birds to renest after nest failure.

Suggested Reading: Banko 1960; Johnsgard 1975.

Canada Goose
Branta canadensis

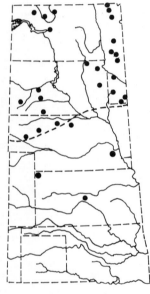

Breeding Status: The giant race of the Canada goose (*B. c. maxima*) originally bred over much of the region concerned, south to central Kansas. Reintroductions at refuges and other localities have reestablished Canada geese as breeding birds in most of these states.

Breeding Habitat: The breeding habitat of the giant Canada goose typically consisted of prairie marshes, especially those in the glaciated portions of the upper Great Plains. Some larger lakes were also used for breeding in early times, with the birds usually nesting on islands.

Nest Location: Muskrat houses probably originally were important nest sites for this race of geese in prairie marshes, but emergent plants such as phragmites and bulrushes were no doubt frequently used. Where terrestrial predators are significant, islands are important nest locations. Nests are often some distance from water, such as in depressions in the prairie or under shrubs. Elevated nest sites are often used if available. Considerable down is normally present.

Clutch Size and Incubation Period: From 4 to about 10 eggs (147 Missouri clutches averaged 5.6). Eggs are white with a smooth surface. Incubation period is 26-29 days, averaging 28 days.

Time of Breeding: Eggs dates in North Dakota range from April 13 to June 14, and dates of dependent young range from June 3 to August 12. Minnesota egg dates range from March 27 to June 2. In northwestern Missouri (Clinton County) the first eggs are laid about March 15-20, and the nesting period (to the hatching of the last egg) averages 73 days, or until early June.

Breeding Biology: Canada geese have strong, permanent pair bonds, and most begin to breed when 2 or 3 years old. Males establish fairly large territories in marshes, usually including the same area and often the same nest site as in previous years. Pair bonds are maintained by mutual displays, especially the triumph ceremony, and unless nest sites are limited or predator pressures are present, the nests tend to be well scattered. The nest is

46

constructed primarily by the female, with the male standing guard and helping to some extent. Copulation occurs on the water, primarily during the egg-laying period, and incubation does not begin until the clutch is complete. Males remain close to the nest and take the major responsibility for guarding it but do not help incubate. Both sexes tend the young, which soon begin to fend for themselves. During the fledging period of about 70 days, both parents undergo a flightless molting period, and thereafter the family may leave the area, with the family bonds persisting through the winter.

Suggested Reading: Brakhage 1965; Johnsgard 1975.

Fulvous Whistling Duck (Fulvous Tree Duck)
Dendrocygna bicolor

Breeding Status: Accidental. The only nesting record for our region is from Morton County, Kansas, in 1971 (*American Birds* 25:873). The nearest regular breeding ground is the Gulf coast of Texas.

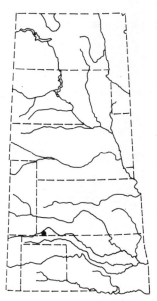

Breeding Habitat: Typical original breeding habitat consisted of freshwater marshes with extensive beds of cattails and bulrushes. Recently the birds have colonized rice fields, particularly those heavily infested with weeds.

Nest Location: In freshwater marshes these birds typically construct their nests in clumps of living or dead bulrushes or in knotweeds, or they build floating nests in open water. Nests in rice fields are usually on levees, over water between levees, or attached to growing plants. On coastal marshes of Texas they typically float over water 3-7 feet deep. No down is present in the nests.

Clutch Size and Incubation Period: From 10 to 16 eggs in nests produced by a single female; addition of eggs by other females often produces larger clutches. Eggs are white with a slightly roughened surface. Incubation period is 24-26 days. Single-brooded.

Time of Breeding: In Texas, egg dates range from May 10 to September 16. Downy young have been seen there as late as October 19, indicating a very long and rather irregularly timed breeding period.

Breeding Biology: Like other whistling ducks, this species is highly monogamous and probably forms lifelong pair bonds. Courtship displays are virtually nonexistent, at least as now understood. The best-known displays are those associated with copulation, which occurs on water and is preceded by mutual head-dipping. A ritualized "step-dance" by both birds follows treading. The female presumably builds the nest (not yet established), but both sexes are known to incubate. Incubation begins when the last egg is laid, and hatching is simultaneous. Both parents tend the young, which require about 65 days to fledge.

Suggested Reading: Meanley and Meanley 1959; Johnsgard, 1975.

Common Mallard
Anas platyrhynchos

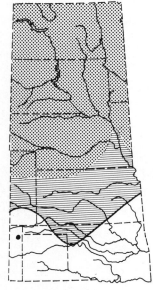

Breeding Status: Breeds over nearly the entire area except the eastern half of Oklahoma, where it is rare and local. Most common in the northern half of the region, with surprisingly few nesting records for Kansas, Oklahoma, and the Texas panhandle.

Breeding Habitat: One of the most widespread and adaptable species of ducks, the common mallard occupies a diversity of water types and surrounding environments. Fairly shallow waters, either still or slowly flowing, and surrounding dry sites of nonforested vegetation seem to be their preferred breeding habitat. They will sometimes breed in forested areas, but never in large numbers.

Nest Location: Nests are on dry ground, usually under relatively tall grass or herbaceous vegetation, and are generally well concealed from above and from all sides. Grasses 1–4 feet tall seem to be the most common nest cover, but weeds such as thistles and nettles are also frequently used. The nest is a shallow depression in the soil, well lined with brown down.

Clutch Size and Incubation Period: From 5 to 15 eggs (118 North Dakota nests averages 9.6). Eggs are creamy white to greenish white with a smooth surface. Incubation period is 24–30 days, averaging 28 days. Normally single-brooded, but renesting is regular after nest loss, and a few cases of double-brooding have been reported.

Time of Breeding: North Dakota egg dates range from April 15 to July 23, and dependent young have been seen from May 22 to September 24. Kansas egg records are for the period April 1 to June 10, with egg-laying most frequent during the first 10 days of May. Oklahoma egg dates are from April 13 to August 1, and young have been seen from mid-May onward. In Texas, eggs have been reported from February to May 17, and small young from April 26 to August 11.

Breeding Biology: Mallards begin social display early in the fall, with many adults probably forming new pair bonds with earlier mates, and those hatched the previous summer beginning courtship for the first time. By spring, nearly all females have formed pair bonds, and on arrival at their breeding grounds pairs spread out across the available habitat. Home ranges of such pairs vary greatly in size but at times may exceed 700 acres; spacing is enhanced by males' evicting other males from the vicinity of their mates. Females choose their nest sites and are abandoned by their mates when incubation gets under way. The newly hatched young are quickly led to water, and the fledging period is 55–59 days in Manitoba. Mallards often try to renest if their first attempt fails; the clutch sizes of renesting efforts tend to be slightly smaller than the original clutches.

Suggested Reading: Girard 1941; Johnsgard 1975.

Mottled Mallard (Mottled Duck)
Anas platyrhynchos fulvigula (*Anas fulvigula*)

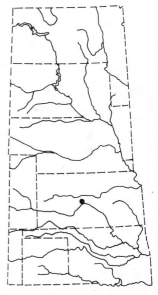

Breeding Status: Accidental. There is one breeding record for Cheyenne Bottoms, Barton County, Kansas, in 1963. The nearest regular breeding area is the Gulf coast of Texas.

Breeding Habitat: The preferred location is much like that of the common mallard, but the species inhabits brackish and saltwater habitats. Coastal marshes with extensive emergent vegetation are the primary breeding habitat, but the birds also breed in coastal prairies, bluestem meadows, and fallow rice fields.

Nest Location: Grass cover, usually fairly tall and dense, seems to be the preferred location for these coastal-nesting mallards. Distance from water is variable, and probably depends on local topography. The down lining of the nest is brown.

Clutch Size and Incubation Period: From 5 to 15 eggs (108 clutches in Texas averaged 10.4). Eggs are creamy white to greenish white. Incubation period is 25–27 days. Probably single-brooded, but up to five nesting attempts have been reported in a single female.

Time of Breeding: In Texas, eggs have been reported from March 18 to July 21, and dependent young have been seen from April 3 to August, except for one remarkable case of a brood observed in December.

Breeding Behavior: In most respects the behavior and breeding biology of this species is mallardlike. But it is possible that the pair bond may be relatively continuous in this population, even though the male does not seem to be present while the brood is reared. The fledging period of 54-60 days is virtually identical to that of mallards, and during this period the male apparently migrates to the coastal areas of Texas to molt.

Suggested Reading: Johnsgard 1975; Engeling 1950.

Black Duck (American Black Duck)
Anas rubripes

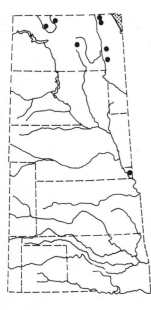

Breeding Status: Rare, limited mainly to Minnesota as a breeder. There are also a few scattered records of breeding in North Dakota, and it has bred at Squaw Creek N.W.R., Holt County, Missouri.

Breeding Habitat: Throughout nearly all of their range black ducks are associated with coastal marshes and eastern forests, and forest seems to represent their primary habitat. In the interior, the birds are found on fairly alkaline marshes, acidic bogs and muskegs, stream margins, and lakes and ponds, especially those near woodlands.

Nest Location: The margins of woody areas appear to be favored for nesting, with more open areas such as marshes or cultivated fields a secondary preference. Plants that serve as cover for the nest frequently are shrubby forms with evergreen or persistent leaves and dense branching patterns that provide excellent overhead concealment. A dry nest substrate, such as dead leaf litter, is an important component.

Clutch Size and Incubation Period: From 6 to 12 eggs (average of various studies about 9). Eggs vary from creamy white to pale greenish or buffy. The incubation period is 26-27 days, beginning with the last egg. Single-brooded, but replacement clutches are frequent.

Time of Breeding: In North Dakota the probable breeding season is from late April to mid-August. In Minnesota, young have been seen from June 22 to August 1.

Breeding Biology: Apart from their ecological preferences, there are virtually no differences in the breeding biology of the black duck and the common mallard. Competition between them is reduced by their generally complementary ranges, but in recent years the extent of range overlap has increased and hybridization between the two types has become more prevalent. Yet there is no good evidence that the black duck is extending its breeding range into the Great Plains, and it will probably continue to be a rare breeder there.

Suggested Reading: Coulter and Miller 1968; Johnsgard 1975.

Gadwall
Anas strepera

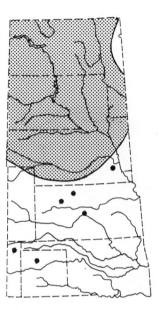

Breeding Status: Breeds over nearly all of the northern half of the region, being relatively common in the Dakotas and western Minnesota, as well as in the Nebraska Sandhills. Although it is regular at Cheyenne Bottoms (Barton County), there is only one other definite breeding record (Trego County) for Kansas. However, it is considered a breeding species at Quivira and Kirwin N.W.R. There is only one breeding record for the Texas panhandle (Potter County), but it is reported as breeding in the area of Clayton, New Mexico.

Breeding Habitat: Breeding occurs on a variety of mostly temporary or semipermanent water areas in the Dakotas, ranging from fresh to subsaline. Shallow prairie marshes that are relatively alkaline are apparently preferred over deeper, more permanent marshes, and those with grassy or weedy islands are also heavily used.

Nest Location: Nests are built on dry ground under a variety of covers, in particular amid broad-leaved weeds. Dry upland sites are preferred to wetter areas, and dense cover is preferred to sparser cover. Cover 1-3 feet in height, especially on islands, is frequently used for nesting; island nesting in dense populations is at times almost colonial. The nest cavity is lined with rather dark grayish down.

Clutch Size and Incubation Period: From 7 to 13 eggs (667 nests in North Dakota averaged 9.9). Eggs are dull creamy white. Incubation period is 25-27 days, averaging 26 days. Renesting is fairly common, and such nests have a slightly smaller average clutch size than initial nesting efforts.

Time of Breeding: Egg dates in North Dakota range from May 18 to August 10, and dates of dependent young are from June 14 to September 24. Three-fourths of the North Dakota egg records are for the month of June.

Breeding Biology: Gadwalls form their pair bond relatively early, during a period of social courtship involving aquatic display as well as aerial chases. Most birds are paired by the time they arrive on their nesting grounds, and pairs establish home ranges that may exceed 50 acres, often overlapping with the home ranges of other pairs. Territorial behavior as such is not significant, and nests are often close together, especially on islands. The female constructs the nest alone and is usually abandoned by her mate about a week or two after incubation has begun. The hen thus raises her brood alone, usually on deepwater marshes unlikely to dry up before fledging, which requires 7–8 weeks.

Suggested Reading: Oring 1969; Johnsgard 1975.

Northern Pintail (Common Pintail)
Anas acuta

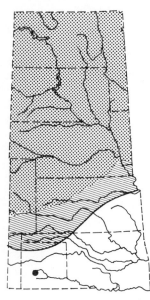

Breeding Status: Breeds over most of the region, becoming more local and rarer southward; there are only a few breeding records south of Kansas, including Cimarron County, Oklahoma, Union County, New Mexico, and Dallam, Lipscomb, and Randall counties, Texas.

Breeding Habitat: Over their vast range, northern pintails are associated with water types ranging from fresh to brackish, and from small temporary ponds to permanent marshes, but they are most abundant where there is open terrain surrounding areas of shallow water. In the Great Plains, stock ponds and similar water areas with little or no vegetative cover are used more by pintails than by most other waterfowl.

Nest Location: Nests are invariably in dry, upland locations, often in dead plant growth of the previous year, at times with very little concealment. Many nests are placed in cover less than a foot high with no concealment on at least one side. Nests are often in shallow natural depressions, rendering them susceptible to flooding by heavy rains. They are lined with dark gray to brownish down.

Clutch Size and Incubation Period: From 5 to 11 eggs (68 nests in North Dakota averaged 7.9). Eggs are white to greenish yellow or grayish, with a smooth shell. Incubation period is 21 days, starting with the last egg. Single-brooded, but renests regularly if the first clutch is lost.

Time of Breeding: North Dakota egg dates range from April 13 to July 6, and dependent young have been reported from May 16 to September 17. Kansas egg dates are from April 21 to June 10, with a peak of egg-laying in early April.

Breeding Biology: Northern pintails form monogamous pair bonds during a prolonged period of social courtship, which continues as the birds migrate north in spring. Most or all females are paired by the time the birds arrive on their nesting grounds, and the pairs tend to become well spaced as they establish large home ranges. Females begin nesting very early, shortly after hillsides are free of snow, and like most ducks they complete their clutches at the rate of one egg per day. Incubation begins with the laying of the last egg, and by that time or shortly afterward the pair bond is broken. When the brood hatches, the female leads them to water, sometimes shifting ponds and moving them nearly a mile from where they were hatched. The fledging period is 47–57 days in South Dakota and averages 41 and 46 days for females and males respectively in Manitoba.

Suggested Reading: Sowls 1978; Johnsgard 1975.

Green-winged Teal
Anas crecca

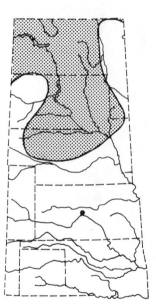

Breeding Status: Breeds over virtually all of the two Dakotas, in southwestern Minnesota, northwestern Iowa, and the northern half of Nebraska. Bred in 1968 in Barton County, Kansas (3 nests, 1 brood).

Breeding Habitat: Green-winged teal breed in greatest numbers where there is a mixture of grassland, sedge meadows, and dry hillsides with low trees, brushy thickets, or open woods adjacent to ponds or sloughs. Grasslands lacking shrubs or thickets are not used as extensively as those with some woody cover, but the breeding ponds are often shallow and transient.

Nest Location: Nests are on dry land, usually well away from water and extremely well shaded by rushes, dense grasses, or shrubs. Low shrubs are apparently the preferred nesting site, especially those that offer excellent overhead concealment. The nest is a shallow excavation, lined with a very dark brown down.

Clutch Size and Incubation Period: Usually from 6 to 12 eggs (25 North Dakota nests averaged 8.6), varying from dull white to olive buff. Incubation period is 21–23 days. Single-brooded, but at least some renesting is known to occur.

Time of Breeding: Egg dates in North Dakota range from May 7 to July 28, and dependent young have been seen from June 20 to September 1.

Breeding Biology: Green-winged teal are highly social and display over a long period of late winter and spring while forming their pair bonds, which are renewed annually. Pair-forming displays are numerous and elaborate and are highly animated. On reaching their breeding grounds, pairs spread out and establish home ranges that center on small ponds. Females select nest sites while accompanied by their mates, which usually remain attached to them until incubation is under way. After the clutch has hatched, the female leads her young to shallow ponds, and they grow very rapidly. They fledge in no more than 44 days, and some Alaska estimates are of as little as 35 days, but fledging is unusually rapid at such high latitudes, where summer daylight allows for continuous feeding.

Suggested Reading: McKinney 1965; Johnsgard 1975.

Blue-winged Teal
Anas discors

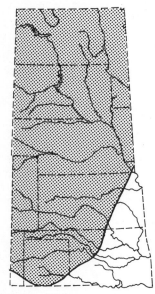

Breeding Status: Breeds commonly over the northern half of the region, and is the commonest surface-feeding duck breeder in much of the area. Progressively more local in eastern Colorado, Kansas, and Oklahoma, being uncommon to rare in the eastern parts of Kansas and Oklahoma. Also breeds locally in the Texas panhandle and northeastern New Mexico.

Breeding Habitat: The highest concentrations of blue-winged teal are in marshes surrounded by native prairies, especially the tall-grass prairies. Relatively small, shallow ponds or marshes are favored over larger and deeper ones during breeding, especially where grassy or sedge meadows are nearby.

Nest Location: Nests are on dry land, often very near water, particularly in sedges or grasses that average about a foot high. Steep slopes and very dense cover are avoided, and nests are often placed about halfway between water and the highest surrounding point of land in gently rolling country. The nests are well lined with grasses and dark down having conspicuous white centers.

Clutch Size and Incubation Period: From 8 to 13 eggs (349 nests in North Dakota averaged 10.3). Eggs are creamy white to pale olive-white with a slightly glossy surface. Incubation period averages about 24 days, starting with the last egg laid. Single-brooded, but renesting efforts are frequent.

Time of Breeding: North Dakota egg dates range from April 26 to July 28, and dependent young have been seen from June 17 to September 13, with most broods seen in July and August. Kansas egg dates are from May 1 to May 30, and in Oklahoma dependent young have been seen from May 26 to August 3.

Breeding Biology: Pair bonds are formed fairly late in blue-winged teal, mainly during the migration northward, but some displays may occur on the nesting grounds. Pairs are relatively tolerant of other pairs and often center their home ranges on very small ponds or even roadside ditches. The female chooses the nest site and builds the nest, while the male waits nearby. After incubation begins the pair bond is dissolved, and males often fly elsewhere to complete their summer molt. Females take their broods to water after hatching and usually raise them in rather heavy brooding cover. The fledging period is about 6 weeks, and females also begin to molt at about the time the young are fledged.

Suggested Reading: Dane 1966; Johnsgard 1975.

Cinnamon Teal
Anas cyanoptera

Breeding Status: Apparently a rare breeding species at the eastern limit of its range in our region, but confusion with the blue-winged teal makes the status of this species difficult to ascertain. There are no definite breeding records for North Dakota, although presumably nesting pairs have been seen. Probable broods have been seen at Lacreek N.W.R. in South Dakota. In Nebraska, cinnamon teal are occasionally present during summer at Crescent Lake N.W.R., and there is at least one breeding record for Garden County. They bred at Cheyenne Bottoms, Barton County, Kansas, in 1969 and are occasionally found at Salt Plains N.W.R., Oklahoma, where a brood probably of this species was seen in 1959. They have been recorded breeding in Dallam, Hutchinson, and Potter Counties, Texas, and in Union County, New Mexico.

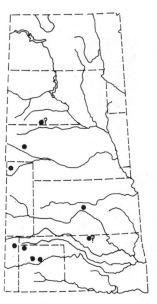

Breeding Habitat: There is no obvious difference in habitat preferences between blue-winged and cinnamon teal, although the latter inhabits more alkaline marshes during the breeding season.

Nest Location: This species usually nests in fairly low herbaceous cover 12-15 inches tall, often consisting of grasses, sedges, or broadleaf weeds. Such cover is favored for nesting if it provides excellent concealment and is close to stands of taller vegetation. Islands that provide low grasses are also preferred. The nest is lined with down nearly identical to that of blue-winged teal.

Clutch Size and Incubation Period: From 8 to 13 eggs (76 Utah clutches averaged 9.3). Eggs are white to pale pinkish buff with a slight gloss. The incubation period is 24–25 days, starting with the last egg. Single-brooded, but a persistent renester.

Time of Breeding: In Utah, clutches are initiated between April 15 and June 24, with a peak in early May and about three-fourths initiated between May 6 and June 2.

Breeding Biology: The social behavior and breeding biology of the cinnamon teal are extremely similar to those of the blue-winged teal, and in a few areas they breed on the same marshes, nesting at the same time and using the same habitats. Nesting densities of cinnamon teal in the middle of their range are appreciably higher than those of blue-winged teal, however, and their home ranges tend to be very small.

Suggested Reading: Spencer 1953; Johnsgard 1975.

American Wigeon (Baldpate)
Anas americana

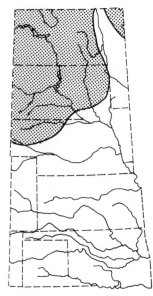

Breeding Status: Locally common in northwestern Minnesota, and uncommon to rare in most of North Dakota and South Dakota. A common to uncommon breeder in the Nebraska Sandhills, but there are no specific nesting records for the more southerly parts of this region. It is thought to nest in Barton County, Kansas (Cheyenne Bottoms), but only one record seems to have been published (*Kansas Ornithological Society Bulletin* 24:36).

Breeding Habitat: American wigeons favor marshes or lakes with abundant aquatic food at or near the surface, but with limited emergent aquatic vegetation. Areas surrounded by sedge meadows are favored, as are those with partly wooded or brushy habitats near the water.

Nest Location: Nests are on dry land, often 100 yards or more from water. The surrounding cover is often of sedges, rushes, mixed prairie grasses, or weeds, but the nests are also sometimes placed near the base of a tree. The actual nest is simply a slight depression in the soil well lined with light grayish down.

Clutch Size and Incubation Period: From 7 to 12 eggs, averaging 8 or 9 in most areas, and 8.2 for 14 nests in North Dakota. The eggs are creamy white with a smooth surface. Incubation period is 23–24 days. Single-brooded, but renesting apparently is frequent.

Time of Breeding: North Dakota egg dates range from May 31 to July 13, and dates of dependent young are from June 26 to September 21.

Breeding Biology: Wigeons form seasonally monogamous pair bonds after a period of social courtship in winter and spring. Males perform fairly simple displays, mainly involving calling, chin-lifting, and raising the folded wings high above the back. After pair-formation, pairs establish a home range on marshes ranging from less than an acre to more than 20 acres in area. There is no territorial defense, although males evict other males from the vicinity of their mates. Nest sites are well hidden, and shortly after incubation begins males abandon their mates. The female thus incubates and rears the brood alone. Broods are reared on relatively open marshes, and fledging occurs at about 70 days of age.

Suggested Reading: Sowls 1978; Johnsgard 1975.

Northern Shoveler
Anas clypeata

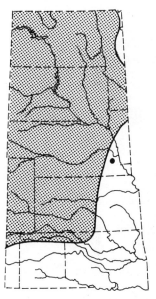

Breeding Status: Breeds commonly in the Dakotas and western Minnesota, is common to uncommon in Nebraska, northwest Iowa, and eastern Colorado, and breeds locally and occasionally in western Kansas, northwestern Oklahoma, and extreme northeastern New Mexico.

Breeding Habitat: Shallow prairie marshes with an abundance of plant and small animal life floating on the surface provide ideal shoveler habitat. Submerged plants whose leaves reach or nearly reach the surface, such as pondweeds, also provide food by supporting an abundant aquatic invertebrate life. Nonwooded shorelines are preferred to wooded ones, and muddy bottoms seem to be preferred.

Nest Location: Shovelers usually build their nests well away from water, in grassy cover that is less than a foot tall and almost never more than 2 feet tall. Broad-leaved weeds and shrubby cover are used secondarily for nesting. The nest is a shallow depression lined with some vegetation and with brownish down having lighter centers.

Clutch Size and Incubation Period: From 8 to 13 eggs (54 North Dakota clutches averaged 10.2). Eggs are buffy, usually with a greenish tint. The incubation period is 22–25 days. Single-brooded, but renesting efforts are frequent.

Time of Nesting: North Dakota egg dates range from May 6 to July 20, and records of dependent young range from June 8 to September 17. There is a record of an Oklahoma nest for June 12, and young have been seen from May 23 to early July. In Texas, dependent young have been reported from early March to June 30.

Breeding Biology: Shovelers begin pair-formation on their wintering grounds and continue it through their arrival on the breeding grounds. Most of the displays are aquatic, but there are also "jump-flights" and aerial chases associated with courtship. The birds are seasonally monogamous (contrary to the early literature) and at least in captivity some birds remate with previous mates while others choose new ones. The pairs spread out over the breeding habitat and have been described as territorial by some workers, while others have simply reported that they occupy overlapping home ranges from 15 to 90 acres in area. The females do all the incubation, and the males abandon them during the incubation period. The fledging period is about 6–7 weeks.

Suggested Reading: Poston 1969; Johnsgard 1975.

Wood Duck
Aix sponsa

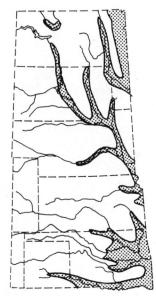

Breeding Status: Largely restricted to wooded rivers east of the 100th meridian, but occurring rarely and only locally west to the Souris and Missouri rivers of North Dakota, central South Dakota, the eastern thirds of Nebraska and Kansas, and the eastern half of Oklahoma, with breeding extending locally along the Red River into the Texas panhandle (to Randall County).

Breeding Habitat: The wood duck breeds in floodplain forests along rivers, creeks, and oxbows and around wooded lakes but is generally associated with slow-moving rivers, sloughs, or ponds where large trees are found. Forests providing acorns or other mast are desirable, and water areas with an abundance of flooded shrubs or trees and depths no greater than 18 inches are especially favored.

Nest Location: Nests are in natural or artificial cavities. Favored trees are at least 16 inches in diameter, having openings at least 3½ inches wide and interior cavities at least 8 inches in diameter. The cavity must be well drained and the entrance well protected from the weather. High cavities with small entrances are preferred, as are locations over water. Likewise, trees growing in clusters or groves are favored over isolated trees, and open stands of trees are preferred over dense stands.

Clutch Size and Incubation Period: From 12 to 16 eggs, averaging about 14 in clutches produced by a single female, often more in "dump nests" produced by several females. Eggs are creamy white with a smooth surface. Incubation period is 25–37 days, averaging 30 days. Normally single-brooded.

Time of Breeding: Egg dates in North Dakota are from May 7 to June 14, and dates of dependent young are from June 3 to September 8. Kansas egg dates are from March 21 to May 10,

58

with mid-April a probable peak of egg-laying. Texas egg dates are from March 1 to July 15, and dependent young have been seen from March 14 to August 25.

Breeding Biology: Pair bonds are established each year, after a prolonged period of courtship displays. No definite territorial behavior exists, but males assist females in seeking out suitable nest sites, which may take days. Competition for nest sites is frequent, and thus collective "dump nests" are locally prevalent. The female does the incubation, and males normally desert their mates before hatching. The female raises the brood, which fledges at about 60 days of age. Renesting after loss of the first clutch is fairly frequent, and a second brood may be raised on rare occasions.

Suggested Reading: Grice and Rogers 1965; Johnsgard 1975.

Redhead
Aythya americana

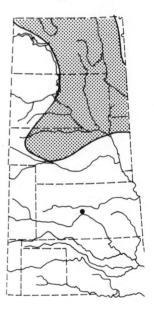

Breeding Status: Breeds in western Minnesota and the Dakotas, mainly east and north of the Missouri River, as well as in northwestern Iowa and the Sandhills of Nebraska. There are only a few nesting records from Kansas, for Barton County in 1928 and 1961–64.

Breeding Habitat: Redheads are similar to canvasbacks in their habitat needs but occur on more alkaline marshes. They usually nest in marshes at least an acre in area, with about 10–25 percent of the surface open water and emergent vegetation 20–40 inches tall.

Nest Location: Nests are built in emergent vegetation, in water about a foot deep and in vegetation 20–40 inches tall. They are placed within 50 yards of open water, often less than 5 yards away. Cattails and hardstem bulrushes are favored nesting sites. The nest bowl is lined with down that varies from white to medium gray.

Clutch Size and Incubation Period: From 8 to 15 eggs (74 North Dakota nests averaged 10.2), but parasitic and "dump nesting" often makes clutches abnormally large. The eggs are usually creamy white, rarely greenish to buffy. Incubation period is 24–28 days, starting with the last egg.

Time of Breeding: North Dakota egg dates range from May 5 to August 10, and dates of dependent young range from June 14 to October 17. Most egg records are for July, and those of dependent young are for August. In northwestern Iowa, nest dates range from May 1 to mid-July, with most before June 1. Brood records are from June 1 to early September, most broods hatching between June 8 and June 30.

Breeding Biology: Redheads have seasonal pair bonds, established each winter and spring. Their displays and associated behavior are much like those of canvasbacks, and the two species often associate. On reaching their nesting grounds, pairs establish home ranges that typically include nest-site potholes and waiting-site potholes, often shared with other pairs. Nest parasitism by redheads is high in most areas, and they drop eggs in the nests of a large variety of other marsh birds, although not all females are parasitic nesters. Males abandon their mates early in incubation and often fly elsewhere to molt. In Iowa the young fledge at 70–84 days of age, and shorter fledging periods have been reported for Canada. In Iowa there is also a moderate amount of renesting, but little or none occurs in Canada.

Suggested Reading: Low 1945; Johnsgard 1975.

Ring-necked Duck
Aythya collaris

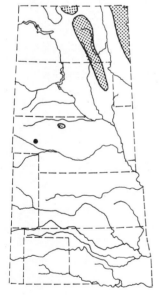

Breeding Status: Breeding is confined mainly to northwestern and west-central Minnesota and the Turtle Mountains of north-central North Dakota, with local breeding on the Drift Plain of east-central North Dakota and scattered breeding records for northeastern South Dakota and the Sandhills of Nebraska.

Breeding Habitat: The primary habitat of ring-necked ducks is sedge-meadow marshes and bogs, with freshwater marshes used secondarily. Acidic or freshwater areas are preferred to brackish ones, and the birds especially frequent ponds that support water lilies and are surrounded by shrubby cover.

Nest Location: Nests are very frequently on floating islands in bogs or on hummocks of vegetation in open marshes. Rarely, nests are on dry land well away from water, but most are close to water of swimming depth and within 15 yards of water open enough for landing and taking off. Both sedges and brushy cover are used for specific nest cover, and small clumps are apparently used more often than larger ones. Nests are lined with sooty brown down having white centers.

Clutch Size and Incubation Period: From 6 to 14 eggs, with an average of about 10. The eggs are olive buff and smooth. The incubation period ranges from 25 to 29 days, averaging about 27 days after the laying of the last egg. Single-brooded, but a persistent renester.

Time of Breeding: North Dakota egg dates range from May 30 to July 20, and dates of dependent young are from June 29 to September 7. Most of the brood records are for August.

Breeding Biology: Pair bonds in ring-necked ducks start to become established on the wintering grounds through social display that begins in January and February, but some displays persist until the birds arrive on their nesting grounds. Display patterns are much like those of redheads and canvasbacks, in spite of plumage differences. Pairs become spaced out over the breeding grounds but show little aggression when they come into contact, and nests are often close together on islands. The pair bond is usually broken near the end of the incubation period, and females raise their broods alone, on ponds often largely covered by water lilies. By the end of the fledging period of 7–8 weeks the female has begun her flightless period and family bonds terminate.

Suggested Reading: Mendall 1958; Johnsgard 1975.

Canvasback
Aythya valisineria

Breeding Status: Breeding is mostly confined to the glaciated areas of the Dakotas and western Minnesota, north and east of the Missouri River. There is also local breeding in the Nebraska Sandhills, but only one breeding record farther south, from Barton County, Kansas, in 1962.

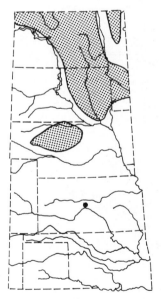

Breeding Habitat: Canvasbacks are most abundant on shallow prairie marshes that are surrounded by cattails, bulrushes, and similar vegetation, with open water for landing and taking off and little or no wooded area around the shoreline.

Nest Location: Nests are constructed over water, among emergent vegetation that is 1–4 feet tall and composed of bulrushes (usually hardstem bulrush) or cattails, with phragmites used less often. Nests are usually 10–15 yards from areas of open water that are at least 50 feet square. The nest bowl is usually well lined with pearly gray down having inconspicuous white tips.

Clutch Size and Incubation Period: From 6 to 16 eggs (26 North Dakota nests averaged 9.9), with larger clutches of "dump nests" not infrequent. The eggs are grayish olive to greenish. The incubation period is about 24 days. Single-brooded, but renesting occurs frequently.

Time of Breeding: North Dakota egg dates range from April 28 to July 15, and dates of dependent young are from May 29 to

September 5. More than 90 percent of the egg dates are for May and June, and dates of dependent young are for July and August.

Breeding Biology: Canvasbacks renew their pair bonds annually, and courtship is usually intense as the birds are returning to their nesting grounds. Several aquatic displays, including cooing calls and head-throw displays, are conspicuous then. As pairs form, they separate from the flocks and seek out nesting areas in smaller and shallower ponds than those used for courting. In densely populated areas a substantial amount of nest parasitism occurs among canvasbacks and between canvasbacks and redheads. Although parasitic redheads are prone to lay their eggs in canvasback nests, the latter usually lay eggs only in the nests of other canvasbacks. Thus mixed-species broods sometimes occur, but parasitized nests are less successful than nonparasitized ones. The fledging period is 8-9 weeks.

Suggested Reading: Hochbaum 1944; Johnsgard 1975.

Lesser Scaup
Aythya affinis

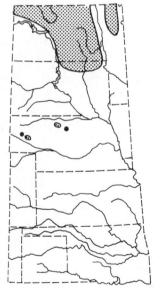

Breeding Status: Breeds locally in northwestern Minnesota and in the Dakotas, mainly east and north of the Missouri River. Occasionally breeds in the Nebraska Sandhills (Garden, Morrill, Cherry, and Brown counties) and probably breeds sparingly in eastern Kansas, though no definite records exist.

Breeding Habitat: Prairie marshes, "potholes," and ponds or lakes in partially wooded parklands are the major habitat of lesser scaups. Favored ponds are usually slightly to moderately brackish and vary from semipermanent to permanent. Those supporting high populations of amphipods ("scuds") and aquatic insects are most heavily used.

Nest Location: Nests are on dry land but usually fairly close to shore. Islands on lakes, especially those with grassy or weedy cover, are especially preferred. Nests are usually within 50 yards of water, and the cover is 1-2 feet tall. In some boggy areas the nests are placed on floating sedge mats. The nest cup is lined with very dark brown down with inconspicuous white centers.

Clutch Size and Incubation Period: From 7 to 12 eggs (25 North Dakota nests averaged 9.4). Eggs are greenish to olive buff with a smooth but not glossy surface. The incubation period is 21-26 days, with an average of 24.8 days after the last egg is laid. Single-brooded, but a regular renester (up to three renesting attempts have been reported).

Time of Breeding: North Dakota egg dates range from May 21 to August 10, with more than 80 percent of the records in June. Dependent young have been recorded from July 16 to September 6, predominantly in August.

Breeding Biology: Lesser scaup form pair bonds that persist for a single season, during a prolonged period of winter and spring social display. Pairs establish relatively large but poorly defined home ranges, often centering on marshes 2-5 acres in area that include some deep water. Females build their nests alone and are abandoned by their mates shortly after they begin incubation. Although scaup often nest in or near gull colonies, the gulls sometimes prey severely on ducklings, and much brood disruption is typical. Females often desert their ducklings early to molt, and large broods consisting of ducklings from several families are frequent in some areas. The fledging period is about 47-50 days, relatively short for diving ducks.

Suggested Reading: Trauger 1971; Hines 1977.

Common Goldeneye
Bucephala clangula

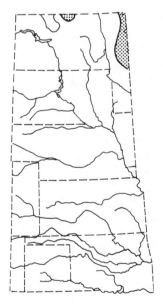

Breeding Status: Limited as a breeder to forested portions of western Minnesota and the Turtle Mountains of north-central North Dakota. Nonbreeders may often be seen during the summer elsewhere in the region.

Breeding Habitat: Goldeneyes are generally associated with deep marshes and lakes that have adjacent stands of trees, especially hardwoods, that provide nest cavities. Although the coniferous forest region of North America is the center of their distribution on this continent, they probably nest mainly in aspens or poplars.

Nest Location: Nests in North Dakota are usually in elms, but oaks and sometimes box elders are also used if they are hollow and have cavity entrances 2-35 feet above the ground. In New Brunswick, the birds nest in a variety of hardwood trees, most often those in open stands at the edges of marshes or fields, and prefer those with lateral cavity openings rather than vertical entrances. An internal cavity diameter of 6-10 inches is preferred, but the entrance diameter is apparently not critical. The down lining is very pale gray with white centers.

Clutch Size and Incubation Period: From 7 to 16 eggs (13 North Dakota nests averaged 11.2). The eggs are olive green with a smooth surface. Incubation period is about 30 days, varying from

27 to 32. Single-brooded, with renesting incidence unknown but probably low.

Time of Breeding: North Dakota egg records range from May 2 to late June, and dependent young have been seen from June 18 to July 29.

Breeding Biology: Goldeneyes become mature when 2 years old and form pair bonds in their second winter. Their social displays are conspicuous and elaborate and continue until the birds reach their nesting areas. Thereafter pairs spread out and seek nesting sites, females usually returning to the place where they nested previously. No definite nest territory is established, although males defend their mates until the pair bond is broken early in incubation. Competition for nest sites sometimes results in dump nesting and nest desertion, and raccoons are serious nest predators. Nests with small entrances (3.5 by 4.5 inches) may be more successful than those with larger ones where predators are serious problems. After hatching, the young remain in the nest for about a day, then leave as a group by jumping to the ground. The broods feed in relatively open water and often form mixed broods or even motherless aggregations. They fledge in about 55–60 days and are abandoned by their mothers a week or two before fledging.

Suggested Reading: Carter 1958; Johnsgard 1975.

Bufflehead
Bucephala albeola

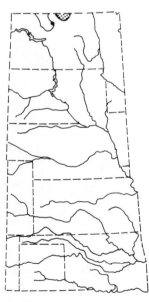

Breeding Status: Restricted mainly to the Turtle Mountains of North Dakota, where it is uncommon, and also breeds rarely at the J. Clark Salyer N.W.R., McHenry County. Nonbreeding birds often summer elsewhere in the region.

Breeding Habitat: Buffleheads are associated with ponds or lakes in or near open woodland, where nesting sites are available and where there is an abundance of aquatic invertebrates. Moderately deep and fertile lakes, with open shorelines and sparse reedbeds that have nest sites nearby, are especially favored.

Nest Location: Nests are almost always in cavities made by woodpeckers, primarily flickers. Cavities that are 3–20 feet above the ground or water level are used, especially those in dead trees that are either standing in water or are very close to it. The birds will also nest in artificial cavities, with entrances about 3 inches wide and internal diameters of about 7 inches. The nest is lined with a down that is pale grayish to brownish with indistinct lighter centers.

Clutch Size and Incubation Period: From 5 to 16 eggs (averaging about 9 in initial clutches and 7 in late or renesting efforts). Eggs are creamy to olive buff with a slightly glossy surface. The incubation period ranges from 29 to 31 days and is usually 30 days. Single-brooded; renesting is apparently uncommon and possibly does not occur.

Time of Breeding: North Dakota egg dates are from June 6 to June 13, and dependent young have been seen from June 15 to August 15.

Breeding Biology: Buffleheads form seasonal pair bonds during a prolonged period of courtship that extends from winter through the spring migration. It is not known how often males remate with mates of the previous year, but females have a strong tendency to return to the place where they nested previously, and they often nest in the same cavity. Competition for nest sites among buffleheads and with other hole-nesting birds such as starlings and tree swallows makes nest-site availability an important facet of their biology. Males abandon their mates and often leave the area shortly after incubation gets under way, and the females leave their nests occasionally during incubation to feed. The young remain in the nest 24–36 hours after hatching, then at their mother's signal they jump down to the ground and leave as a group, usually during the morning. The fledging period is about 50–55 days, during most of which the female keeps the young within a brood territory. However, some brood transfers and formations of multiple broods have been reported.

Suggested Reading: Erskine 1972.

White-winged Scoter
Melanitta fusca

Breeding Status: Virtually extirpated. At present an accidental or very rare breeder in northern North Dakota, with the most recent records for Des Lacs N.W.R., in 1952 and 1953. At one time breeding was common at Devils Lake and Stump Lake, Nelson County.

Breeding Habitat: In North Dakota, large and brackish lakes surrounded by dense brushland or woodlands were the favored breeding habitats. Lakes having boulder-covered islets that support herbaceous vegetation and shrubs or trees are used in more northerly parts of the species' range and are probably the primary habitat type.

Nest Location: In North Dakota, most nests were on islands and under shrubby cover, sometimes a considerable distance from

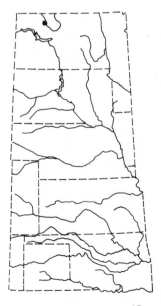

65

water. Studies in Finland indicate that junipers and other bushes are the usual cover and provide excellent concealment. Somewhat exposed nests were common only near gull colonies, and often the nest is protected by rock overhangs or by heavy stems and branches of junipers. The nest cup usually contains some plant materials as well as a lining of deep grayish to olive brown down with small white centers.

Clutch Size and Incubation Period: From 6 to 14 eggs (21 nests in North Dakota averaged 10.0). Eggs are creamy buff with a smooth but not glossy surface. The incubation period is 26–29 days, with an average of 27.5. Single-brooded, with no evidence of renesting.

Time of Breeding: North Dakota egg dates range from June 15 to July 25, and sightings of dependent young are from July 28 to September 3.

Breeding Biology: White-winged scoters reestablish their pair bonds annually after becoming sexually mature in their second year. They arrive on their northern breeding grounds rather late, and there is about a month's delay between arrival and nesting in Alberta and Manitoba. Males remain with their mates through the egg-laying period but desert them soon afterward. Rarely, males have been seen defending broods from the attacks of gulls. At least in some areas, there are extensive brood mergers shortly after hatching, and groups of 100 or more ducklings sometimes develop. The fledging period is not definite but is probably between 63 and 77 days.

Suggested Reading: Rawls 1949; Johnsgard 1975.

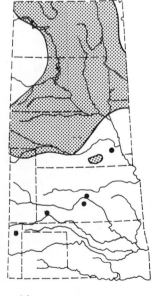

Ruddy Duck
Oxyura jamaicensis

Breeding Status: Breeding occurs locally over most of western Minnesota and northwestern Iowa, in the Dakotas in the glaciated areas to the east and north of the Missouri River, and in the deeper marshes of the Nebraska Sandhills and rainwater basin in Clay County. Breeding in Kansas is very local, with definite nesting records from Barton, Stafford, and Grant counties, and possible breeding birds have been seen in Finney and Kearny counties. Reported as nesting in Union County, New Mexico, and may also nest occasionally in Harper and Beaver counties, Oklahoma.

Breeding Habitat: Permanent or semipermanent prairie marshes having stable water levels and an abundance of emergent vegetation, especially cattails and hardstem bulrushes, are prime ruddy duck habitat. The marshes must have some open water close to the nesting cover provided by the emergent vegetation or be connected with open water by muskrat channels.

Nest Location: Hardstem bulrush stands are optimum nesting cover for ruddy ducks, which build platforms or floating nests in water up to 3 feet deep. Dense cover is preferred to sparse cover, and bulrushes are preferred to cattails or other emergent plants. Bulrushes that can be readily bent over to form the nest are preferred; species with stiff, tough stalks are little used. Many nests have canopies and ramps for easy access. Usually there is a little down present as a lining, which is very light colored with white centers.

Clutch Size and Incubation Period: From 3 to 15 eggs (68 North Dakota nests averaged 7.7), but dump nests or parasitically laid eggs often cause inflated clutch sizes. The eggs are white with a chalky and granular surface. The incubation period is 23 days. Single-brooded (perhaps double-brooded at the southern end of its range), with limited evidence of renesting.

Time of Breeding: North Dakota egg dates range from May 29 to August 21, and dependent young have been seen from July 3 to October 14. Most egg dates are for June and July, and most broods dates are for August. In northwestern Iowa egg initiation dates are from May 8 to July 11, with the peak of laying during the first half of June.

Breeding Biology: Ruddy ducks apparently mature in their first year, though not all females are thought to breed as yearlings. Pair bonds are rather weak, and much display is related to territorial advertisement rather than courtship itself. Females show little or no pair-forming or pair-maintaining behavior, although males may remain in the vicinity of their mates after they have begun nesting. Some even persist in remaining with them after the brood has hatched, though they do not "assist" in rearing the brood. The young are highly precocious and independent, so that broods often become scattered before they fledge, about 6–7 weeks after hatching.

Suggested Reading: Low 1941; Johnsgard 1975.

Hooded Merganser
Mergus cucullatus

Breeding Status: Limited as a breeder to the forested portions of western Minnesota and the Souris and Des Lacs rivers of north-central North Dakota, with a few scattered records from elsewhere in the state. There are also records of young birds seen

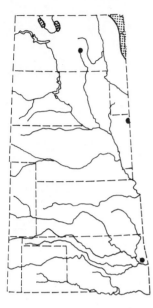

(1954, 1958) at Trumbull Lake, Clay County, Iowa, and a single breeding record for Sequoyah County, Oklahoma (*Bulletin of the Oklahoma Ornithological Society* 10:22).

Breeding Habitat: In North Dakota, hooded mergansers are associated with rivers, creeks, and oxbows bordered by woods and supporting large populations of small fish. In general, they prefer clear streams that have woods nearby to provide nesting cavities, especially streams that have sandy or cobble bottoms rather than mud bottoms and are not very deep or murky enough to interfere with underwater vision.

Nest Location: Cavities or nesting boxes close to water are used more often than those well away from it. The cavity entrance may be near the ground, or up to 60 feet high. Hooded mergansers readily accept nesting boxes that have been set out for wood ducks, so their general requirements for cavity size are probably very similar. Both species often use the same boxes where they occur together, resulting in mixed clutches that may be incubated by either species. The down lining of hooded merganser nests is pale gray with white centers (as in wood ducks), but the associated breast feathers are narrow and off-white rather than wide and white.

Clutch Size and Incubation Period: From 7 to 13 eggs (5 North Dakota clutches averaged 8.7). Eggs are white with a glossy surface and are slightly larger (over 40 mm wide) and more rounded than wood-duck eggs. The incubation period is about 33 days. Single-brooded, with no evidence of renesting tendencies.

Time of Breeding: North Dakota egg dates are for June and early July, with young (or hatched eggs) reported from June 18 to August 7. In southeastern Missouri, records indicate nest initiations from February 9 to April 20, with most initiated between February 24 and March 20.

Breeding Biology: Hooded mergansers first form pairs in their second winter of life and thereafter establish pair bonds annually during a prolonged courtship period. On return to their nesting areas, the females usually find nest sites near their former nest areas and many nest in the very same locations. These locations are often within a few miles of where the females hatched. Pair bonds are probably disrupted early in the incubation period, and incubating females leave their nests two or three times a day to forage. The newly hatched young are taken to rearing areas, usually rivers but sometimes standing-water habitats such as beaver ponds. They fledge in about 70 days.

Suggested Reading: Morse, Jakabosky, and McCrow 1969; Johnsgard 1975.

Common Merganser
Mergus merganser

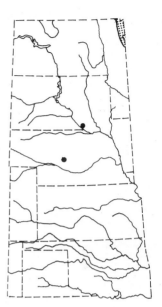

Breeding Status: Nearly extirpated. Previously bred in the vicinity of Devils Lake, and probably also the Turtle Mountains, North Dakota. Currently the species breeds in north-central and north-eastern Minnesota, and there is a breeding record for Roseau County. Summer birds in Clearwater and Otter Tail counties suggest possible recent breeding there. There was a sight record of a brood in Custer County, Nebraska, in 1968 (*Nebraska Bird Review* 37:45), and another 1968 sighting of a brood near Pickstown, Charles Mix County, South Dakota.

Breeding Habitat: Common mergansers are mostly associated with forest-lined lakes and ponds near rivers that support high fish populations and are clear enough to allow for visual foraging.

Nest Location: In our region nesting is mostly confined to cavities in hardwood trees, but in areas without trees large enough to provide nest cavities the birds will nest under boulders, in buildings, or under dense brush. Concealment and darkness in the nest cavity are prime requirements. The birds will also nest in artificial cavities such as nest boxes with entrances 4 ¾ inches in diameter, an internal cavity about 10 inches wide, and a depth of about 20 inches below the entrance. Nests are usually lined with some down, which is nearly white, unlike the dark down of red-breasted mergansers.

Clutch Size and Incubation Period: From 7 to 16 eggs, averaging 9 or 10. The eggs are pale cream to buffy and lack a glossy surface. Incubation period is 32–35 days. Single-brooded.

Time of Breeding: Eggs in North Dakota nests are reported from May 26 to June 25. In Minnesota, dependent young have been seen from June 3 to August 11.

Breeding Biology: During fall and winter, these mergansers usually stay in small flocks that sometimes feed cooperatively, but as spring approaches much time is spent in social display and in establishing pair bonds, and flock sizes decrease. Females remain fairly gregarious while looking for nest sites and often nest close together. Probably some dump nesting occurs in locations where nest cavities are limited. The males usually leave their mates before hatching but on rare occasions have been seen with broods. The young are led to water a day or two after hatching, and the brood is usually raised in shallow rivers. At times the female carries part of her brood on her back, especially when they are frightened. The fledging period is 60–70 days.

Suggested Reading: White 1957; Johnsgard 1975.

FAMILY CATHARTIDAE (AMERICAN VULTURES)

Turkey Vulture

Turkey Vulture
Cathartes aura

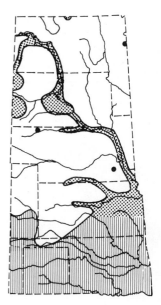

Breeding Status: Breeds from eastern New Mexico, the Texas panhandle, and most of Oklahoma, southern Colorado, and northward across eastern Kansas, becoming progressively less common and generally restricted to the deeper river valleys and wooded canyons such as the Pine Ridge area of Nebraska, the Black Hills of South Dakota, and the North Dakota Badlands. Breeds rarely in north-central Minnesota (*Loon* 49:87).

Breeding Habitat: At the northern part of its range, this species is associated with brushy woodlands adjoining more open grasslands or croplands, but farther south the birds have a more general distribution. Areas providing thermal updrafts or declivity winds along cliffs or mountains are favored for foraging, and the cliffs also provide nest cavities.

Nest Location: In the northern areas, cliff overhangs, rocky cavities, and even badger holes have reportedly been used for nesting, while farther south the birds often nest in bank hollows, caves, and tree cavities or among rocks. The nest sites are always dark and well concealed, and sometimes the birds nest in abandoned buildings. No actual nest is constructed; the eggs are simply laid on the substrate.

Clutch Size and Incubation Period: Usually 2 eggs, rarely 1 or 3. The eggs are dull to creamy white with various-sized spots and blotches of pale and brighter brown. The incubation period lasts 38–41 days. Single-brooded.

Time of Breeding: Kansas egg records are for the period April 21 to June 10, and the peak of egg-laying is probably about May 1. Oklahoma egg dates are from April 28 to June 20, and young have been seen there from May 9 to July 4.

Breeding Biology: Turkey vultures are monogamous, but little is known of their pairing behavior or their age of sexual maturity. However, the nests are well scattered even where nest sites are restricted, and the pair often uses a cave or other possible nest site as a roost for some time before laying their eggs there. Both sexes participate in incubation, and the incubating bird usually takes morning and afternoon breaks to preen and sit in the sunshine. Injury-feigning at the nest has been reported, and young birds will often disgorge their food or bite when approached. The young are relatively precocial and soon move to the mouth of the nesting cavity to sun themselves. The fledging period is surprisingly long, from 70 to 80 days.

Suggested Reading: Coles 1944; Brown and Amadon 1968.

Black Vulture
Coragyps atratus

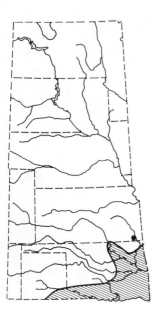

Breeding Status: Now confined as a breeding species to eastern Oklahoma and adjacent Texas. Formerly bred in southeastern Kansas, but there is only a single old nesting record, from Labette County.

Breeding Habitat: Black vultures are mostly confined to tropical and subtropical areas, in both wooded and semiopen country, and in particular concentrate in areas where garbage or offal is plentiful.

Nest Location: Nests are in sites similar to those used by turkey vultures, such as caves, dense thickets, and hollow trees. They rarely nest more than 15 feet above the ground and sometimes use low cliffs that have eroded recesses. Sometimes nests are close together, especially where nest sites are limited. The eggs are simply deposited on the ground; no actual nest is built, but loose materials are accumulated in the general vicinity of the eggs.

Clutch Size and Incubation Period: Usually 2 eggs, occasionally 1 or 3. The eggs are pale grayish green to whitish, usually with spots or blotches of brown on the larger end. The incubation period is 32–39 days starting with the last egg. Single-brooded.

Time of Breeding: Oklahoma egg records are from April 15 to June, and young have been reported from June (unfledged) to July 6 (fully fledged). In Texas, eggs have been reported from January 28 to July 29, and dependent young from April 4 to August 21.

Breeding Biology: Black vultures are monogamous, presumably permanently. Courtship chases and diving through the air have been reported, and males have also been seen strutting with outstretched wings near females. The eggs are laid at intervals of a day or two. They are incubated equally by both sexes, by being placed between the toes and thus aligned side by side and parallel with the body's axis. After hatching the young are brooded almost continuously for several weeks and are fed by letting them take regurgitated food from the parents' open mouths. By 80 days they can fly short distances, and they leave the nest after about 90 days.

Suggested Reading: Stewart 1974; Brown and Amadon 1968.

FAMILY ACCIPITRIDAE (HAWKS, EAGLES, AND HARRIERS)

Golden Eagle

White-tailed Kite
Elanus leucurus

Breeding Status: Extirpated. Probably once bred in eastern and central Oklahoma, but apparently never common. One breeding record, from near the present location of Davis, Murray County. The nearest current breeding area is coastal Texas.

Breeding Habitat: Open and fairly dry country, such as savanna or agricultural lands with scattered trees and a permanent source of water provides the preferred habitat for this subtropical to tropical species.

Nest Location: Nests are built in trees, often near marshes. The trees may be almost any species but must be moderate in height and near a food source, often mice. The nest is a platform of twigs, about 20 inches in diameter and lined with grasses.

Clutch Size and Incubation Period: From 3 to 6 eggs, typically 4. Eggs vary from pure white to heavily marked with reddish purple blotches. The incubation period is 30–32 days, starting with the first egg. Frequently double-brooded.

Time of Breeding: In Texas, egg records are from March 18 to August 21. The single Oklahoma record of eggs is for May 9.

Breeding Biology: These kites are very social, even during the breeding season. At the onset of the breeding season the male establishes a perching place where he regularly sits and soon attracts a female, but pairs evidently do not establish exclusive territories. Males soon attempt to copulate and select a nesting site and at times may even begin building a nest. However, females invariably choose their own sites and do all or nearly all of the nest construction, which requires 7-28 days. During this phase the male performs distinctive "flutter-flights" while uttering a chittering sound. The nest itself is primarily defended by the female, and unlike most raptors, the female does all the incubation. During this time the male does the hunting and transfers food to her while in the air near the nest. Females also do all the feeding of the young, which fledge in 35–40 days. The young often return to the nest to roost at night, but eventually they are driven from the nest area by the adults, which often begin to renest even before the first brood has fledged. This is done by the male initiating copulations and nest-construction attempts.

Suggested Reading: Dixon, Dixon, and Dixon 1957; Brown and Amadon 1968.

Swallow-tailed Kite
Elanoides forficatus

Breeding Status: Extirpated. Formerly bred in eastern Nebraska, Kansas, Oklahoma, and the adjacent Red River Valley of Texas. Also once bred in Iowa and Minnesota, apparently north to Itasca Park, Clearwater County. Still found in Texas, but extremely rare.

Breeding Habitat: The species is now largely confined to lowland cypress and mangrove swamps, but it also occurs in pine forests, near freshwater marshes, and along river-bottom forests with adjacent grasslands. The presence of very tall trees for nesting and adjacent relatively wet open land for hunting is probably crucial.

Nest Location: The nests are 60–130 feet above the ground in the tops of tall trees, often pines. Trees at the edges of clearings are typically used, and the nests are 15–20 inches in diameter, constructed mostly of twigs and lined with Spanish moss and lichens.

Clutch Size and Incubation Period: Two eggs, occasionally 3, and rarely 4. The eggs are whitish with large blotches of bright brown. Incubation period is 28 days, probably starting with the first egg. Single-brooded.

Time of Breeding: Texas records of eggs extend from March 10 to June 7. In Kansas the eggs are probably laid in May. Egg dates in Minnesota are from May 15 to June 16.

Breeding Biology: One of the most graceful fliers of all birds, these kites prey on animals they catch in flight, ranging in size from large insects to nestling birds and arboreal snakes. In Florida the birds arrive already paired and frequently soar over their nesting area, at times performing swoop-flight displays. Nest-building begins soon after arrival, with the female doing most of the building, and both birds gather material by breaking off twigs of dead branches while in flight. Copulation is mostly limited to the nest-building period and occurs on horizontal tree branches. The male sometimes feeds the female during this time as well as during incubation. Most of the incubation is done by the female, and when males take over they usually bring in Spanish moss to line the nest. After hatching, the female does most brooding while the male brings in food, including frogs, lizards, and nestling birds. The young fledge in 36–39 days and do not return to the nest.

Suggested Reading: Snyder 1974; Brown and Amadon 1968.

Mississippi Kite
Ictinia mississippiensis

Breeding Status: Nests in south-central and southwestern Kansas southward through the western half of Oklahoma and the Texas panhandle. There also are recent nesting records for Otero County, Colorado (*Colorado Field Ornithologist* 11:5), and the species is now resident along the Arkansas and Cimarron rivers. Formerly ranged to northwestern Kansas and possibly to Iowa.

Breeding Habitat: Studies in Kansas indicate that this species prefers open barren terrain, although in the southeastern states it inhabits forests. In such open and dry country, groves of cottonwoods provide perching and nesting sites, and water is supplied by wells and small reservoirs.

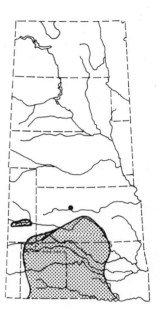

Nest Location: Nests are usually at moderate heights (25–50 feet), in a variety of tree species such as cottonwood, willow, elm, and black locust in Kansas. The nests are rather small, only 10–18 inches across, and are placed in forks or crotches or branches that vary from 1 to 10 inches in diameter. The nests are made of twigs about the size of pencils and are lined with fresh green leaves.

Clutch Size and Incubation Period: Usually 2, sometimes 1 or 3 eggs, which are bluish white. The incubation period is about 30 days. Single-brooded, but renesting occurs if the first clutch fails.

Time of Breeding: Egg records in Kansas are April 20 to June 10, with a probable peak of laying during the first week of May. Egg records in Oklahoma are from May 19 to June 24, and dependent young have been seen from July 24 to September 8 (barely fledged). Texas egg records are from May 3 to July 12, and young in the nest are reported from May 23 to August 27.

Breeding Biology: Mississippi kites are highly social, and the flocks arriving in the Great Plains in spring consist of already mated birds. The birds apparently do not establish territories and often nest close to one another. Sometimes the previous year's nest is used again, but even then nest-building is a protracted and leisurely process. The male gathers most of the material, by snipping twigs off branches while perched or breaking them off with the talons while in flight. He brings these twigs to the female, who incorporates them into the nest. Nesting activities in a colony are usually well synchronized, with incubation starting about the same time. Both sexes incubate, and both gather food for the young when they hatch. The young leave the nest after 34 days but continue to follow their parents and may be fed by them for some time.

Suggested Reading: Fitch 1963; Parker 1975.

Goshawk
Accipiter gentilis

Breeding Status: A rare breeder in the Black Hills of South Dakota and in northern Minnesota, with breeding records for Clearwater and Roseau counties.

Breeding Habitat: Goshawks are limited to wilderness or near-wilderness areas of forest, especially montane forest. They breed in both coniferous and deciduous forests but seem to prefer nesting in broad-leaved trees.

Nest Location: Nests are usually at least 30 feet above the ground and sometimes up to 75 feet and are placed in a crotch or at the base of limb against the tree trunk. Pines, beeches, birches, and poplars seem to be preferred, and nests are relatively large, up to 3 or 4 feet in diameter. They are rather flat platform structures, usually lined with bark or sprigs of coniferous vegetation. There may also be some down and feathers in the nest.

Clutch Size and Incubation Period: From 2 to 5 eggs, usually 3. Eggs are dirty white to pale bluish with a rough surface. They are laid at about 3-day intervals. The incubation period is generally reported as 36–38 days, but about 41 days elapse between the laying of the first egg and the nearly simultaneous hatching of the young, so not all eggs develop at the same rate. Single-brooded.

Time of Breeding: There are apparently only a few actual nest records for the Black Hills, with building observed in late March and young seen in late July. Minnesota egg dates are from April 12 to May 10, and a young bird a few days old was collected in Roseau County on May 23.

Breeding Biology: Goshawks are believed to pair for life, even though they may spend the winter period in somewhat different areas. As the nesting season approaches, the female returns to her old nest site and begins calling to attract her mate. She may also attract the male by performing aerial displays. The male is likewise known to display in flight, by flying in an undulating fashion with alternating dives and swoops, perhaps as a territorial advertisement. The pair occupies a large home range from 6 to 15 miles in diameter, encompassing the nesting tree. Often the birds use an old nest, with the female simply helping the male refurbish it, but if a new nest is built the male constructs it entirely alone, while the female watches from a nearby perch. Copulation occurs through the nest-building and egg-laying period, which may take 2 months. The female does most of the incubation, with the male periodically bringing her food. By the time the young are 35 days old they move out of the nest onto nearby branches, and they fledge when they are about 45 days of age. However, they are not completely independent of their parents until they are about 70 days old.

Suggested Reading: Schnell 1958; Brown and Amadon 1968.

Sharp-shinned Hawk
Accipiter striatus

Breeding Status: Breeding status somewhat uncertain, but uncommon to rare in North Dakota, with most records for the Turtle Mountains and Little Missouri drainage. Rare in western Minnesota, with an early (1921) nesting record for Murray County. Fairly common in the Black Hills of South Dakota, but generally rare elsewhere in the state. The only nesting records for Nebraska are for Sioux County, but the species probably also nests in the Missouri River floodplain forest. It is an uncommon breeder in northwestern Missouri (Squaw Creek N.W.R.) and probably is a rare breeder in northeastern Kansas, but there are only two breeding records (Cloud and Pottawatomie counties). There is a single old nesting record for Oklahoma, in Cimarron County. No nesting records exist for the Texas panhandle or adjacent New Mexico, but it may breed locally in east-central Colorado (Kingery and Graul, 1978).

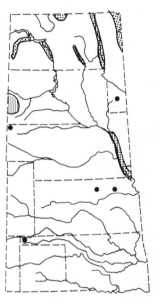

Breeding Habitat: During the breeding season this species is associated with fairly dense forests, particularly mixed woods with some coniferous trees.

Nest Location: Nesting usually occurs in a dense grove of trees, and the nest is 20–60 feet above the ground on a large limb against the tree trunk. It is about 2 feet in diameter, and usually a new nest is constructed each year. At least in Utah, nests are typically in stands of trees with dense foliage, and foliage density appears to be the reason the birds choose conifers over broad-leaved trees. The nest may be unlined or lined with pieces of bark.

Clutch Size and Incubation Period: From 3 to 5 eggs (34 Utah nests averaged 4.3). Eggs are white to bluish white with bright brown blotches. Eggs are laid on alternate days, but incubation does not begin until the clutch is complete, so that hatching occurs 30 days after the last egg is laid. Single-brooded.

Time of Breeding: Little information is available, but the birds in Kansas probably lay in April and May, and in the Black Hills the nesting season probably is from May through July. The only egg date for northern Texas is April 30, and for Oklahoma, July 1.

Breeding Biology: Like other hawks, these are monogamous, and the female is appreciably larger than her mate, foraging on some-

81

what larger prey, primarily birds. In Utah the birds appear at their nest sites as much as a month before egg-laying and probably spend much of that time constructing new nests, since old ones are rarely used even if they are still intact. However, a crow nest or squirrel nest is sometimes modified for use. After the clutch is complete both sexes assist in incubation, and the young hatch almost simultaneously. They grow rapidly, with the males fledging at 24 days of age and the somewhat larger females at 27 days.

Suggested Reading: Platt 1976; Brown and Amadon 1968.

Cooper Hawk
Accipiter cooperii

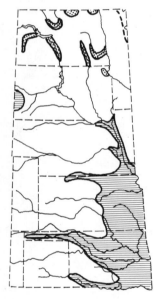

Breeding Status: Occupies a range similar to that of the sharp-shinned hawk, but more common in the southern part of the region, occurring uncommonly over the eastern halves of Kansas and Oklahoma. Rarer and more local toward the west. In the northern part of the region generally uncommon to rare and restricted to wooded river valleys and wooded uplands such as the Black Hills and the Turtle Mountains and Pembina Hills in northern North Dakota. It may also breed in southeastern Colorado (Kingery and Graul, 1978).

Breeding Habitat: The Cooper hawk breeds in mature forests, particularly hardwood forests. It rarely if ever breeds in the same areas as sharp-shinned hawks or goshawks.

Nest Location: Nests are in large trees, 20–60 feet above the ground, usually in the crotch of a deciduous tree or against the trunk of a conifer. The nest is normally about 2 feet across and is constructed of twigs with a lining of bark. Sometimes an old crow or squirrel nest is used for a nest base, but the previous year's nest is rarely used.

Clutch Size and Incubation Period: From 3 to 6 eggs, usually 4 or 5. Eggs are pale blue to dirty white, sometimes with a few pale spots. The eggs are laid on alternate days. Incubation begins with the third egg, and the first three eggs hatch after 34 days, with the rest hatching by the 36th day. Single-brooded, but renests regularly.

Time of Breeding: North Dakota egg dates range from May 5 to May 28, and young have been seen from June 29 to August 22 (recently fledged). In Kansas, egg dates are from March 21 to May 30, with a peak about April 25. Oklahoma egg dates are from April 16 to June 30, and in Texas eggs have been reported from April 1 to May 30.

Breeding Biology: In New York, Cooper hawks arrive in their nesting areas in March, and the male establishes a territory about 100 yards in diameter. From this area he calls and feeds any female that might appear. As a pair is being formed they perform courtship flights, either alone or together. Such flights may be seen for a month or more. During that time the male selects a nest site; rarely he uses an old nest, but more frequently a new location is chosen. The male gathers most of the material and does most of the actual nest building, and he also continues to feed his mate during this period. The female incubates while the male provides food for her and guards the nest briefly while she is eating. At the time of hatching the female carries the eggshells away from the nest and drops them and may even help the young birds out of the shell. For the first 3 weeks after hatching the female rarely leaves the nest, and thus all foraging is done by the male. The young birds fledge at slightly more than a month of age, the females about 4 days later than males, but they remain dependent on their parents for food until they are about 2 months old.

Suggested Reading: Meng 1951; Bent 1937.

Red-tailed Hawk
Buteo jamaicensis

Breeding Status: Pandemic throughout the region. This species is a common resident throughout the eastern half of the region, especially in more wooded areas, but occurs uncommonly and locally elsewhere in western areas where woodlands are scarce.

Breeding Habitat: A combination of extensive open habitat for visual hunting and clumps or groves of tall trees for nesting are the general need of this species. It occupies both coniferous and hardwood forest areas, but especially the latter.

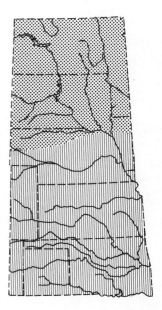

Nest Location: Nests are 15-70 feet high in tall trees. In North Dakota, cottonwoods, elms, and oaks are commonly used, while in Kansas, cottonwoods, honey locusts, Osage oranges, sycamores, and walnuts serve. Nests have also been seen on rock pinnacles in the Black Hills. The platform-like nests are about 30 inches in diameter, constructed of sticks and twigs, and sometimes old nests are used again. They are usually lined with bark and sprigs of green vegetation.

Clutch Size and Incubation Period: From 2 to 4 eggs (28 North Dakota nests averaged 2.8, and 20 Kansas nests averaged 2.6). Eggs are bluish white to dirty white with varying amounts of brown spotting. The eggs are laid at intervals of several days, and incubation begins almost immediately. The incubation period is 28–32 days, usually about 30. Single-brooded.

Time of Breeding: North Dakota egg dates range from April 19 to June 11, and nestling records are from June 23 to July 22. Kansas egg records are from February 21 to April 11, with a peak of egg-laying about March 5. Oklahoma egg records extend from March 16 to April 17, and unfledged young have been reported from May 2 to June 15.

Breeding Biology: Red-tailed hawks pair monogamously and arrive at their nesting areas already mated. Nonetheless, courtship flights are common in early nesting phases, with the birds dramatically soaring and swooping together and occasionally locking talons in flight. Copulation often follows such flights. The nest is built by both birds well before egg-laying, and after it is completed the female stays near it while the male feeds her and brings nest-lining materials. Both sexes help incubate, but the female assumes most of the responsibility and is fed by her mate during this period. The young are hatched at intervals of several days and grow rapidly. By the time they are a month old they may climb out onto adjoining branches, and they can fly at about 45 days. After leaving the nest they are fed progressively less by their parents and become relatively independent in about a month.

Suggested Reading: Austin 1964; Brown and Amadon 1968.

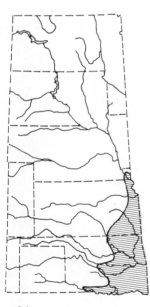

Red-shouldered Hawk
Buteo lineatus

Breeding Status: Breeds from the eastern half of Oklahoma northward to the eastern third of Kansas. It breeds occasionally in northwestern Missouri (Squaw Creek N.W.R.), which is probably close to the northwestern limit of its present range in our region, though there are early records of its breeding in the Missouri Valley of Nebraska. It also breeds in southeastern Minnesota; stragglers sometimes occur in western Minnesota, but with no evidence of breeding.

Breeding Habitat: Relatively moist woodlands, especially floodplain forests, and adjacent open country for foraging, are needed by this species. It is usually not found where the larger red-tailed hawk is common. Like most other raptors, its range is now contracting and its numbers are declining.

Nest Location: Nests are in tall trees, 20–60 feet above the ground, usually close to the trunk. A variety of hardwood trees are used, including elms, sycamores, and oaks in our region, and occasionally coniferous trees such as pines. The platform of twigs is about 2 feet in diameter, and the same nest is occasionally used more than once. The nest is well lined with small twigs, sprigs of vegetation, and usually down.

Clutch Size and Incubation Period: From 2 to 4 eggs (normally 3). Eggs are white to pale bluish with variable brown spotting. They are laid at 2- or 3-day intervals. The incubation period is about 28 days, probably starting with the first egg. Single-brooded, but renesting attempts may be made if the first clutch or two fails.

Breeding Biology: These monogamous birds arrive at their nesting territory already paired. One or both birds spend considerable time in aerial display, performing a series of soaring and diving gyrations that presumably serves as a territorial advertisement and perhaps as a stimulus to mating. The nest is built in a leisurely manner and may be lined with the nests of tent caterpillars as well as with vegetation and down. Both sexes incubate, but the female probably does most, since she is regularly fed by the male. The young begin to leave the nest when about 5 or 6 weeks old, and soon thereafter they are fully fledged.

Suggested Reading: Stewart 1949; Brown and Amadon 1968.

Broad-winged Hawk
Buteo platypterus

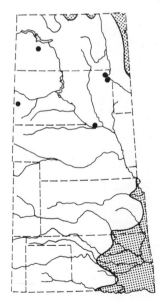

Breeding Status: Breeds in moist woodlands in the eastern half of Oklahoma, the extreme eastern portions of Kansas, and in west-central and northwestern Minnesota. Also breeds locally in North Dakota, particularly in the Pembina Hills and the Turtle Mountains, and recently bred in Dunn County. It is uncommon in northwestern Missouri and has bred in the Missouri Valley of eastern Nebraska, but the present northern limits of its breeding range in this area are not clear. There is a notable record of nesting in the Black Hills in 1977 (*South Dakota Bird Notes* 29:72).

Breeding Habitat: Mature deciduous forests, especially those near water, are the habitat of this species. It forages within the confines of the woods to a greater extent than most other *Buteo* species and is thus seldom seen.

Nest Location: Nests are 20-40 feet above the ground in large trees of a variety of species, usually hardwoods growing near water and having large crotches. They are normally placed in the main crotch, but in pines they are placed against the trunk. Nests are about 15 inches in diameter, poorly constructed of twigs and leaves and lined with bark and sprigs of leaves.

Clutch Size and Incubation Period: From 1 to 4 eggs, usually 2. Eggs are white to bluish white with variable brown spotting. They are probably incubated as soon as they are laid, and the incubation period is probably at least 28 days. Single-brooded.

Time of Breeding: In Minnesota, eggs have been reported from May 9 to June 13, and young from June 17 to July 4. Egg records in Kansas are for the period April 21 to May 30, and in Oklahoma eggs have been seen in late April and dependent young from June 19 to early July. Texas egg dates are for March and April.

Breeding Biology: Broad-winged hawks usually arrive on their nesting grounds about a month before egg-laying, which seems to be timed to coincide with the leafing-out of the trees. They are monogamous, and courtship or territorial display consists of soaring and swooping in the vicinity of the nest. Both sexes build the nest, the female bringing most of the materials and apparently all of the bark lining. During the nestling stage many sprigs of vegetation are also brought. During incubation, males cover the eggs only while the female eats the food her mate brings, and thus they rarely incubate for more than 15 minutes at a time. Only the females brood the young, but males continue to provide food for both their mates and the developing brood. After a week or two the female also begins to hunt for food, but she broods the young at night until they are 3 weeks old. The young begin to leave the nest when about 30 days old and are able fliers during their 6th week of life.

Suggested Reading: Matray 1974; Brown and Amadon 1968.

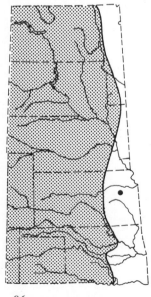

Swainson Hawk
Buteo swainsoni

Breeding Status: Breeds from western Minnesota and virtually all of North Dakota southward to eastern New Mexico, the Texas panhandle, and western Oklahoma, becoming more uncommon eastwardly and rare to absent in Iowa and northwestern Missouri.

Breeding Habitat: This species occupies a habitat approximately complementary to that of the broad-winged hawk, being absent from dense and moist woodlands and most abundant on the open plains. It overlaps in range and habitat with the red-tailed hawk and the ferruginous hawk and probably competes with both these species.

Nest Location: Nests are in isolated trees or bushes or, rarely, on the ground. In any case they usually have a commanding view. Typically they are about 20 feet above the ground, but they may be as high as 60 feet. In North Dakota, they have been found frequently in elms, oaks, willows, cottonwoods, and shrubs, and less frequently in pines, aspens, box elders, ashes, hawthorn, wild plum, and balsam poplars. Other species such as walnut, hackberries, and honey locusts are used farther south. The nest is about 2 feet in diameter, constructed of large sticks and typically lined with bark and fresh green leaves. Normally a new nest is built each year, sometimes near the old one.

Clutch Size and Incubation Period: From 2 to 4 eggs (16 North Dakota nests averaged 3.0 eggs, and five Kansas nests averaged 2.4). Eggs are white to pale bluish white, sometimes unmarked but usually with brown spots. They are apparently incubated from the time they are laid, and incubation lasts about 28 days. Single-brooded, but a new nest may be built and a second clutch laid if the first one fails.

Time of Breeding: North Dakota egg dates range from April 29 to July 15, and nestling records are from July 1 to August 15. In Kansas, egg records are for the period April 11 to June 10, with many clutches completed in late April. Oklahoma egg records are from May 6 to June 27, and young have been recorded between June 11 and August 3 (barely fledged).

Breeding Biology: Swainson hawks are monogamous and arrive on their breeding ground in eastern Wyoming about a month before egg-laying begins. They soon begin nest-building and sometimes use old magpie nests for a base, but infrequently they use their own old nests. Although males rarely assist in incubation they do bring prey to the incubating female. The female also broods the young during the first 20 days after hatching but thereafter spends considerable time hunting. The young fledge in 28 to 35 days.

Suggested Reading: Dunkle 1977; Brown and Amadon 1968.

Ferruginous Hawk
Buteo regalis

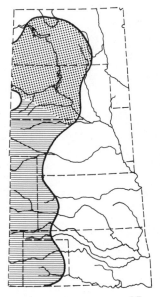

Breeding Status: Breeds throughout most of the western half of the region from the Missouri Coteau of central North Dakota southward through western Nebraska, eastern Colorado, western Kansas, the Oklahoma and Texas panhandles, and eastern New Mexico.

Breeding Habitat: Grasslands with scattered trees, or with clay buttes or bluffs for nesting sites, provide the favored habitat of this distinctive plains-adapted hawk.

Nest Location: In North Dakota, about half of 61 nests were on the ground in prairie vegetation, about a third were in trees, and the rest were on boulders, piles of rocks, haystacks, strawstacks, and miscellaneous locations. Most of the ground nests were on slopes near hillcrests or ridgetops, while the tree nests were in single trees or groves and involved a variety of hardwood species. The tree nests ranged from 10 to 45 feet above the ground and averaged 26 feet. Nests are often 3 feet in diameter and are constructed of large sticks, lined with bark or dried grasses and typically also with chunks of dry dung.

Clutch Size and Incubation Period: From 2 to 6 eggs (35 North Dakota nests averaged 3.9). Eggs are white to bluish white with variable amounts of brown spotting. Incubation lasts 28 days, and probably begins with the first egg, since the young hatch at about 2-day intervals. Single-brooded.

Time of Breeding: North Dakota egg dates range from May 27 to July 16, and dependent young out of the nest have been reported from June 15 to July 27. Kansas egg records are for the period March 11 to April 30. Oklahoma egg dates are from May 9 to June 9, and Texas egg records are from May 6 to June 2.

Breeding Biology: Pairs return to their breeding territory each year and usually use the same nest, so it gradually increases in size. Both sexes bring nesting material in the form of sticks and nest lining, which the female molds to fit her body. Evidently the female does most of the incubation, while the male takes over for a part of the afternoon. After hatching, the male did most of the brooding in a nest observed in Washington, while the female brought in prey, primarily jackrabbits. The adults defended their nest ferociously and were observed attacking a coyote as well as evicting a red-tailed hawk from the area. The young may leave the nest when only about a month old but do not fledge until they are about 44–48 days of age. They start catching live prey only a few days after fledging.

Suggested Reading: Angell 1969; Ohlendorf 1975.

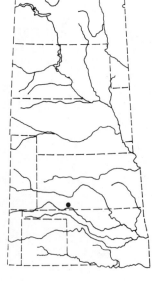

Harris Hawk
Parabuteo unicinctus

Breeding Status: Accidental. There is a single nesting record for the region, for Meade County State Park, Kansas, in 1963. Central Texas is the nearest regular breeding area.

Breeding Habitat: In Texas, the prime habitat of this species is mesquite woodlands containing prickly pear. The birds are also found in the yucca, cactus, and creosote-bush deserts of Trans-Pecos Texas and occur in limited numbers in juniper-oak habitats of Edwards Plateau.

Nest Location: Nests are in a variety of trees, including mesquite, hackberry, and chaparral oaks, and also in yucca and cactus. They are usually less than 30 feet from the ground and are of flimsy construction. The nest is a platform of twigs, sticks, and roots, lined with softer materials such as leaves and grass.

Clutch Size and Incubation Period: From 2 to 4 eggs (50 Arizona nests averaged 2.96). Eggs are white, sometimes slightly spotted with pale brown. The incubation period averages 35 days and ranges from 33 to 36 days. Double-brooded, at least in some areas.

Time of Breeding: The single Kansas nest was built in late March; eggs were seen in late April, and a small chick was seen in late May. Texas egg records are from March 8 to November 14, but young have been seen as early as February 17. This long nesting period in Texas suggests that multiple brooding might occur there, as it does in southern Arizona.

Breeding Biology: Harris hawks are apparently monogamous, but at least in some areas they breed as trios, with a second male present that serves as a nest "helper" and at times even copulates with the female. Courtship flights by a single male, the pair, or even all three birds of a trio occur and involve a swooping dive from a height of more than 100 yards. Nest-building is performed mainly by the female, though sometimes she is assisted by the male, and begins up to a month before egg-laying. Often several nests are built or old nests repaired; these surplus nests may be used as feeding platforms. Incubation may begin with the first egg or be delayed until the last is laid. It is performed by the female and by at least one of the males when trios participate in breeding. After hatching, the female does most of the brooding, while the male (or males) does the hunting. Apparently the excess male does not feed the young directly, but passes food to the bird on the nest, which then feeds it to the young. By the time they are 40 days old, the young begin to perch on branches, and they fledge a few days later. Pairs regularly renest in new sites when their first clutch fails, and in four of five observed cases of double-brooding, a new nest was constructed or an old nest was refurbished for the second clutch.

Suggested Reading: Mader 1975; Brown and Amadon 1968.

Golden Eagle
Aquila chrysaetos

Breeding Status: Breeds in the badlands area of the Dakotas, the South Dakota Black Hills, western Nebraska, and the high plains

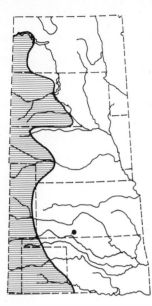

and Rocky Mountain piedmont of eastern Colorado, eastern New Mexico, and the Texas panhandle. In Oklahoma limited to the rough parts of Cimarron County, and now extirpated from Kansas, with a single probable nesting record before 1891.

Breeding Habitat: Golden eagles are found in arid, open country, often with associated buttes, mountains, or canyons that offer remote nesting sites and large areas of natural vegetation for foraging.

Nest Location: Eleven nest sites in North Dakota included 8 on cliff ledges or crevices and 3 in large trees, including 2 cottonwoods. The cliff nests were mostly 20–100 feet above level ground, while the tree nests were 30–75 feet high. The nests are large, often more than 10 feet across, constructed of sticks up to 2 inches in diameter, and lined with various soft materials. Frequently a pair will have several nest sites, sometimes rotating them in different years.

Clutch Size and Incubation Period: From 2 to 3 eggs (8 North Dakota nests averaged 2.4). Eggs are white with brown to reddish brown spotting. The incubation period is generally believed to be 43–45 days, but there are some questionable estimates of 35 days. Single-brooded, but renesting has been reported.

Time of Breeding: Egg dates in North Dakota are from April 29 to May 14, and nestlings have been seen from April 29 to July 22. In the Black Hills the nesting season probably extends from May through August, and young have been seen in July and August. Completed nests or nests with eggs have been seen in Oklahoma between March 12 and April 12, and young have been reported from May 1 to June 8. Texas egg records span the period February 16 to October 11.

Breeding Biology: Golden eagles are monogamous, and pairs occupy large home ranges (averaging about 35 square miles in California) that provide them and their young with an adequate food supply. Aerial displays are most common before the nesting season but may occur at other times too. They consist of soaring and swooping by one or both members of the pair. Both members of the pair work on the massive nests, and up to 12 alternate nests may be maintained. The eggs are laid at intervals of 3–4 days, and incubation begins almost immediately. The female does most of the incubation, but the male begins to assist in brooding soon after the young have hatched, especially during the afternoon. By about 50 days of age the young are feathered, and they fledge at about 65–70 days. However, they remain dependent upon their parents for at least some food for as much as 3 months after fledging.

Suggested Reading: Beecham and Kochert 1975; Ohlendorf 1975.

Bald Eagle
Haliaeetus leucocephalus

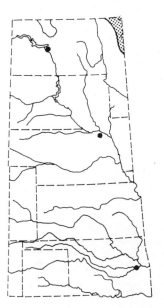

Breeding Status: Now limited as a regular breeding bird to northwestern Minnesota, with nesting records for Becker, Clearwater, and Marshall counties. There are two single recent nesting records for North Dakota (McLean County, 1975) and Nebraska (Cedar County, 1973). There are also some old nesting records for the Texas panhandle (Potter and Armstrong counties), and the birds still breed rarely in southern Texas. There was an Oklahoma nesting attempt in 1976 on the R. S. Kerr Reservoir (*Bulletin of the Oklahoma Ornithological Society* 11:4).

Breeding Habitat: Breeding is largely confined to forested regions in the vicinity of lakes or larger rivers that support a good supply of fish.

Nest Location: In Minnesota, bald eagle nests are usually in upland areas, high in the crowns of living red pines or white pines. The recent nests in North Dakota and Nebraska were in large cottonwoods. They are built of large branches picked up from the ground or broken off dead trees. The nest gradually increases in size with each year's use and is generally about 4–7 feet in diameter and up to 10 feet thick in old nests. It is lined with aquatic vegetation such as cattails and bulrushes, or with other soft, leafy materials.

Clutch Size and Incubation Period: From 1 to 3 eggs, usually 2 (3 North Dakota nests averaged 1.7). Eggs are dull white with a rough surface. The incubation period is probably 34–45 days; published estimates vary widely. Single-brooded.

Time of Breeding: In Minnesota, eggs are laid from about March 16 to April 3. They hatch about April 22 to May 10, and young may be seen in the nest from about July 16 to August 12. In Texas, eggs have been reported from November 6 to June 20, and well-grown young from March 1 to June 17.

Breeding Biology: After maturing and acquiring the adult plumage at 4 or 5 years of age, eagles pair monogamously and remain paired permanently. They perform aerial displays, one of which involves locking talons and tumbling downward through the sky for several hundred feet. In Minnesota these flights occur in March, or during the nest-building period. Copulation occurs at the same time, and egg-laying soon follows. Both sexes assist in incubation, and the young hatch at intervals of several days. The female and young are brought food by the male, which in Minnesota consists primarily of bullheads and suckers rather than important game fish. As the birds grow, both parents gather food for them, but rarely do more than two eaglets survive to fledging. This occurs at about 70 days of age, but the young birds follow their parents for some time afterward, until they are evicted from the area by the adults.

Suggested Reading: Dunstan, Mathisen, and Harper 1975; Sherrod, White, and Williamson 1976.

Marsh Hawk
Circus cyaneus hudsonius

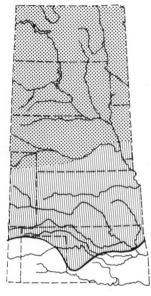

Breeding Status: Breeds nearly through the region, but rare or absent in the Black Hills of South Dakota, and also virtually absent from the Staked Plain of Texas and adjacent New Mexico, probably breeding along the northern tier of Texas counties and also at Muleshoe N.W.R. Widespread in Oklahoma, but no breeding records exist for the southeastern parts of that state.

Breeding Habitat: This species inhabits open country habitat, particularly native grasslands, prairie marshes, and wet meadows, and croplands that are close to natural grasslands.

Nest Location: Nests are in grassy vegetation, ranging from upland situations to wetland habitats including emergent plants such as cattails, bulrushes, and whitetop standing in water up to 2½ feet deep. North Dakota nest sites also include locations in shrubby willows along wet meadows or swamps and in patches of upland shrubs such as wolfberry, silverberry, and rose. The nest is constructed of sticks, twigs, and grasses and is up to 30 inches in diameter, without specific lining materials.

Clutch Size and Incubation Period: From 4 to 6 eggs (39 North Dakota nests averaged 5.0). Eggs are white to pale bluish white, usually unmarked, but sometimes with pale brown spots. The incubation period is 24–30 days, usually beginning before the clutch is complete. Single-brooded, but renesting is frequent.

Time of Breeding: North Dakota egg dates range from April 26 to June 25, and nestlings have been recorded from June 15 to July 15. Kansas egg dates are from April 11 to May 20, with a peak of egg-laying about May 5. Oklahoma egg dates are from April 26 to June 9, and nestling dates range from June 4 to June 30.

Breeding Biology: Males migrate separately from females and arrive on the nesting grounds first. They display aerially by performing a series of spectacular dives and swoops, especially in the presence of females. Later the pair may display in this way and also by locking talons in flight. The nest is constructed mainly by the female, though the male may help gather materials. Frequently the birds are semicolonial, with up to six nests concen-

92

trated in a square mile. The eggs are laid at intervals of several days, and the female may begin to incubate at almost any time during the prolonged egg-laying period. Males feed their incubating mates, and on the basis of a group of six nests studied in Manitoba, sometimes provide food for two females. The young hatch at staggered intervals and while they are very small are brooded continuously by the female while the male brings in food. Later the female also hunts, but she usually receives by aerial transfer the food the male brings in. She is the only parent to feed the young directly. Where males are tending two nests the females must do more hunting by themselves, and starvation of young nestlings is frequent. The young fledge at about 5 weeks, males a few days sooner than females.

Suggested Reading: Watson 1977; Brown and Amadon 1968.

FAMILY PANDIONIDAE
(OSPREYS)

Osprey

Osprey
Pandion haliaetus

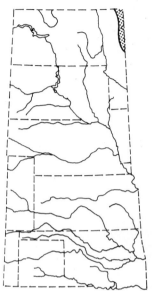

Breeding Status: Restricted as a breeding species to northwestern Minnesota, with nesting records for Becker and Clearwater counties. There is also a 1973 record of attempted nesting in McLean County, North Dakota, and an old nesting record for Cass County, Nebraska. Previously ospreys had a wider nesting distribution in Minnesota, including most or all of the state.

Breeding Habitat: Ospreys are associated with clear rivers and lakes in forested areas of our region but also occur along coastlines throughout much of the world.

Nest Location: In Minnesota, nests are 30–90 feet above the ground, usually at the tops of dead or partially dead lowland conifers, but sometimes on artificial structures such as powerline poles. They consist mostly of dead sticks and branches. Unlike eagle nests in the region, they are usually rounded rather than cone-shaped and are generally smaller; the cup is typically lined with lichens.

Clutch Size and Incubation Period: From 1 to 4 eggs, usually 3. Eggs are white with grayish and bright brown markings. They are laid at intervals of 1–3 days. Incubation begins with the first egg; its duration averages about 37 days. Single-brooded.

Time of Breeding: In Minnesota, eggs are laid between May 6 and 15, hatching occurs about June 15, and young are present in the nest until the middle of August.

Breeding Biology: In Minnesota, ospreys arrive in late April as the ice is melting from their nesting grounds, and males soon begin courtship flights. These swooping and soaring flights may serve to attract females but also continue for a time after pair bonds are established or reestablished. Nest-building or repair of the old nest starts very soon, the male bringing most of the larger sticks and the female bringing in the lining materials as well as doing the final shaping of the nest. From the time she arrives until the young are nearly fledged, the female catches few if any fish and thus relies on the male for virtually all her food. Mating occurs on the nest site or a nearby branch and continues during the egg-laying period. Both sexes incubate, but the female undertakes most of the responsibility and does all the nighttime incubation. The eggs hatch at intervals of up to 5 days, which results in considerable differences in the sizes of the young. For the first month of brooding the female rarely leaves the nest, and the male does all the hunting. As the young approach fledging at about 55 days of age, the female may also help in hunting. After fledging the young continue to use the nest for roosting and as a feeding platform, but they soon attempt to catch fish on their own. They do not mature sexually until they are three years old.

Suggested Reading: Green 1976; Dunstan 1973.

FAMILY FALCONIDAE (CARACARAS AND FALCONS)

American Kestrel

Prairie Falcon
Falco mexicanus

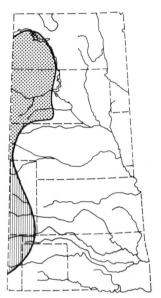

Breeding Status: Breeds locally and uncommonly in the western parts of North and South Dakota, northwestern Nebraska (Dawes and Sioux counties), eastern Colorado, extreme western Oklahoma (Cimarron County), and northeastern New Mexico (Quay County). Originally also bred on the Texas panhandle (Randall County), but there is no twentieth-century record for this area.

Breeding Habitat: This species is associated with large expanses of open grassland or sagebrush scrub with adjacent cliffs, bluffs, or rock outcrops suitable for nesting.

Nest Location: Prairie falcons invariably nest on tall bluffs or escarpments, usually those having ledges with overhanging rocks. The cliffs usually range from 30 to 300 feet high but at times are much higher. The ledges should have gravel or other loose materials for a nest site. The nest is only a shallow depression scooped in the gravel or sand, with no lining of vegetation. It often has a southern exposure and typically overlooks a large hunting area.

Clutch Size and Incubation Period: From 3 to 6 eggs (55 Colorado and Wyoming nests averaged 4.5). Eggs are pinkish buff with red to brownish spotting. The incubation period is 29-31 days, starting when the clutch is nearly complete. Single-brooded, with infrequent renesting after loss of clutches early in incubation.

Time of Breeding: Nests with dependent young have been seen in North and South Dakota in June and July, and a nest with 4 young and 1 egg was reported in the Black Hills for June 13. In northern Colorado and Wyoming, clutch completion records extend from April 12 to May 9, hatching from May 12 to June 8, and fledging from June 21 to July 19. Nests with eggs or both eggs and young have been reported in Oklahoma in late May.

Breeding Biology: Prairie falcons probably first nest when two years old; yearlings normally wander during the breeding season. The birds arrive on the Wyoming and Colorado nesting grounds in late February or early March, and the male engages in aerial courtship for about a month while the pair examines potential nest sites. Frequently, nest sites of the previous year are used, even if the female is mated to a new male. The male begins to do most of the hunting for the pair during the courtship period, and the female does nearly all the incubation. Only when the female is eating food brought by her mate does he incubate, but the male performs the major role in nest defense. The young begin to acquire their flight feathers at about 30 days and fledge at about 40 days of age. Evidently a large portion survive their first autumn, but there is an overall mortality rate of about 80 percent by the end of the first year of life.

Suggested Reading: Enderson 1964; Brown and Amadon 1968.

Peregrine Falcon
Falco peregrinus

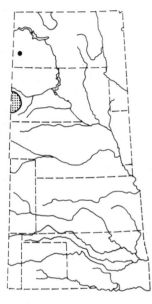

Breeding Status: Extremely rare and currently limited to the Black Hills of South Dakota. Previously more widespread, including the Missouri slope of southwestern North Dakota (most recent breeding record is 1954, Billings County; earlier records for Williams, Stark, Slope, and Dunn Counties). It may also have once bred in Dawes County, Nebraska, and also formerly bred in Kansas (Woodson and Ellis counties). There are no breeding records for Oklahoma or the Texas panhandle.

Breeding Habitat: In the Black Hills this species is associated with deep canyons having high, vertical limestone or sandstone cliffs.

Nest Location: Nest sites vary considerably in various parts of this species' wide range, but in the Great Plains nests are typically on tall, steep cliffs. In general these falcons choose the steepest and most inaccessible locations and, unlike prairie falcons, definitely prefer tall cliffs to lower ones. Limestone cliffs with eroded recesses are favored, especially those offering protected ledges that are flat and grassy, at least 18 inches wide, with sheer rock above and below. The ledge must also offer a potential nest scrape where a depression can be made in sand or gravel. A few nests in tree hollows have been reported, notably in eastern Kansas.

Clutch Size and Incubation Period: From 2 to 5 eggs, usually 3 or 4 (23 North Dakota clutches averaged 3.6). Eggs are creamy white to pinkish with heavy brown spotting. Incubation begins with the second or third egg and lasts 28-29 days. Single-brooded, but renesting is at least locally prevalent.

Time of Breeding: Egg dates in North Dakota range from May 1 to May 29, and flightless young have been observed from June 28 to August 2. The estimated breeding season in the Black Hills is May through July, and in Kansas the eggs were laid in February and March. Texas records are from March 20 (eggs) to July 21 (young).

Breeding Biology: The use of modern pesticides has nearly eliminated this species from much of North America, but it still breeds in a few remote areas. After a pair has returned to its nesting area, the male, or both birds, performs courtship flights

102

consisting of diving and swooping, and sometimes passing food in the air. Mating is frequent during this period and the period of egg-laying, with the eggs laid at intervals of about 2–3 days. The female does most of the incubation, with the male bringing prey to his mate and also occasionally relieving her. The female typically eats away from the nest, on a nearby "plucking post." For the first 2 weeks after hatching the female does nearly all the brooding and feeding of the young, but later both adults hunt extensively and simply drop their prey into the nest, letting the young birds compete for it and tear it up. Usually only 2 or 3 young fledge from each brood; fledging occurs about 35–42 days after hatching. Since the advent of modern pesticides, virtually no young are fledged at most nests, since the thin-shelled eggs laid by pesticide-poisoned females fail to hatch or else the young do not survive to fledging.

Suggested Reading: Hickey 1969; Brown and Amadon 1968.

Merlin (Pigeon Hawk)
Falco columbarius

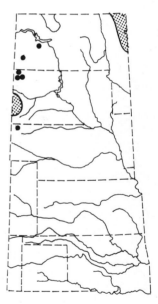

Breeding Status: Currently a rare breeding species in the area, with possible nesting in northwestern Minnesota (especially Kittson and Roseau counties), southwestern North Dakota (recent definite records only for Slope and Dunn counties), the Black Hills region of South Dakota (nesting record for Pennington County in 1948), and northwestern Nebraska (nest records for Dawes County, 1975 and 1978).

Breeding Habitat: Merlins prefer a habitat that includes both tree groves and adjoining open areas of fields, grassland, and brushy vegetation. Tree-lined coulees or ravines in otherwise open country are the typical habitat in the Great Plains.

Nest Location: Old crow or magpie nests are favored nest locations in the northern states and prairie provinces; such nests may be rather close to the ground in coniferous or deciduous trees. Nests have been reported in ponderosa pines in North Dakota, South Dakota, and Nebraska, and in an elm in North Dakota, the birds most often using old magpie nests.

Clutch Size and Incubation Period: From 2 to 7 eggs, usually 5 or 6. The eggs are light buff with heavy stippling of brown, reddish brown, and purple tones. The incubation period is 28–32 days, starting before the clutch is completed. Single-brooded.

Time of Breeding: North Dakota egg records are for mid-May, and dependent young have been seen from June 21 to July 10. The estimated nesting season in the Black Hills is from May

103

through July (newly hatched young seen on June 11), and the Nebraska nest had a probable incomplete clutch on June 13.

Breeding Biology: Typically, males return to the breeding ground before females and begin calling while flying from one perch to another. Little actual nest-building is done, since the birds typically take over an already constructed nest of another species. The eggs are laid at 2-day intervals, and the female begins to incubate when the clutch nears completion. Probably the only time the male takes over incubation is during the short periods the female is off the nest eating food he has brought. The eggs hatch at intervals, resulting in marked size differences among the young, which develop rapidly. Early in the brooding period all the food is brought in by the male, passed on to the female, then torn up and divided among the young. Later the female assists in hunting and bringing in food, which mostly consists of small birds. The young fledge in 25–30 days but remain in the vicinity of the nest for some time. They initially begin to hunt by catching insects but soon learn to chase and capture young birds.

Suggested Reading: Lawrence 1949; Brown and Amadon 1968.

American Kestrel (Sparrow Hawk)
Falco sparverius

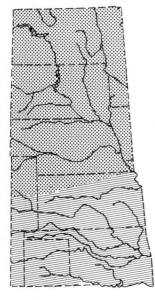

Breeding Status: Pandemic and relatively common throughout most of the region, but least common in heavily wooded areas and also relatively uncommon where trees are too small or too scattered to provide suitable nesting sites.

Breeding Habitat: This most abundant of the American falcons occurs especially where scattered trees or tree groves adjoin large areas of open country, including grasslands, croplands, or badlands.

Nest Location: Nests are in cavities, typically in trees, but occasionally in rock crevices, old magpie nests, or even large birdhouses. When natural cavities of trees are used (including flicker cavities) they usually are 8–30 feet above the ground. Kestrels can use woodpecker cavities with entrances as small as 3 inches and occupy a wide variety of hardwood trees having such cavities.

Clutch Size and Incubation Period: From 3 to 5 eggs (8 North Dakota nests averaged 4.7 eggs, and 5 Kansas nests averaged 4.2). The eggs are whitish to light cinnamon, usually with small brown spots. They are laid at intervals of 2–3 days, and incubation lasts 29–30 days, starting before the last egg is laid. Single-brooded, but possible double-brooding has been found in Colorado (*Wilson Bulletin* 89:618).

Time of Breeding: North Dakota egg dates are from May 9 to June 21, and dependent young have been seen from July 29 to August 30. In Kansas, egg-laying dates are from March 21 to May 20. Egg dates in Oklahoma are from April 18 to May 26, and young have been seen from May 27 to August 1.

Breeding Biology: American kestrels are perhaps the most sociable of the falcons, and until pair-formation may associate in small groups. During the courtship period males perform aerial dive displays, whining vocalizations, and courtship feeding, which reaches a peak shortly before egg-laying. Courtship feeding serves to maintain the pair bond and also provides food for the female and her young. Sometimes the female begs for food in flight by performing a distinctive "flutter-glide" display. Copulations reach a peak just before egg-laying, but courtship feeding peaks during the egg-laying period and continues through the brooding period. The female does nearly all the incubating, and the young often hatch at intervals of about a day, somewhat less than the egg-laying interval. During the fledging period of approximately 30 days, the male continues to do most of the food gathering while the female broods and directly feeds the young. But after about 20 days the adults bring in entire prey animals and place them in the nest for the young birds to tear apart and feed themselves. The family typically remains together for some time after fledging.

Suggested Reading: Willoughby and Cade 1964; Balgooyen 1976.

FAMILY TETRAONIDAE (GROUSE AND PTARMIGANS)

Greater Prairie Chicken

Blue Grouse
Dendragapus obscurus

Breeding Status: Extirpated. This species was collected in the Black Hills of South Dakota in the 1850s and was reported seen in 1874, but there have been no records since. An effort was made to reintroduce it in 1969 and 1974, when blue grouse from Colorado were released, but the success of this attempt is unknown. (During the 1960s two other galliform species were released in western South Dakota. These include the California quail [*Lophortyx californicus*], released in 1961, and the chukar partridge [*Alectoris chukar*], released in 1960 and 1964 in Harding and Washabaugh counties. There is no evidence that the former species has survived, but the latter was observed as recently as 1973 in Harding County.)

Breeding Habitat: Although blue grouse winter among coniferous cover in Colorado, they breed at lower altitudes; they are especially associated with habitats dominated by aspens, with a variety of shrubs and grasses.

Nest Location: Nests are typically beside logs or under low tree branches and are fairly well concealed. The nest is simply a scrape, with little or no lining.

Clutch Size and Incubation Period: From 6 to 10 eggs are normally laid; they are buffy and are almost covered with brownish spots and dots. The incubation period is 26 days. Single-brooded.

Time of Breeding: Breeding dates are not available for our region, but in Colorado the published egg dates are for the month of June. Although at low altitudes eggs may hatch as early as June 1, hatching normally is 3–4 weeks later at higher altitudes.

Breeding Biology: Blue grouse are promiscuous, with the males establishing exclusive "hooting territories" in early spring. Within these rather large and well-separated territories in wooded areas, the males strut, perform other displays, and utter associated calls to attract the females for mating. The males play no further role in reproduction; the females incubate the eggs and rear the brood alone. The chicks grow surprisingly rapidly. They can fly short distances at only 5–6 days of age, and at 2 weeks old they can fly about 200 feet. By the time the young are about 2 months old the brood begins to break up, and juveniles slowly make their way toward wintering areas at higher elevations.

Suggested Reading: Johnsgard 1973; Bailey and Niedrach 1965.

Spruce Grouse
Dendragapus canadensis

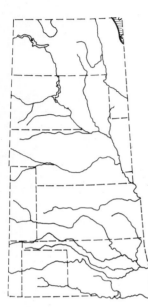

Breeding Status: A local resident in northwestern Minnesota, with breeding evidence only for Roseau County.

Breeding Habitat: In Minnesota, observations indicate that spruce grouse usually breed in forest cover consisting of at least 75 percent evergreen species, and that most of these are upland forest types including cedar and black spruce. In general, open stands having an abundance of ground cover and shrubs are preferred to dense and mature stands, probably in part because the shrubs bear berries that are important late-summer foods.

Nest Location: Nests are always well concealed, often under low-hanging branches, in brush, or in deep mosses. The forest cover is typically open, mature coniferous forest or mixed coniferous and deciduous forest. The nest is a shallow scrape, lined with leaves, needles, and some feathers.

Clutch Size and Incubation Period: From 4 to 10 eggs (averaging about 6 in Nova Scotia). The eggs are buffy to cinnamon, with large brown spots. The incubation period is 21–23 days, starting with the completion of the clutch. Single-brooded; with renesting apparently rare.

Time of Breeding: Broods in Minnesota have been seen between July 7 and August 1; thus most egg-laying probably occurs in early June.

Breeding Biology: As the breeding season approaches, males become relatively sedentary and localized in territories that range from about 3 to 20 acres of forest. They use a variety of stationary and aerial displays to advertise these territories, primarily strutting conspicuously and performing short and noisy flights. In Minnesota these flights and postures have been observed in May, and no doubt they correspond to the period of female receptivity and fetilization. No pair bonds are formed, and the male probably fertilizes as many females as he can attract through his advertising displays. The female may nest within the male's territory, but he does not defend her during the incubation or brooding period. Until the chicks fledge at about 10 days of age the brooding female is highly aggressive and defensive toward her brood, performing either threatening movements resembling strutting or a sneaklike display. Sometimes males accompany females with broods, but apparently only because of a continued sexual attraction.

Suggested Reading: Robinson and Maxwell 1968; Johnsgard 1973.

Ruffed Grouse
Bonasa umbellus

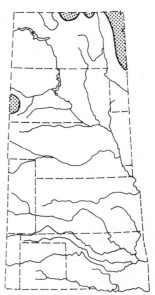

Breeding Status: Breeds in west-central and northwestern Minnesota (Roseau, Marshall, Beltrami, Clearwater, Otter Tail, and Norman counties), in the Turtle Mountains and Pembina Hills of North Dakota, and in the Black Hills of South Dakota. Once also bred in the Missouri Valley of Kansas, Missouri, Nebraska, and Iowa, but now extirpated there.

Breeding Habitat: Ruffed grouse are associated with deciduous forests or with mixed forests containing aspens, poplars, and birches. Open stands or those containing small clearings with considerable berry-bearing shrubs are more valuable to grouse than stands lacking a well-developed understory.

Nest Location: In a study involving more than 1,200 nests in New York, the base of a large hardwood tree seemed to be the most consistent choice for a nest site. Other locations include tree stumps, logs, bushes, or brush piles. The nests are usually close to a forest opening, on fairly level ground, and offer a combination of visibility, protection, and an escape route. Most nest sites are rather well lighted, with an open forest canopy and relatively open shrub cover nearby.

Clutch Size and Incubation Period: From 9 to 15 eggs (averaging 11.5 in New York for initial nesting attempts). The eggs are buffy, sometimes with a few small brownish spots. The incubation period is 23-24 days, beginning with the last egg. Single-brooded. Renesting efforts are somewhat frequent in New York, although second clutches tend to be smaller.

Time of Breeding: Minnesota egg dates are from May to June 17, and dependent young have been seen from June 6 to July 16. The nesting season in the Black Hills is from May to early August, with egg dates as early as May 2 and broods seen as late as August 16.

Breeding Biology: As the breeding areas become free of snow, male grouse establish territories that usually include a clump of aspens and one or more "drumming logs" from which they display daily. The drumming behavior of this species is a ritualized form of flight display; the bird does not leave the ground, and the sound generated by wing-beating attracts females to the log. When a female (or another male) appears, the male begins an elaborate strutting behavior that is a ritualized form of threat, leading to copulation if the intruder is a female or to fighting if it is another male. After mating the female selects a nest site that is usually near a clump of male aspens that provides a food source during incubation. After hatching the young grow rapidly. They can fly short distances after 10-12 days, but they remain with their mother until they are about 4 months old, when the juveniles begin to disperse.

Suggested Reading: Johnsgard 1973; Bump et al., 1947.

Pinnated Grouse (Greater and Lesser Prairie Chickens) *Tympanuchus cupido pinnatus and T. c. pallidocinctus*

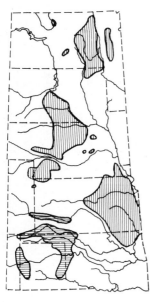

Breeding Status: This grassland species once bred throughout the region, but the greater prairie chicken is now localized and restricted to west-central and northwestern Minnesota (Clay, Norman, Wilken, Polk, and Mahnomen counties), eastern North Dakota (mainly Stutsman, Grand Forks, and Barnes counties), south-central South Dakota (mainly Lyman, Tripp, and Gregory counties), the eastern Sandhills area of Nebraska, eastern Kansas, and adjacent northeastern Oklahoma. The lesser prairie chicken (considered by the A.O.U. to be a separate species) occurs in sandy areas of southwestern Kansas, western Oklahoma, southeastern Colorado, the Texas panhandle, and northeastern New Mexico.

Breeding Habitat: Greater prairie chickens are associated with native grasslands and with combinations of native grasslands and grain croplands, where the proportion of croplands is fairly low. The lesser prairie chicken differs somewhat in that it needs brushy vegetation such as sagebrush, shinnery oaks, and wild plums for summer shade, winter protection, and supplemental foods.

Nest Location: Greater prairie chicken nests are typically in grassy, open habitats such as ungrazed meadows or hayfields, usually in fairly dry situations, but sometimes are in brushy vegetation and occasionally in open woods or the edges of woods. Nests of the lesser prairie chicken are usually between clumps of bunchgrass under shrubby vegetation no more than 15 inches tall. The nest is a shallow scrape, usually lined with leaves and grasses.

Clutch Size and Incubation Period: From 9 to 14 eggs (7 North Dakota clutches of the greater prairie chicken averaged 11.4, and 7 Oklahoma clutches of the lesser prairie chicken averaged 10.7). The eggs are buffy to olive, usually with small darker spots. The incubation period is 23–26 days. Single-brooded, but with some renesting.

Time of Breeding: North Dakota egg dates are from April 28 to July 1, with young seen from May 31 to July 27. Kansas nesting dates are from May 1 to June 10 for the greater prairie chicken. Oklahoma egg dates for the lesser prairie chicken are from May 16 to June 8, with hatching dates from late May to mid-June.

112

Breeding Biology: Male prairie chickens establish individual territories in early spring on communal "booming" or "gobbling" grounds and perform their distinctive displays every day for several months. Females are attracted to birds holding central territories, the "master cocks," and such birds are able to mate with most females. After fertilization the female lays her clutch, and incubation begins at about the time the last egg is laid. Until they are about a week old the chicks are brooded much of the time, but they are highly precocial and can fly in less than 2 weeks. Families gradually disintegrate when the young are about 6–8 weeks old.

Suggested Reading: Johnsgard 1973; Schwartz 1945.

Sharp-tailed Grouse
Tympanuchus phasianellus

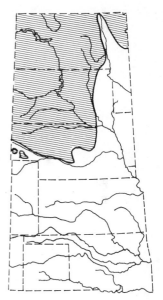

Breeding Status: Resident breeder locally in northwestern Minnesota, most of North Dakota excluding the Red River Valley, the western three-fourths of South Dakota, and the Nebraska Sandhills and adjoining plains areas. A remnant population may occur in northeastern Colorado (Yuma County), but the species is evidently extirpated from Kansas, where it once occurred at least as far east as Ellis County, and also from northwestern Oklahoma. There was also a remnant population in Colfax County, New Mexico, that is now probably extirpated.

Breeding Habitat: The Great Plains race of this species is adapted to a grassland habitat where trees are rare or absent. Native grassland vegetation interspersed with from 5 to 30 percent bushy cover is a preferred habitat type in North Dakota and probably elsewhere through this region.

Nest Location: In North Dakota, 10 of 22 nests were in unused prairie vegetation, 10 were in unused alfalfa or sweet clover, and grain stubble and hayfields accounted for the others. Vegetation at nest sites is usually at least 12 inches tall. Studies in Michigan indicate a preference there for nesting near shrubby or woody cover and in sites varying from open to 75 percent shaded. The nest is a shallow scrape lined with grasses, leafy materials, and a few feathers.

Clutch Size and Incubation Period: From 7 to 18 eggs (29 North Dakota nests averaged 11.9). The eggs are buffy to brownish, usually with a few small darker spots. The incubation period is 23–24 days, starting with the completion of the clutch. Single-brooded, with infrequent efforts at renesting.

113

Time of Breeding: North Dakota egg dates range from April 29 to July 28, and dependent young have been seen from May 22 to September 11. In Minnesota, eggs have been reported from May 10 to June 22. In South Dakota the nesting season is from May through July, and broods are seen from mid-June to mid-August.

Breeding Biology: By late February or March, when snow melts from hilltops, male sharp-tailed grouse begin to assemble and establish or reestablish territories in a communal display area, or "lek." Older, more experienced males tend to occupy the more central and desirable territories, sought out by females when they arrive for fertilization. Most of the males' elaborate "dancing" behavior is thus directed toward the other males and consists of ritualized hostile behavior, though a few displays and calls are reserved for females. Females visit the leks only long enough to be fertilized, and the males take no further part in reproduction. The young hatch simultaneously and soon leave the nest to begin feeding on small insects. They can fly short distances by the time they are 10 days old and may move a quarter-mile in a day even before fledging. They are nearly independent by the time they are 6–8 weeks old and often disperse considerable distances at that time.

Suggested Reading: Lumsden 1965; Sisson 1976.

Sage Grouse
Centrocercus urophasianus

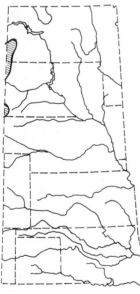

Breeding Status: A local resident in southwestern North Dakota, adjacent northwestern South Dakota, and also southwestern South Dakota. Probably extirpated as a breeding species from adjacent northwestern Nebraska, but displaying birds have recently been reported from Sioux County.

Breeding Habitat: Sage grouse are never found far from sagebrush, both during the spring display and during the summer nesting period. They usually breed in semiopen stands of sage, with a diversity of other species, including grasses, broad-leaved weeds, and herbaceous legumes.

Nest Location: Nests are nearly always under sagebrush. Stands that have 20–30 percent canopy coverage seem to be preferred, and the average height of sage plants used for nest cover in Montana was about 16 inches, or significantly taller than those in adjacent areas. The nests are shallow scrapes, well lined with grasses and sage leaves.

Clutch Size and Incubation Period: From 5 to 9 eggs (averages in various states range from 6.8 to 7.5). The eggs are generally pale olive buff with small brownish spots. Incubation begins with the final egg and lasts 25–27 days. Single-brooded, with very limited renesting.

Time of Breeding: In North Dakota eggs have been reported in mid-May, and broods are reported from late May to late July.

Breeding Biology: From early spring onward, large groups of male sage grouse assemble on traditional "strutting grounds" in open sage country, where they compete for territories and where females later come for fertilization. Typically a single "master cock" dominates each display ground and accounts for most of the matings there. The male displays are complex and highly stereotyped but include stepping, wing-brushing movements, and a series of rapid inflations and deflations of the esophageal "air sacs," with associated plopping sounds. Females are attracted to such groups of males just before their egg-laying period, and most mating occurs at about sunrise. After copulation, the female leaves the strutting ground and probably does not return unless her clutch is destroyed. She usually nests some distance from the display ground, and abut 10 days are needed to complete a clutch of 8 eggs. Males do not take part in incubation or nest defense, and the chicks hatch in a highly precocial condition. Their mother quickly moves them to moist areas where insect food is plentiful, and they fledge rapidly, in less than 2 weeks. However, they usually remain with their mother for most of the late summer, gradually becoming more independent. Eventually the birds are forced to move to their wintering areas, which are usually at lower elevations and may be 50 miles or more from the nesting areas.

Suggested Reading: Patterson 1952; Johnsgard 1973.

115

FAMILY PHASIANIDAE (QUAILS, PHEASANTS, AND PARTRIDGES)

Ring-necked Pheasant

Bobwhite
Colinus virginianus

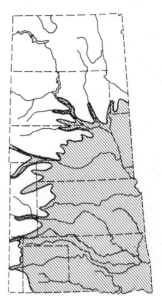

Breeding Status: Resident from southwestern Minnesota, southeastern South Dakota, and eastern Nebraska southward through the region, with extreme western limits along the rivers of eastern Wyoming, Colorado, and New Mexico.

Breeding Habitat: For breeding bobwhites need a diversity of cover types, including grassy nesting cover, cultivated crops or a similar source of food, and brushy cover or woodlands with a brushy understory. A source of water and a place for dusting are also important.

Nest Location: Nests are in open herbaceous cover with nearly bare ground, with the vegetation short and sparse enough so the birds can readily walk through it. The nest is a shallow scrape, well lined with grass, with vegetation often arched over the top to conceal it from above.

Clutch Size and Incubation Period: From 8 to 20 eggs (22 Kansas nests averaged 12.8). Eggs are white with a smooth surface. The incubation period is 23-24 days. Possibly double-brooded in the south, and a persistent renester.

Time of Breeding: Nesting records in Kansas are from May 1 to September 20, with a peak of egg-laying in late May. Eggs in Texas have been reported from March 15 to November 9, indicating a remarkably long 9-month breeding period.

Breeding Biology: Bobwhites mature in their first year, sometimes breeding in captivity when only 5 months old. Near the end of winter, coveys tend to break up and males establish singing posts, from which they begin to utter their familiar *bob-white* whistle. This whistle is not a territorial defense signal but is simply an indication of sexual availability. When a pair is formed, the whistling becomes soft and infrequent or even terminates. Both sexes seek the nest site and build the nest. Eggs are laid about one per day, and incubation begins when the clutch is complete. Males rarely help with incubation but will regularly take over if their mate is killed. In captivity, at least, double-brooding has been observed in bobwhites, with the male taking over the care of the first brood while the female begins a second clutch. The young grow rapidly, and in less than 2 weeks they can fly short distances. The family bonds remain intact through the summer and fall and provide the basis for coveys. Families may also be joined by unsuccessful nesting pairs during covey-formation, since six or more birds are needed to form the heat-conserving circular roosting formation that the birds habitually assume.

Suggested Reading: Robinson 1957; Rosene 1969.

Scaled Quail
Callipepla squamata

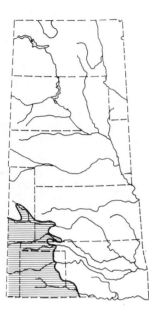

Breeding Status: Resident in southeastern Colorado, southwestern Kansas, the panhandles of Oklahoma and Texas, and New Mexico. Unsuccessfully introduced in Nebraska.

Breeding Habitat: In our region the scaled quail is generally found near sandy soils and associated sand sagebrush vegetation. Secondarily it occupies arid grasslands with cactus or yuccas and to a limited extent uses the piñon pine and juniper habitat type. Shrubby cover for escape and for protection from mid-day heat is an important aspect of this species' needs. Water for drinking is probably not needed if succulents are present, but populations tend to be higher where there is surface water.

Nest Location: Nests are among shrubs or in some other protected and shady location, rarely in open situations among rocks. The nest scrape is lined with dried grasses and few feathers and is often arched over with overhead grasses, which effectively conceal it.

Clutch Size and Incubation Period: From 5 to 22 eggs (39 Oklahoma nests averaged 12.7). Eggs are white and smooth-surfaced. Incubation period is 23–24 days. Probably sometimes double-brooded, and a persistent renester.

Time of Breeding: Oklahoma egg records extend from May 8 to September 6, those for New Mexico from April 15 to September 22, and those for Texas from March 11 to September 9. All of these reflect the influence of renesting or double-brooding.

Breeding Biology: Scaled quail spend the colder months in large coveys that average about 30 birds but sometimes exceed 100. As the breeding season approaches, males in the coveys begin fighting and mated pairs become intolerant, so that the coveys break up. Unmated males soon take up calling sites and begin uttering loud *whock* calls. The calls attract unmated females, and the calling ceases as soon as a pair bond is formed. Nest-building soon follows, and the male remains nearby as the female begins incubating. Rarely do males assist in incubation, but at least one case is known where the male has taken over the brooding responsibilities for a young brood and thus allowed the female to begin a second clutch. The young feed on a variety of insects and can probably fly within 2 weeks. Family bonds remain intact, and there is a gradual fall merging of broods and pairs to form the rather large coveys characteristic of the species.

Suggested Reading: Schemnitz 1961; Johnsgard 1973.

Ring-necked Pheasant
Phasianus colchicus

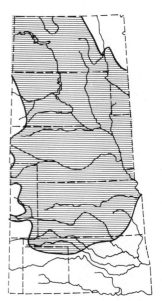

Breeding Status: Introduced. Resident throughout most of the region except for northwestern Minnesota, most of Oklahoma, the Texas panhandle, and New Mexico. Increasingly local toward the south and west and largely limited to irrigated areas or other sites with permanent surface water.

Breeding Habitat: A combination of small grain croplands and adjacent edges, such as weedy ditches, sloughs, wooded areas or shelterbelts, or uncut hayfields provide optimum breeding habitat for this species.

Nest Location: In North Dakota, most nests have been found in roadside ditches, in alfalfa or sweet clover fields, and in heavy grasses, with smaller numbers in cropland, shelterbelts, pastures, and the like. In Nebraska, roadside ditches with an abundance of early-maturing rather than warm-season grasses provide valuable nesting cover. Alfalfa, cool-season grasses, and winter wheat accounted for about 80 percent of all nests found in a study there. The nest is usually well concealed and lined with leaves, grasses, and a few feathers.

Clutch Size and Incubation Period: From 6 to 15 eggs (469 Nebraska nests averaged 9.4). Eggs are uniformly olive to brownish with a slight gloss. The incubation period is 23–25 days, starting with the last egg. Single-brooded, but renesting is frequent.

Time of Breeding: North Dakota egg dates are from late April to July 29, with young being reported from May 19 to September 2. In south-central Nebraska, estimated nest-initiation dates range from mid-April to late July, and the average estimated date of hatching is June 11, but the hatching period extends from late April to mid-August.

Breeding Biology: Pheasants spend the winter in small groups wherever there is food and cover, but by early spring the males begin to spread out and establish crowing areas. Although these are not typical territories, the male displays in these ill-defined areas and attracts a variable number of females. This harem is maintained only until the females are fertilized, after which they leave the male and establish a nest. The male plays no further role in reproduction, and the female must both incubate and defend the nest. Nesting losses during incubation are often substantial because of predation and hay-cutting, but renesting efforts help compensate for these high losses. The chicks grow rapidly, and when only a week old they can fly a few feet. The female typically remains with her offspring for 6–8 weeks, or about the time that young males begin to acquire their adult plumage.

Suggested Reading: Baxter and Wolfe 1973; Baskett 1947.

121

Gray Partridge
Perdix perdix

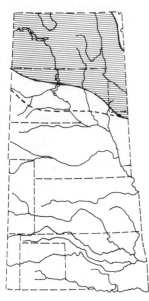

Breeding Status: Introduced resident over nearly all of North Dakota, western Minnesota, and the northern third of South Dakota, extending southward to extreme northern Iowa. Previously extended to central Iowa and northern Nebraska (Holt and Knox counties) but now extirpated from the latter state.

Breeding Habitat: In the northern Great Plains, gray partridges are associated with grainfields and adjoining edge habitats such as weedy borders, shelterbelts, and abandoned farmsteads. Surface water may not be essential if enough succulent vegetation is available, but the birds need a supply of grit, and probably also a dusting area.

Nest Location: Hayfields and grainfields accounted for more than half the nest locations in three studies, and alfalfa appears to be a particularly favored cover plant. In a North Dakota study, nearly all of 23 nests were in sweet clover, along roadsides, or in heavy grasses. In hayfields the nests are usually close to the edge of the field and rarely more than 100 feet from the edge. The nests are shallow scrapes lined with dead leaves and grass.

Clutch Size and Incubation Period: From 6 to 20 eggs (470 Wisconsin nests averaged 16.4). Eggs are olive-colored, without spotting. The incubation period is 23–24 days, starting with the last egg laid. Single-brooded, but renesting is frequent.

Time of Breeding: The nesting period in North Dakota is from mid-May to mid-August, with a hatching peak in late June and early July. In the nine-year period from 1955 to 1963, 70 percent of more than 8,000 birds analyzed had hatched between June 16 and July 20, with about 1 percent hatching before June 8 and fewer than 1 percent after August 25.

Breeding Biology: Gray partridges remain in winter coveys in North Dakota until about March, when aggression among the males and competition for mates cause social disruption. Mates of the past year frequently pair again, but many females mating for the first time change their mates several times before a pair bond is firmly established and the pair leave the covey for nesting. Unmated males establish crowing posts from which they regularly call, but there is no territoriality as defense of a specific area against conspecifics. The female builds the nest, with the male standing guard, and lays eggs at about one per day. Incubation is by the female alone, but the male may at times sit beside her, especially at the time of hatching. The precocial young fledge when about 2 weeks old, but family bonds remain intact and form the basis for fall coveys.

Suggested Reading: McCabe and Hawkins 1946; Johnsgard 1973.

FAMILY MELEAGRIDIDAE
(TURKEYS)

Wild Turkey

Wild Turkey
Meleagris gallopavo

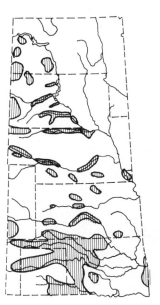

Breeding Status: Originally resident over much of the region but extirpated from most areas, with indigenous populations persisting from southern Kansas southward. Reintroductions have repopulated nearly all the states in this region except for western Iowa and western Minnesota, where the status of introductions in several counties (Clay, Becker, Otter Tail) is in doubt.

Breeding Habitat: Breeding habitats vary greatly among the several subspecies; in the southern Rio Grande race the birds are found in very arid habitats dominated by short grasses but including the scattered trees necessary for roosting as well as a water supply and succulent vegetation. The Merriam race of the Badlands, Black Hills, and Pine Ridge country is typically associated with red cedar and ponderosa pines, running water, and rugged topography. The eastern race, native to eastern Oklahoma, eastern Kansas, and adjacent Missouri, is adapted to a variety of hardwood trees, of which oaks and other mast-bearing species are most important.

Nest Location: Nests are in well-concealed locations in forested areas, often under a log or a bush or at the base of a tree. The nest is lined with dead leaves.

Clutch Size and Incubation Period: From 8 to 15 eggs, buffy with reddish or brownish spots. The incubation period is 28 days, starting with the last egg. Single-brooded.

Time of Breeding: Egg dates in Oklahoma are from May 5 to July 18, and dependent young have been seen from May 8 to August 6. Texas egg dates are from February 15 to July 25.

Breeding Biology: Turkeys spend the winter in small flocks consisting of adult males or larger groups of hens and family units. When the "gobbling" season begins the males may establish individual gobbling or strutting areas, but groups of brothers typically associate, displaying in synchrony and allowing the most dominant of the brothers to fertilize any female that is attracted. Additionally, single highly aggressive males may dominate entire local populations, in a manner equivalent to the "master cock" situation in grouse. Females have brief contact with males until they are fertilized, then they establish their nests. Within a week the chicks can make short flights, and they soon begin to roost in trees. When young males reach the age of 6-7 months they may break away from their families and begin to establish brother unions that will persist for their entire lives.

Suggested Reading: Watts and Stokes 1971; Lewis 1973.

FAMILY GRUIDAE
(CRANES)

Whooping Crane

Whooping Crane
Grus americana

Breeding Status: Extirpated. Previously bred locally in North Dakota, Minnesota, and northern Iowa. The most recent known nesting was during 1915 in North Dakota and before 1900 in Minnesota and Iowa.

Breeding Habitat: Breeding on the Great Plains occurred in large prairie marshes and their adjoining wet meadows.

Nest Location: Nests are on piles of rushes, cattails, and similar vegetation in shallow water or on damp ground near water. They are often in water 12 to 18 inches deep and elevated about a foot above the water. There sometimes is a moat of open water immediately around the nest, probably because the birds use the emergent vegetation in the vicinity to build the nest.

Clutch Size and Incubation Period: Normally 2 eggs, buffy to olive with spots and blotches of brown tones. The incubation period is about 30 days. Single-brooded.

Time of Breeding: The relatively few egg dates available for this region are from May 4 (Iowa) to June 3 (North Dakota).

Breeding Biology: At present wild whooping cranes nest only in Wood Buffalo National Park, Northwest Territories, Canada. There, crane pairs defend the same territories year after year and occupy nonoverlapping home ranges of about a square mile. Nests are thus well isolated from one another, and both sexes defend the nest fiercely. Incubation is likewise shared, with several changeovers per day. As in sandhill cranes, incubation begins with the first egg, and thus hatching is staggered over a period of about 24 hours. Also like sandhill cranes, the young birds often fight and thus the younger, weaker bird sometimes is killed. The fledging period is apparently similar to that of the greater sandhill crane, about 2 months.

Suggested Reading: Allen 1952.

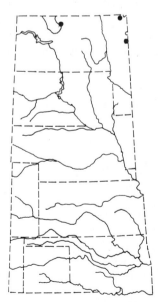

Greater Sandhill Crane
Grus canadensis tabida

Breeding Status: Virtually extirpated. The greater sandhill crane formerly bred rather widely in North Dakota and locally in South Dakota, as well as in Iowa and western Minnesota. The only definite North Dakota breeding record since 1916 was obtained in 1973 in McHenry County. Recent breeding evidence for northwestern Minnesota is from Roseau County (1954 and 1966) and Clearwater County (1972).

129

Breeding Habitat: Cranes require extensive areas of minimal human disturbance for their nesting. They have large territories that vary with population density but often exceed 100 acres, usually consisting of wet meadows that provide water, sites for feeding, nesting, and roosting, and brood-rearing areas. The extensive prairie marshes that once offered these features are now mostly drained, and human disturbance at the remaining ones is too severe for cranes.

Nest Location: Studies in Idaho indicate that nests are usually either in shallow water (averaging about 8 inches deep) or on the shoreline fairly near water (averaging about 15 feet away). In decreasing order of usage, they nest in wet meadow–marsh edge areas near shore, islands, dry upland meadows, marsh area far from shore, and artificial dikes. Old nest sites are rarely used in following years, but the nest is often placed near the old site. Nests on dry land are small and simply constructed, whereas those on water are more bulky, constructed of any vegetation easily available in the vicinity.

Clutch Size and Incubation Period: Normally 2 eggs (rarely 1 or 3). Eggs are olive with darker olive or brown spotting. The incubation period is 30 days, starting with the first egg. Single-brooded, with renesting frequent.

Time of Breeding: Egg dates in North Dakota are from May 1 to July 9, and dependent young have been seen from June 6 to July. Iowa egg dates are from May 2 to May 27, and in Minnesota eggs have been found from April 29 to May 21.

Breeding Biology: Cranes are monogamous, probably pairing for life after reaching reproductive maturity at about four years of age. Upon returning to their breeding areas, pairs establish territories as early as 2–4 weeks before nest-building gets under way. Nest-building is done by both sexes and may take from a day to a week or more. Eggs are laid at 2-day intervals, and both sexes participate in incubation, with the female apparently always doing the nighttime incubation. The eggs typically hatch 24 hours apart, and the chicks begin to feed immediately, with the first-hatched often taken away from the nest by one adult while the other remains to hatch the second chick. Perhaps because the young "colts" are very aggressive toward each other, they are often brooded separately. Fledging occurs at 67–75 days of age, and the family soon migrates as a unit.

Suggested Reading: Drewien 1973; Littlefield and Ryder 1968.

FAMILY RALLIDAE (RAILS, GALLINULES, AND COOTS)

Yellow Rail

King Rail
Rallus elegans

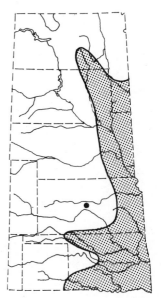

Breeding Status: Breeds in southwestern Minnesota and adjacent South Dakota (possibly to southeastern North Dakota), southward through eastern Nebraska (where rare), western Iowa (generally rare), northwestern Missouri (uncommon), eastern Kansas (locally common), Oklahoma (rare to uncommon), and adjacent Texas.

Breeding Habitat: In our region the king rail is generally associated with freshwater marshes. In an Iowa study the birds were found on shallow marshes up to 4 feet deep, with abundant shoreline and emergent vegetation of grasses and sedges. They are often associated with muskrats, whose runs open up the vegetation and provide passageways for the rails.

Nest Location: Nests are in rather dense emergent vegetation. In Iowa, 6 nests were in such vegetation, including 4 in lake sedges and 2 in river bulrushes, and all were in water 4–18 inches deep. The nests are basketlike structures of dead herbaceous vegetation, with an overhead canopy of emergent plants.

Clutch Size and Incubation Period: From 8 to 14 eggs (34 Ohio nests averaged 10.9). Eggs are pale buff with a few darker brown spots. The incubation period is 21–22 days, starting near the end of the clutch. Apparently single-brooded, except perhaps in the Deep South. Renesting probably occurs frequently after nest loss.

Time of Breeding: Egg records for northwestern Iowa are from May 13 to June 23. Kansas nesting records are from May 1 to July 20, with most eggs laid in early June. Oklahoma egg records are from April 29 to July 6, and Texas egg records are from February 6 to August 12.

Breeding Biology: The onset of the breeding season in king rails is marked by the males establishing territories and beginning their low-pitched mating call, *chuck-chuck-chuck*, which attracts unmated females. Males evict other male rails, even of such small species as soras, from their territories. They also choose the nest site and do most of the nest-building. Usually several brood nests are also built and later are used for brooding the chicks. Both sexes incubate, with most of the young hatching simultaneously. The young grow rather slowly and remain close to their parents for more than a month. They do not fledge until they are 9–10 weeks old, and during this fledging period the adults molt and become flightless for a time.

Suggested Reading: Meanley 1969; Tanner and Hendrickson 1956.

Virginia Rail
Rallus limicola

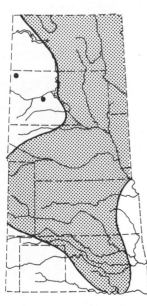

Breeding Status: Although rather elusive, this species is evidently a fairly common breeder in wetlands throughout most of the region concerned, with the probable exception of the Texas panhandle and adjacent New Mexico.

Breeding Habitat: Marshes with extensive stands of emergent vegetation such as taller grasses (cattails, phragmites), bulrushes, and sedges are the primary breeding habitat of this species. Habitat needs of Virginia rails and soras appear to be virtually identical. However, at least in Iowa, Virginia rails tend to nest in cattails and eat more insects and duckweeds, while soras favor whitetop or sedges for nesting and include a larger proportion of seeds in their diets.

Nest Location: Nests are built over wet ground or shallow water in stands of emergent vegetation. When nests are built over water, the water is rarely more than 10 inches deep. In Minnesota, all of 17 nests found in one study were in cattails, usually within 20–30 feet of open water or other vegetational edges. In Iowa, lake sedge was found to be the most important cover for 27 nests, and a Virginia study also indicated a preference for sedges and grasses over cattails for nesting. The nest is typically lined with fine grassy material and has an overhead canopy of live emergent plants.

Clutch Size and Incubation Period: From 6 to 13 eggs (28 Iowa clutches averaged about 8). Eggs are buffy to white with a few brown spots near the larger end. The incubation period lasts 17–20 days, with an average spread of 3.3 days between the hatching of the first and last egg. Probably single-brooded, but some renestings have been reported.

Time of Breeding: North Dakota egg dates are from June 12 to August 1. In northwestern Iowa the eggs are found over a period of about 50–60 days, with most hatching between June 6 and July 12. In Kansas, eggs are evidently laid in May and June, and in Oklahoma eggs (or females about to lay eggs) have been reported from late April to June 13.

Breeding Biology: Shortly after returning to their breeding grounds, males establish territories, which they proclaim by uttering their distinctive *ticket, ticket* calls and maintain by evicting other male Virginia rails, though they reportedly tolerate sora rails. They probably construct their nests in a few days, but like other rails they may also build several "dummy nests" that are later used as brood nests. Males perform bill-nibbling and courtship feeding of their mates and perhaps do most of the nest-building as well. Eggs are laid approximately daily, and incubation (by both sexes) begins near the end of the clutch, resulting in a slight scattering of hatching periods. The young leave the original nest soon after hatching and can fly in 6–7 weeks. When they are about 60 days old the parents begin to peck at them and evict them from their territories.

Suggested Reading: Tanner and Hendrickson 1954; Kaufmann 1971.

Sora
Porzana carolina

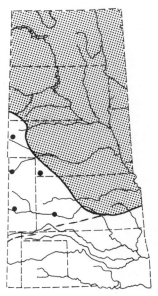

Breeding Status: Locally common in marshes over nearly all of North Dakota, western Minnesota, and western Iowa and the eastern portions of South Dakota and Nebraska. Local and uncommon in Kansas and northwestern Missouri and apparently only a migrant in Oklahoma and northern Texas. The southern and western breeding limits are uncertain; some reportedly occur in the Wyoming and Colorado plains, but there is no good evidence that the species breeds in western Kansas.

Breeding Habitat: Much like the Virginia rail, the sora prefers marshlands that have extensive stands of dense emergent vegetation, especially tall grasses and grasslike plants, and fresh to slightly saline waters. The birds feed mostly at the surface on plant seeds rather than probing for invertebrates as is typical of Virginia rails.

Nest Location: Where Virginia and sora rails nest in the same marshes, sora nests tend to be in deeper water, averaging from about 9 to 12 inches in depth. The nest is elevated several inches above the water level and is often hidden in cattails, bulrushes, or sedges. It is basketlike, with a deep cup and sometimes a lateral runway to water.

Clutch Size and Incubation Period: From 6 to 13 eggs (29 Minnesota nests averaged 9.9). Eggs are a rich buffy color with some darker spotting and are darker overall than those of Virginia rails. They are laid daily, and incubation begins at varied stages of clutch completion. The incubation period averages about 19 days, but the spread of hatching is from 3 to 13 days, averaging about 7. Considered single-brooded, but there is some evidence of double-brooding.

Time of Breeding: North Dakota egg dates are from May 20 to July 30, and Minnesota records extend from about May 10 to July 10. Both ranges indicate a long nesting period and suggest renesting or double-brooding.

Breeding Biology: Territorial male soras are more aggressive than Virginia rails, evicting individuals of that species as well as of their own. The *whinney* is the male advertisement call and peaks

135

at the time egg-laying gets under way. Nest-building is probably by both sexes, and several "dummy nests" are usually constructed near the primary nest. Both sexes incubate, and as the first chicks hatch they are tended by one parent while the other incubates the remaining eggs. Compared with Virginia rail young, soras are fed and brooded for a relatively short time, which perhaps facilitates second broods in some circumstances. The young attain their full juvenile plumage by 6 weeks and can fly when only about 36 days old. By this time in late July the adults have become flightless and are replacing their wing and tail feathers.

Suggested Reading: Pospichal and Marshall 1954; Tanner and Hendrickson 1956.

Yellow Rail
Coturnicops noveboracensis

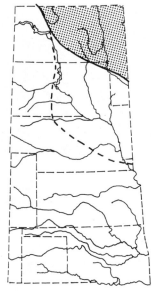

Breeding Status: A local and elusive species, with few known nesting records in north-central North Dakota (Benson County, scattered summer records for Bottineau, McLean, and Mountrail counties), and southwestern Minnesota (breeding records from Becker, Mahnomen, and Murray counties, and several other summer records). It probably also breeds in southeastern South Dakota, and it possibly breeds in Nebraska, where there are two summer records (*Nebraska Bird Review* 41:24).

Breeding Habitat: In North Dakota, yellow rails are limited to fenlike areas or boggy swales associated with springs. Often they are quaking surface mats of emergent vegetation such as cattails, bulrushes, sedges, and associated species. Yellow rails sometimes nest in the same marshes as sora and Virginia rails but occupy the densest areas of sedges, while the other species occupy cattails and bulrushes.

Nest Location: Nests may be built over wet ground or over water up to 4 inches deep, usually in dense emergent vegetation consisting of grasses and grasslike plants. The nest is usually under a canopy of dead grass and fairly close to a spring-fed brook. It is a coiled cup of dead grass lined with bits of grasses, sedges, and mosses.

Clutch Size and Incubation Period: From 8 to 10 eggs (5 North Dakota nests averaged 9.4). Eggs are buffy to pinkish with numerous small brown spots. The incubation period is 17 days, beginning with the last egg.

Time of Breeding: Egg dates in North Dakota are from May 25 to June 19, with most egg-laying probably occurring in the first 10 days of June, and hatched eggs have been seen as early as June 16.

Breeding Biology: During spring, males establish territories in dense marshes, patrolling them frequently and uttering their distinctive clicking notes (*tic-tic, tic-tic-tic*). Males are immediately evicted, but females are approached with a wing-spreading display. After pair bonds are formed, the mates preen each other and copulations are frequent. Nest-building begins nearly a month before incubation, with both sexes participating, and several extra brood nests are often constructed. The female does the final lining of the nest and apparently performs all the incubation, leaving the nest only to feed for brief periods. She also finishes building a brood nest after the clutch hatches; the chicks hatch nearly synchronously. The female broods and feeds her young both in and out of the nest for about 17 days, after which they are brooded only at night. They are nearly independent by their 3rd week of life but do not fledge until they are about 35 days old. Evidently the male plays no active role in defending or caring for the brood.

Suggested Reading: Stalheim 1975; Terrill 1943.

Black Rail
Laterallus jamaicensis

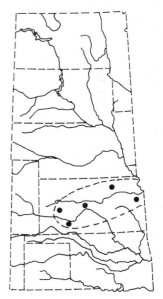

Breeding Status: The breeding range of this tiny and elusive species is most uncertain. It is considered hypothetical in North Dakota, an accidental migrant in Minnesota (no summer records), a rare migrant in Iowa and South Dakota, a rare migrant and summer resident in Nebraska, an uncommon summer resident in Kansas, and a rare migrant and possible breeder in Oklahoma. There is one summer record from the Texas panhandle. Thus only Kansas (Finney, Meade, Riley, Barton, and Franklin counties) can be definitely considered breeding range for this region on the basis of current evidence.

Breeding Habitat: Marshy meadows, heavily overgrown with sedges and grasses, are the breeding habitat of this species in the interior, although it also nests in salt-grass marshes just above the tideline of coastal areas.

Nest Location: Nests are on damp ground, in dense grass or sedge vegetation, or above water on a mat of grasses. The nest is typically arched over with interwoven grasses and has a lateral entrance. The surrounding vegetation is usually 18-24 inches high, and the deep nest cup sometimes contains a few black feathers.

Clutch Size and Incubation Period: From 6 to 13 eggs, probably averaging about 8. The eggs are creamy to buffy with large reddish spots at the larger end. The eggs are laid daily, but the incubation period is unknown. Hatching reportedly is synchronous, and the young have been reported to leave the nest the day they hatch. Probably single-brooded.

Time of Breeding: In Kansas, eggs are laid at least during June, and the species is present in the state between March 18 and September 26. Texas egg dates are May 9 and June 5.

Breeding Biology: Very little is known of the breeding biology of this species. The male's best-known call is a metallic *kik-kik-kik-ker* or *kik-kik-ker*, while the female is said to use a more cuckoo-like *croo* note in response to her mate. Virtually nothing is known of the specific aspects of behavior associated with nesting, but presumably they are much like those of the yellow rail.

Suggested Reading: Bent 1926; Todd 1977.

Purple Gallinule
Porphyrula martinica

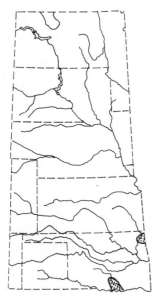

Breeding Status: Breeds locally in Oklahoma, at least in Bryan County, and possibly also in Marshall, Johnston, Carter, Delaware, and Grady counties.

Breeding Habitats: In our region this species is limited to marshes with extensive growth of water lilies, lotus, and other aquatic vegetation. It also occurs in tropical swamps, rivers, lagoons, rice plantations, and similar habitats through much of the western hemisphere.

Nest Location: Nests are built over relatively deep water, sometimes on floating islands, in pickerelweed, or in woody vegetation. The nest is well concealed from above by arched-over vegetation and has a flat lateral runway leading downward to the water.

Clutch Size and Incubation Period: From 5 to 10 eggs, usually 6-8. The eggs are pale buff with a few small brown spots. The incubation period is 20-23 days, starting before the clutch is completed. Probably double-brooded in our area.

Time of Breeding: In Oklahoma, eggs have been seen from May 15 to July 18 and young recorded from May 15 to August 18. In Texas, egg dates are from April 9 to August 12, and barely fledged young have been seen as late as September.

Breeding Biology: Few studies have been done on the behavior and biology of this species, but it resembles the common gallinule in being highly territorial and in advertising courtship and feeding territories with repeated *kuk* or *keek* notes. Nest-building begins a few weeks after the birds arrive and establish territories; the male probably does most of the nest-building. Both sexes incubate, and mates perform a nest-relief ceremony of presenting a leaf to the incubating bird, which incorporates it into the nest before departing. The young hatch over a period of about 4 days and are brooded actively for about a week, after which they are brooded only at night. One or more brood nests is typically present. The fledging period is uncertain but is probably about 6–7 weeks, and adults also apparently undergo a flightless period during late summer.

Suggested Reading: Meanley 1963; Trautman and Glines 1964.

Common Gallinule (Moorhen)
Gallinula chloropus

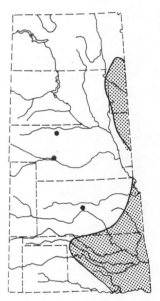

Breeding Status: Breeds in west-central and southwestern Minnesota, northwestern Iowa, eastern Kansas, and the eastern half of Oklahoma. It is an occasional summer resident in northwestern Missouri and rare in Nebraska, with several breeding records, but its range limits are not certain.

Breeding Habitat: The favored habitat of this species is freshwater ponds and marshes with an abundance of emergent vegetation. Unlike the purple gallinule, it does not need floating vegetation.

Nest Location: Nests are in water, suspended above water, or on land surrounded by water. Deepwater nests usually have a ramp up the side, whereas those in shallow water or on land do not. In Iowa, 17 of 19 nests were in cattails, the others in bulrushes. The nest is constructed of emergent and aquatic plants and has a well-developed cup of finer vegetation.

Clutch Size and Incubation Period: From 5 to 10 eggs (13 Iowa clutches averaged 7.1; in England first clutches average about 6 eggs, and renests or second clutches are somewhat smaller). The eggs are buffy with small brown dots or spots. The incubation period is 21–22 days, starting (in first clutches) with the next-to-last egg, or (in later clutches) midway through the laying period. A regular renester and sometimes double-brooded.

Time of Breeding: In Kansas eggs are laid in May and June, and in Iowa nests are also initiated between mid-May and late June. Oklahoma egg dates are from May 15 to July 18, and young have been seen from July 2 to August 8.

Breeding Biology: Common gallinules are highly territorial birds and in some areas maintain winter core-areas that later expand to become breeding territories. Within the territories the birds build three kinds of structures: display platforms, egg nests, and brood nests. Up to five temporary display platforms are built early in the breeding season, and one or two egg nests are constructed a week or two before egg-laying. The male gathers most of the nest materials, and the female incorporates them into the nest. Eggs are laid daily, and both sexes incubate. The young of the first brood hatch nearly synchronously and are fed by their parents within an hour after hatching. Up to 5 brood nests are built after the brood hatches. The young are tended by both parents for varying periods; in one case a pair began a new nest only 26 days after hatching their first brood. The chicks fledge at 60–65 days of age and tend to disperse soon afterward.

Suggested Reading: Frederickson 1971; Wood 1974.

American Coot
Fulica americana

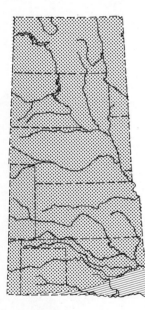

Breeding Status: Pandemic throughout region. A common to abundant breeder in the Dakotas, Minnesota, western Iowa, and Nebraska, becoming more uncommon and local in Kansas and Oklahoma and occasional to rare in the Texas panhandle.

Breeding Habitat: Wetlands with open water and emergent vegetation interspersed are favored, especially those that are fairly shallow and rich in submerged aquatic plants. Coots sometimes also forage in wet meadows and on grassy shorelines of lakes or ponds.

Nest Location: In North Dakota, hardstem bulrush is the predominant species of emergent vegetation used for nesting cover. Cattails and other bulrush species are also frequently used, and in an Iowa study cattail cover accounted for more than 250 of 320 nests studied. Nests are built over water ranging from 5 to nearly 60 inches deep and are floating platforms anchored to the surrounding vegetation.

Clutch Size and Incubation Period: From 5 to 15 eggs (502 North Dakota nests averaged 8.8, and 281 Iowa nests averaged 9.0). The eggs are buffy with small brown spots. The incubation period is 23–27 days, with onset of incubation ranging from the first egg to the last egg, and hatching of the young is usually staggered. Usually single-brooded, but renesting is frequent and double-brooding has been reported in Utah and California.

Time of Breeding: North Dakota egg dates are from April 29 to August 13, and young have been seen from May 22 to September 15. Nest initiation in an Iowa study ranged from early May to late

June, and Kansas egg records span the period May 11 to June 30, with a peak in late May. Oklahoma egg dates are from May 15 to June 23, with young seen as late as July 15.

Breeding Biology: Coots are monogamous, with a potential life-long pair bond, and spend much of their time in advertising and defending territories. These are established soon after arrival on the breeding grounds, and although the male patrols the territory at first, later it is defended by both members of the pair. Pairs also construct display platforms for copulation and, as the egg-laying period approaches, construct one or more egg nests as well as brood nests later on. Both sexes participate in incubation, with the male most often incubating at night. Unlike gallinules, coots seem to have no specific nest-relief ceremony. Hatching is typically staggered over several days. Apparently the male takes the major responsibility for brooding the young birds, although the female may take the first-hatched chicks and leave the male to incubate and tend the later hatchlings. The young begin to beg shortly after hatching and soon begin to follow the adults during their foraging. After a month or so they are nearly independent, but they beg occasionally almost to the time they fledge, at about 75 days of age. If the adults begin a second clutch they may expel the young of the first brood from the area while they are still fairly young.

Suggested Reading: Fredrickson 1970; Gullion 1954.

FAMILY CHARADRIIDAE (PLOVERS AND LAP-WINGS)

Killdeer

Piping Plover
Charadrius melodus

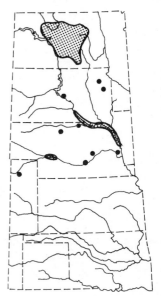

Breeding Status: Breeds uncommonly and locally in North Dakota, mainly on the Missouri Coteau, sporadically in west-central Minnesota (Otter Tail and Douglas counties), occasionally in South Dakota (Union, Clay, Yankton, Hughes, and Codington counties), and rarely farther south. There are many old nesting records for Nebraska (Niobrara, Platte, Loup, and Missouri rivers), and recent ones for Douglas, Hall, Holt, Saunders, and Washington counties, as well as at Lake McConaughy, Keith County. There is also a 1949 breeding record for Washington County, Colorado, but no breeding records exist for Kansas or farther south.

Breeding Habitat: In North Dakota, piping plovers are associated with sparsely vegetated shorelines of shallow lakes and impoundments, especially those that have salt-encrusted areas of gravel, sand, or pebbly mud. Sand dunes with little or no vegetation are also used for nesting.

Nest Location: Nests are simply hollows in the sand, sometimes lined with pebbles, or scrapes in gravel or pebbly mud.

Clutch Size and Incubation Period: From 2 to 4 eggs (typically 4 in first clutches, sometimes 3 in renesting efforts). Eggs are buffy with dark brown spots. Incubation ranges from 27 to 29 days, starting with either the third or the last egg. Single-brooded, but renesting usually occurs if the clutch is lost in the first half of the breeding season.

Time of Breeding: North Dakota egg dates range from May 19 to July 5, and dependent young have been seen from June 26 to July 27. Nebraska egg records are from June 10 to July 14.

Breeding Biology: Piping plovers are monogamous, but mate-changing in successive breeding seasons is fairly frequent, even when the original mate is still available. Eggs are laid every other day, and incubation responsibilities are about equally divided by the two sexes. In most nests the eggs all hatch on the same day, and within 2–3 hours the young have dried off and are able to leave the nest. They are brooded by the adults until they are about 20 days old, and although they can run very well they tend to crouch and freeze when approached. Adults of both sexes feign injury when their brood is threatened. Until they fledge at 30–35 days of age, the young remain within 400–500 feet of the nest.

Suggested Reading: Wilcox 1959; Stout 1967.

Snowy Plover
Charadrius alexandrinus

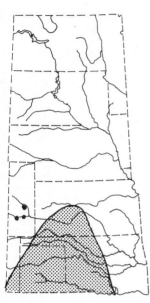

Breeding Status: A local summer resident in central Oklahoma, the Texas panhandle, and south-central to southwestern Kansas, largely limited to saline flats and sandy riverbeds. There are two nesting records for eastern Colorado (Kiowa County, 1939), and the species also nests in the vicinity of Roswell, New Mexico.

Breeding Habitat: The barren salt plains area of northern Oklahoma represents prime breeding habitat for this arid-adapted species, and sandy riverbeds or barren shorelines of reservoirs are used secondarily.

Nest Location: Nests are on rock, gravel, or sandy substrates and consist of a slight hollow lined with bits of debris. Occasionally the nests are clustered in loose colonies, and the birds sometimes nest near tern colonies.

Clutch Size and Incubation Period: Usually 3 eggs, sometimes 2. The eggs are sand-colored or buffy with small black spots or lines. Incubation lasts 23–29 days, averaging 26 days. Possibly double-brooded.

Time of Breeding: Kansas egg dates range from May 25 to June 20, with a peak of laying about June 10. Oklahoma egg records are from April 29 to July 11.

Breeding Biology: After arriving on their breeding areas and establishing territories, males begin to advertise with various calls and displays including "scraping," a ritualized nest-building behavior. One of the other male displays is a slow "butterfly flight" with a trilling call. Although the birds commonly breed around salt water, they can drink no more saline water than other shorebirds and must obtain liquid by eating insects or other succulent foods. Thermal extremes are also common in their often vegetationless and highly reflective environment. Thus, during hot weather parental activity increases, the birds spending most of their time standing over the eggs rather than sitting on them. Both sexes incubate. The eggs are laid about 3 days apart, but hatching is synchronous. Both sexes also defend the eggs and young, performing effective "broken-wing" behavior. In one study the young fledged within 41 days.

Suggested Reading: Purdue 1976; Boyd 1972.

Killdeer
Charadrius vociferus

Breeding Status: A pandemic summer resident throughout the region, locally common around marshes, streams, and other water areas.

Breeding Habitat: Killdeers breed wherever there are wetlands that either have exposed ground nearby or have ground with very little vegetative cover. They seem to prefer gravelly, stony, or sandy areas over muddy or silty substrates, probably because they offer camouflage for the eggs and the incubating bird.

Nest Location: Nests are often some distance from water, in a surprising variety of locations. Of 13 North Dakota nests, 3 were on garden plots, 2 on bare fields, 2 on heavily grazed native prairie, 2 on exposed sand or gravel, 2 on bare lake shorelines, and 1 each in a stubble field and an abandoned farmyard.

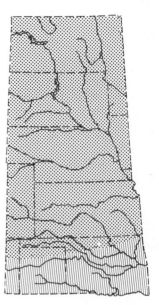

Clutch Size and Incubation Period: Nearly always 4 eggs, rarely 3 or 5. The eggs are buffy with extensive black or dark brownish spotting and blotching. The incubation period is 24–26 days, starting with the laying of the last egg. Sometimes double-brooded, especially toward the south.

Time of Breeding: North Dakota egg dates range from April 18 to June 21, and dependent young have been seen from May 19 to July 25. In Kansas, egg dates are from March 21 to June 30, with a double peak of nest dates suggesting double-brooding. Oklahoma egg dates are from March 30 to July 28, and dependent young have been seen from April 16 to September 19, also indicating double-brooding. Texas egg dates are from March 3 to July 17.

Breeding Biology: Although some birds are paired at the time they arrive on their nesting areas in southern Canada, most arrive unpaired. Males advertise their territories in a variety of ways, such as uttering the familiar *killdeer* calls while flying with slow, deep wingbeats, and by sham-nesting or "scraping" displays resembling nest-building behavior. Such scraping displays are performed not only by unmated males but also before copulation, during hostile encounters, and during actual nest construction. Once pair bonds are formed, the pair remains together and both sexes defend their territory, although they may do some foraging outside the defended area. Both sexes also incubate the eggs and care for their young, but males tend to be more aggressive toward humans, while females vigorously evict other killdeers from the nest vicinity. The familiar injury-feigning display, or "broken-wing act," is primarily directed toward potential mammalian predators; large grazing mammals such as horses and cattle are more likely to be threatened or even attacked. Evidently the male undertakes most of the brooding duties, which last about 3 weeks. Fledging occurs by the time the young are 40 days old.

Suggested Reading: Phillips 1972; Bunni 1959.

Mountain Plover
Eupoda montana (*Charadrius montanus*)

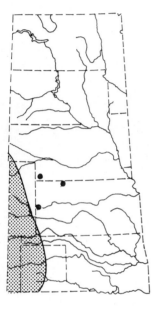

Breeding Status: Summer resident from extreme southwestern Nebraska (nested 1974, Kimball County) southward through eastern Colorado, possibly extreme western Kansas (early nesting records for Greeley and Decatur counties, possibly nested in Hamilton County in 1964), extreme western Oklahoma (Cimarron County), the western panhandle of Texas (breeding record for Swisher County), and adjacent New Mexico.

Breeding Habitat: Mountain plovers are essentially limited to the short-grass plains but at times occur on sandy semiarid flats supporting some brush and cacti. In northeastern Colorado they are restricted to flat, heavily grazed areas dominated by grama grass, buffalo grass, and similar plains vegetation.

Nest Location: The nest is simple and initially consists of a scrape on bare ground, often fairly close to roadways, placed in the open rather than under vegetation. A simple lining of dried grass or other debris may be added as the clutch is completed, but this may be deposited by wind rather than purposely added by the bird.

Clutch Size and Incubation Period: From 2 to 4 eggs (105 of 133 Colorado nests had 3 eggs). Eggs are dark greenish buff with black spots and lines. The incubation period is about 25 days. Multiple-brooded, the female sometimes changing mates between clutches.

Time of Breeding: Oklahoma egg dates are for May 17 and June 30, and in Texas eggs or downy young have been seen in late May. Colorado egg dates extend from April 17 to June 15, and young have been reported from June 16 to July 23.

Breeding Biology: In northeastern Colorado, mountain plovers arrive in late March and soon disperse over their breeding grounds. Males commonly reestablish their old territories, whereas females also return to the same general area but may visit several territories before choosing mates. Territorial males advertise with calls and an aerial "falling-leaf" display, and occasionally with a slow "butterfly flight." As in other plovers, "scraping" is the most frequent courtship display of the male, which produces several potential nest sites throughout his territory.

148

Although monogamous pair bonds are soon formed, social relationships become complex when egg-laying begins. At least some females begin a second clutch with new mates within about 2 weeks of completing their first clutches, leaving their first mates to attend to the original clutches. Evidently the female often incubates the second clutch herself, but current evidence indicates that only one sex is involved in incubation and brooding duties for each clutch and brood.

Suggested Reading: Graul 1974; Laun 1957.

FAMILY SCOLOPACIDAE (WOODCOCKS, SNIPES, AND SANDPIPERS)

Marbled Godwit

American Woodcock
Philohela minor

Breeding Status: Summer resident in west-central and northwestern Minnesota, with nesting records for Pennington and Becker counties and a summer record from Pope County. Nesting in Iowa is fairly frequent (*Iowa Bird Life* 46:65) and has occurred in Brookings County, South Dakota (*South Dakota Bird Notes* 25:6). There is a 1972 nesting record for Sarpy County, Nebraska (*Nebraska Bird Review* 42:43). There are also three recent Kansas nesting records for Jefferson, Woodson, and Douglas counties (*Kansas Ornithological Society Bulletin* 26:22, 27:9, 28:22). There is a 1973 breeding record for Payne County, Oklahoma (*American Birds* 27:789), and since that time there have been a surprising number of Oklahoma nestings. Outside its Minnesota range, these scattered woodcock nestings must be considered as extra-limital records, although there is some evidence of a recent western extension of the breeding range (*American Birds* 32:1122).

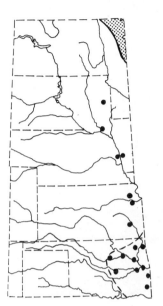

Breeding Habitat: Woodcocks are generally confined to young forests with scattered openings on rather poorly drained land, especially soils supporting a large population of earthworms that can be readily obtained by probing. Nesting cover is usually of hardwood or mixed hardwood and conifer trees but may also be dominated by brushy growth.

Nest Location: Nests are usually within 500 feet of a male's territory and are typically less than 50 yards from the edge of woody cover. Of more than 200 nests studied in Maine, nearly half were in mixed hardwoods and conifers, and most of the rest were in pure alder or other hardwood cover. The nest is usually at the base of a small tree or shrub and is simply a slight depression in the soil with little or no vegetation lining it.

Clutch Size and Incubation Period: Nearly always 4 eggs, rarely 3. The eggs are pinkish buff to cinnamon with darker brown spotting. They are laid daily, incubation beginning with the last egg and lasting 20–21 days. Single-brooded.

Time of Breeding: Minnesota egg dates are from April 27 to June 23. In Kansas the few records suggest that eggs are probably laid from March to May, and young have been seen in April and May. In Oklahoma, young have been seen in early to mid-April.

Breeding Biology: Shortly after returning to their breeding grounds, males begin their distinctive dawn and dusk display flights from territorial "singing grounds." These consist of a series of calls, a hovering flight with "twittering" wing noise, and a zigzag flight back to earth accompanied by a series of liquid trilling notes. The male attempts to copulate with any females that are attracted to such singing grounds, and it is probable that no pair bond is established. The female locates her nest in the

153

general vicinity of the singing ground but is not protected by the male, and she does all the incubation and brooding alone. Female woodcocks are noted for being extremely "tight" sitters and if finally forced off the nest will perform strong injury-feigning displays. The young are soon led from the nest and begin to feed on earthworms as early as 3 days after hatching. They can fly short distances by the time they are 3 weeks old, and most broods probably break up between 6 and 8 weeks after hatching.

Suggested Reading: Sheldon 1967; Godfrey 1975.

Common Snipe (Wilson Snipe)
Gallinago gallinago (*Capella gallinago*)

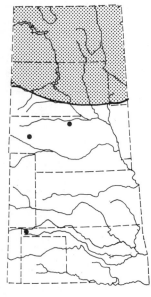

Breeding Status: A summer breeding resident throughout Minnesota, North Dakota, and most probably South Dakota (no actual nesting records), becoming more local and uncommon southward. There are no recent breeding records for northwestern Iowa, and in Nebraska breeding is known only from Garden County and, recently, Rock County (*Nebraska Bird Review* 38: 17). There are no breeding records for Kansas and only a 1910 record for Cimarron County, Oklahoma. It is likewise not known to breed in eastern New Mexico or northern Texas.

Breeding Habitat: In North America this species is primarily associated with peatland habitats such as bogs, fens, and swamps, which in our region are generally confined to Minnesota. Farther south, snipes also breed in marshy habitats along ponds, rivers, and brooks, where mucky organic soil and rather scanty vegetation are to be found.

Nest Location: Nests are usually in rather wet locations, usually on a hummock in cover provided by mosses, grasses, or heather. When nests are built in grasses, the previous year's growth is interwoven to form a canopy. A lining of fine, dry grasses is also typical.

Clutch Size and Incubation Period: Typically 4 eggs, rarely 5 or 3. The eggs are light to dark brown with heavy spotting or blotching of darker brown. Incubation begins with the last egg and lasts 17–20 days, usually about 18 days. Single-brooded, but renesting in some areas is probable.

Time of Breeding: In North Dakota the breeding season lasts from early May to mid-July, with eggs seen from May 20 to June 27 and young reported from June 9 to July 15. In Minnesota, eggs have been seen as early as May 10, and the single Oklahoma egg record is for June 3.

154

Breeding Biology: During migration the males fly in advance of the females and arrive on their breeding grounds up to 2 weeks before them. Males immediately establish territories and begin advertising them with several displays, especially "winnowing," an aerial display in which the bird dives at a 45° angle with the tail fanned horizontally and the wings quivering. The vibration of the outer tail feathers produces the distinctive tremulous sound, with the wings used as "dampers" to prevent excessive vibration. After females arrive there is a good deal of chasing, and the female may mate with several males before forming an association with one, which happens when she selects the nest site and begins to lay. The pair bond lasts only until the chicks are hatched, and during incubation the male may also court other females. Only the female incubates, but the male returns to the nest at the time of hatching and collects the first active chicks, leaving the last two or three to be cared for by the female. The chicks grow rapidly and can flutter short distances at 2 weeks, but they cannot make protracted flights until they are about 3 weeks old. When about 6 weeks old they begin to gather with other young in "wisps" that may number in the hundreds and begin to migrate south before the adults.

Suggested Reading: Tuck 1972.

Long-billed Curlew
Numenius americanus

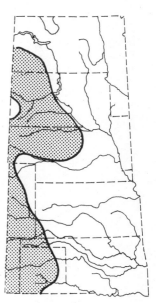

Breeding Status: Breeds in southwestern North Dakota (primarily Bowman and Slope counties), western South Dakota, western Nebraska (the Sandhills area), eastern Colorado (particularly Baca County), southwestern Kansas (Stanton and Morton counties), extreme northwestern Oklahoma (Cimarron County), the western panhandle of Texas, and eastern New Mexico.

Breeding Habitat: This species is most often associated with short-grass plains, grazed mixed-grass prairies, or combinations of short-grasses, sage, and cactus, often on gently rolling terrain.

Nest Location: Favored nest sites are damp, grassy hollows in prairie vegetation or long slopes near lakes or streams. The nest is simply a slight hollow lined with a varying amount of grasses or weeds. At times the birds nest in loose colonies, and they frequently place their nests beside dried cow dung, presumably for better concealment. In the Nebraska Sandhills, the proximity of potential upland nesting areas to moist meadows for foraging was found to be the most important criterion for nest sites.

Clutch Size and Incubation Period: Usually 4 eggs, sometimes 5, and rarely more in multiple clutches. The eggs are mostly olive buff with variable spotting of darker browns. The incubation period is probably 27–28 days. Single-brooded.

Time of Breeding: The probable breeding season in North Dakota is from late April to early August, with a peak from early May to early July. Kansas egg records are for May and June, and Oklahoma egg records extend from May 10 to July 1, with young birds seen as early as June 8. Texas egg records are from May 15 to June 10, and newly fledged young have been reported from June 1 to June 16.

Breeding Biology: In the Nebraska Sandhills, long-billed curlews arrive by early April, usually in flocks of fewer than 12 birds. The rest of the month is spent in prenesting activities, including establishing core areas and foraging areas. Core areas typically consist of rolling sands and are advertised by extended flight displays and calling above the ultimate nest site. Meadows adjacent to nesting locations are used for foraging and are advertised by similar flight displays. The foraging area is a part of the defended territory, and other curlews are forcibly excluded. Both sexes incubate, and both sexes care for the brood. The fledging period is not precisely known, but in a Nebraska study the last fledging occurred about a month after the end of the hatching period in mid-June. By early August, all the adults and juveniles have departed from the area.

Suggested Reading: Bicak 1977; Fitzner, 1978.

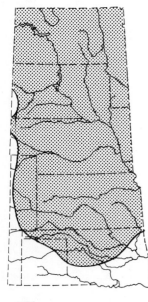

Upland Sandpiper (Upland Plover)
Bartramia longicauda

Breeding Status: Breeding occurs over nearly all of the region except for the southernmost portion, including eastern New Mexico (where it may have bred formerly), the Texas panhandle (where it is sometimes seen in summer but no breeding records exist), and southern Oklahoma. In Oklahoma it formerly bred south to Washita and Comanche counties and more recently (1963) to Oklahoma County.

Breeding Habitat: Breeding occurs on native prairies, especially mixed-grass and tall-grass, on wet meadows, on hayfields, on retired croplands, and to a small extent on fields planted in small grains. Throughout the area this species' abundance has declined as the extent of land in native prairies has decreased in recent decades.

Nest Location: In North Dakota, all of 183 nests in one study were in grassland, mostly native prairie. The nest is simply a slight depression in the ground, usually well hidden in thick grass, with grasses arched overhead to provide protection. It is lined with dried grasses to form a rather deep cup.

Clutch Size and Incubation Period: Typically 4 eggs, rarely 3 or 5 (all of 189 North Dakota nests had 4 eggs). The eggs are creamy to pinkish buff with reddish brown spotting on the rounded end. The incubation period averages 24 days, starting with the last egg and ranging from 21 to 28 days. Single-brooded, but renesting is probable.

Time of Breeding: North Dakota egg dates are from May 15 to July 22, and dependent young have been seen from June 14 to August 2. Egg dates in Kansas are from April 21 to June 10, with a peak of egg-laying in early May. Oklahoma egg dates are from June 1 to June 24, and young have been reported from June 12 to July 5.

Breeding Biology: In North Dakota, the first spring arrivals appear about 2 weeks before the start of nesting and are usually paired birds. Territorial birds perform a flight display consisting of circling with quivering wingbeats while uttering a musical purring or chattering call and finally diving abruptly back to the earth. In North Dakota nesting begins almost simultaneously, and the eggs are laid at approximately daily intervals. Both sexes incubate, and adults typically feign injury when discovered on the nest. There is a fairly long interval between the first pipping and the hatching of the last egg, which may vary from less than 24 hours to about 3 days. The chicks are brooded by both parents, and by the time they are 30 days old they appear to be full grown and presumably are fledged.

Suggested Reading: Higgins and Kirsch 1975; Stout 1967.

Spotted Sandpiper
Actitis macularia

Breeding Status: A breeding summer resident that is locally common throughout the region north of Kansas but also breeds in low densities in Kansas, has bred twice (1910, 1911) in Cimarron County, Oklahoma, breeds locally in Union County, New Mexico, and has bred once in Deaf Smith County, Texas.

Breeding Habitat: This species uses water areas with exposed or sparsely vegetated shorelines or islands, ranging from moving-water habitats such as streams to stillwater ponds and lakes. The

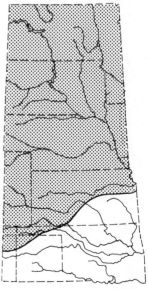

physical and chemical characteristics of the water are evidently secondary to the shoreline features.

Nest Location: Nests are on the ground in rather open terrain, often some distance from water. Cover above the nest varies from grasses 6–30 inches tall to weeds or bushes, and the nest itself is a slight depression lined with dried grasses.

Clutch Size and Incubation Period: Usually 4 eggs, sometimes 3 or 5 (15 North Dakota nests averaged 3.9). The eggs are buffy with heavy spotting of dark brown. The incubation period is 20–22 days, usually 21, and starts with the last egg laid. Some females are sequentially polyandrous and may lay several clutches.

Breeding Biology: Male and female spotted sandpipers arrive on their breeding grounds at about the same time, and pair bonds are formed extremely rapidly during a period of intense aggression, especially among females, which are larger and more aggressive than males. Females establish territories, and pairs are formed by males entering such territories and being either accepted or expelled by unmated females. When a male leaves the shoreline area and enters nesting cover with a female, a bond has been formed, and the female may lay her first egg within 5 days of the male's arrival. Eggs are laid at approximately daily intervals, and by the time she lays the third egg the female begins to show a resurgence of sexual activity, with increased singing and territoriality. Although some females remain monogamous and assist with incubation, others allow their first mates to undertake incubation duties and accept a second mate. Successive mating with as many as four mates in a single season has been found, and typically the female helps incubate the final clutch. The young birds leave the nest as soon as their feathers dry and reportedly are able to fly as early as 13–16 days after hatching.

Suggested Reading: Hays 1973; Oring and Knudson 1973.

Willet
Catoptrophorus semipalmatus

Breeding Status: Breeds locally in prairies and wetlands in the northern part of the region, including most of North Dakota but especially east of the Missouri River, glaciated portions of South Dakota, and the Nebraska Sandhills. There are no breeding records for the region south of Nebraska. It is possibly a casual summer resident in extreme western Minnesota, but there are no recent nesting records.

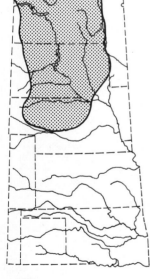

158

Breeding Habitat: In North Dakota, willets are found in a variety of habitats, including fresh to highly saline water areas, streams, and seasonal to semipermanent ponds and lakes, but with highest densities on brackish or subsaline lakes and semipermanent ponds.

Nest Location: Nests are in prairie vegetation, often 100 to 200 yards from the nearest water. Of 12 North Dakota nests, 8 were in native prairie, 3 were in cropland fields, and 1 was in tame hayland. The nests are usually in thick grass, with the grass blades bent down to help provide a nest base and other grass added for lining. Some nests have also been found in almost wholly exposed locations.

Clutch Size and Incubation Period: Typically 4 eggs, rarely 5 (15 North Dakota nests all had 4). The eggs are grayish to olive colored with darker brown spots and blotches. The incubation period is 22–29 days, occasionally starting before the clutch is complete, which results in staggered hatching. Single-brooded, but renesting has been reported.

Time of Breeding: North Dakota egg dates are from May 10 to June 21, and young have been seen from June 11 to July 30.

Breeding Biology: Willets arrive on their breeding grounds several weeks before egg-laying and the group includes paired birds as well as unpaired ones. Courtship is relatively social, and this flocking tendency conflicts with the territorial behavior of males, which tends to space the population. Aerial displays are common, consisting of calling the distinctive *pill-willet* call while moving the wings through a narrow arc. Sparring fights on the ground between males are also common. Precopulatory display consists of standing behind the female and similarly vibrating the open wings, thus displaying the white areas on them. After pair bonds are formed the male follows the female about, often spreading his tail and exhibiting the white feathers, while the female apparently chooses the actual nest site. The nests are usually well spaced, 200 feet or more apart, and the eggs are laid at intervals of 1 to 4 days. Presumably incubation begins before the clutch is completed, since estimates of it range from 22 to 29 days, but the eggs all hatch about the same time and the young are highly precocial. Little is known of posthatching biology, but apparently the parents abandon their offspring before fledging and leave the region.

Suggested Reading: Tomkins 1965; Stout 1967.

Marbled Godwit
Limosa fedoa

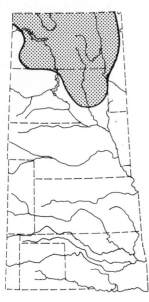

Breeding Status: A summer resident and breeder over most of North Dakota, primarily east of the Missouri, and adjacent prairie region of western and northwestern Minnesota, extending southward into the glaciated portions of South Dakota. Formerly more widespread and previously believed to breed in Nebraska, but no actual records exist.

Breeding Habitat: Godwits use a variety of wetland habitats for breeding in North Dakota, including intermittent streams, ponds, and lakes ranging from fresh to strongly saline. Semipermanent ponds and lakes appear to be the preferred habitat, followed by seasonal ponds and lakes, then miscellaneous wetlands.

Nest Location: Nests are usually in native grassland vegetation, sometimes considerable distances from water. Often the nests are in grassy cover only a few inches high and consist of a simple depression in the ground, lined with dead grasses. In higher grass cover the nest may be concealed by grasses interwoven to form a canopy overhead.

Clutch Size and Incubation Period: Typically 4 eggs, rarely 3 or 5. The eggs are buffy to olive with dull brown spotting and blotching. The incubation period has not been reported, but in related species it is slightly more than 3 weeks.

Time of Breeding: North Dakota egg dates range from April 17 to June 22, and dependent young have been seen from June 7 to July 18. Minnesota egg dates are from May 24 to June 21, and dependent young have been seen from June 8 to June 26.

Breeding Biology: Remarkably little is known of the breeding biology of this species. Females are appreciably larger than males and have considerably larger bills. In a nest found by A. C. Bent, the female was incubating, but more recent observations in North Dakota by T. Nowicki indicate that the male incubates during the day and the female at night. Incubating birds are surprisingly tolerant and have been known to allow themselves to be picked up from the nest. However, humans in the nesting area are sometimes attacked from the air by all the godwits in the vicinity, as many as 50 birds. The young are led to water after hatching and soon begin feeding with groups of adults. The fledging period has not been established, but in the related Hudsonian godwit it is about 30 days.

Suggested Reading: Bent 1907; Roberts 1932; Nowicki 1973.

FAMILY RECURVIROS-
TRIDAE (AVOCETS AND
STILTS)

American Avocet

American Avocet
Recurvirostra americana

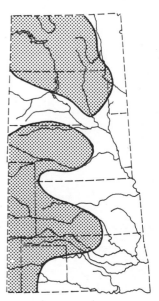

Breeding Status: A summer resident in the western portions of the region, extending locally eastwardly to extreme western Minnesota (Lyon, Stevens, Otter Tail, Lac qui Parle, and Big Stone counties), eastern South Dakota, central Nebraska, central Kansas (Finney, Barton, and Stafford counties), northwestern Oklahoma (Harper, Woods, and Alfalfa counties), and the western panhandle of Texas.

Breeding Habitat: In North Dakota, breeding is usually limited to areas of shallow water with exposed and sparsely vegetated shorelines, most often associated with alkaline to subsaline water areas. Of 253 pairs studied, the largest number were found on strongly saline alkali ponds and lakes, and very few occurred on freshwater ponds and lakes.

Nest Location: Nests are found on mud flats, sandbars, and islands, often only slightly above the water surface and with little or no associated vegetation. The nest is a simple scrape, with a lining of materials found in the immediate vicinity, and is most extensively lined in areas subject to flooding. Nests are often in loose colonies near favored foraging areas.

Clutch Size and Incubation Period: From 2 to 4 eggs (17 North Dakota nests averaged 3.1), typically 4 in completed clutches. The eggs are buffy to olive buff with many darker spots. The eggs are laid at approximately daily intervals, and the incubation period averages about 24 days but varies from 22 to 29 days. Single-brooded, but replacement clutches have been reported.

Time of Breeding: North Dakota egg dates range from May 12 to July 5, and dependent young have been seen from June 9 to July 27. Kansas egg records are from May 11 to June 20, with a modal egg-laying date of June 5. Oklahoma egg dates are from May 15 to June 21, and Texas egg records extend from May 6 to July 13.

Breeding Biology: In Oregon, avocets arrive on their breeding areas 15–20 days before egg-laying, to establish territories and perform precopulatory courtship. They apparently form pairs in late winter, without associated elaborate posturing. Copulation is preceded by a rather simple breast-preening ceremony that may be initiated by either bird. Pairs form close bonds and forage together as well as defend their territory as a unit. Both sexes develop incubation patches and begin to incubate their clutch as soon as it is completed. Early in incubation the male spends more time on the nest than the female, but the female is more attentive later on. The eggs hatch over a 1 or 2 day period, and the young soon become very active, feeding themselves from the outset.

163

They fledge in 4–5 weeks and thereafter the families begin to form flocks.

Suggested Reading: Gibson 1971; Hamilton 1975.

Black-necked Stilt
Himantopus mexicanus

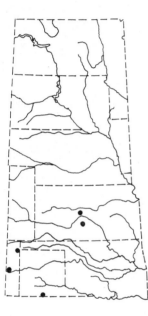

Breeding Status: A casual breeder in the southwestern portions of the region. Reported breeding in Union County and eastern San Miguel County, New Mexico. Apparently a rare summer visitor in eastern Colorado, western Kansas, western Oklahoma, and the panhandle of Texas, but with known breeding records only from central Kansas (Cheyenne Bottoms, 1976, 1978; Quivera N.W.R., 1978), and Hale County, Texas, in 1978.

Breeding Habitat: In inland sites this species breeds around shallow alkali ponds and lakes, but it also is found coastally around brackish and freshwater ponds, on rice plantations, and in other habitats.

Nest Location: Nests are in small colonies, usually of about 6–10 nests, often in grass hummocks and always close to foraging areas. The nest may even be surrounded by water, on a floating platform of sticks and vegetation. At times the eggs are laid in a simple scrape with no lining.

Clutch Size and Incubation Period: Normally 4 eggs, sometimes 3 and rarely 5. The eggs are buffy to sandy with blackish blotches. The incubation period is 25–26 days. Single-brooded, but probably a renester.

Time of Breeding: New Mexico egg records are from May 12 to June 1. In Texas, eggs have been found from April 16 to June 28, and downy chicks have been reported as late as August 1.

Breeding Biology: Like avocets, stilts form pair bonds gradually and without associated elaborate displays, through the persistent association of a female with a particular male, in spite of initial aggressiveness by the male. Stilts defend territories on their breeding grounds better than avocets do and advertise them by aerial displays. Copulation in stilts is preceded by slight ritualized breast-preening by both sexes, apparently identical to that of avocets. Nest-building is probably done by both sexes, and materials are added to the nest through incubation. In periods of rising water the nest may be raised considerably by such added materials, and both sexes apparently share incubation about equally. Incubation begins when the last or penultimate egg is laid, and (in the laboratory) lasts 25 days. The eggs hatch relatively synchro-

164

nously, and the young remain in the nest no more than 24 hours. They are probably brooded for at least a week, and are independent at about 4 weeks.

Suggested Reading: Hamilton 1975; Stout 1967.

FAMILY PHALAROPOD-IDAE (PHALAROPES)

Wilson Phalarope

Wilson Phalarope
Steganopus tricolor

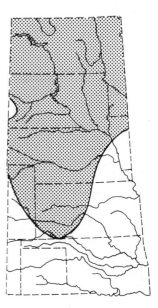

Breeding Status: A summer resident in suitable habitats over most of the region, fairly common over most of North Dakota, western Minnesota, the lowland areas of South Dakota, and the Nebraska Sandhills. It is rare and local in northwestern Iowa, local in eastern Colorado, and local in central and western Kansas, with specific breeding records for Barton and Meade counties, and summer records for Finney, Kearny, and Seward counties. There are no breeding records from farther south than Kansas.

Breeding Habitat: The presence of wet meadows apparently is the major habitat criterion for this species, which is found near fresh to highly saline water and is associated with watery environments ranging from ditches or river edges to seasonal, semipermanent, or permanent ponds and lakes.

Nest Location: Nests are well hidden in wet meadows and sometimes also occur in grassy swales or on hummocky areas of shallow marshy habitats. The nests are scrapes in the ground, lined with dead grass built up into a cup about 2 inches thick. When placed over water they may be built up to a level about 6 inches above the water.

Clutch Size and Incubation Period: Normally 4 eggs, occasionally 3. The eggs are buffy with a varying amount of darker spotting. The incubation period is 16–22 days, probably averaging about 20 days. Incubation is normally by the male, but there is no proof that females produce more than one clutch.

Time of Breeding: Egg dates in North Dakota range from May 26 to July 8, and dependent young have been seen from June 7 to July 17. Kansas egg dates are for May and June.

Breeding Biology: Although female phalaropes are appreciably larger and more brightly colored than males, recent studies have cast doubt on the idea that they are regularly polyandrous. Pair bonds apparently are formed after the birds arrive on the breeding areas, during a period of behavior that is intensely aggressive but little indicative of typical territoriality. The female probably makes the nest scrape after the pair is formed, but the male adds the nest lining. Eggs are laid about 48 hours apart, and presumably the female plays no further role in parental care. The male incubates, and he leads his brood from the nest to foraging areas only a few hours after they hatch. The fledging period has not been reported, but in the closely related northern phalarope it is less than 3 weeks.

Suggested Reading: Höhn 1967; Kangarise, 1979.

FAMILY LARIDAE (GULLS AND TERNS)

Common Tern

California Gull
Larus californicus

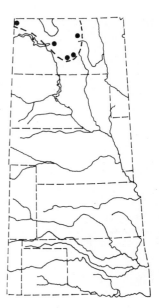

Breeding Status: Limited as a breeder to North Dakota, where it has bred in recent years in Stutsman, Kidder, Ramsey, McLean, and Divide counties. Also breeds just to the west of the limits of this book in Weld County, Colorado.

Breeding Habitat: In North Dakota, this species uses much the same habitat as does the ring-billed gull—barren islands on brackish or alkaline lakes—and the two species sometimes nest in mixed colonies.

Nest Location: Nesting is colonial; individual nests consist of scrapes lined with grasses, sticks, or weeds. In one North Dakota colony, the California gulls nested in the more central and elevated parts of an island, while the ring-billed gulls occupied the area near the water's edge. Likewise in Alberta, California gulls tend to nest on elevated and boulder-strewn areas, while ring-billed gulls occupy more level terrain. Also, California gulls tend to space their nests almost randomly, while ring-billed gulls show a tendency to aggregate.

Clutch Size and Incubation Period: Usually 3 eggs, but 2-5 have been reported. The eggs are white to buffy with darker spotting. The incubation period is 23-27 days. Single-brooded.

Time of Breeding: North Dakota egg dates are from May 3 to June 24, and flightless young have been seen from June 8 to July 28.

Breeding Biology: California gulls arrive from their wintering grounds along the Pacific coast some weeks before the onset of nesting. Territorial establishment and courtship activities begin as soon as they arrive, even if the nesting areas are still covered by snow. Eggs are laid at an average interval of 2 days, so that most clutches are completed in 4-5 days. Egg-laying within colonies is highly synchronized, and in a sample of 100 nests nearly all the eggs were laid within 2 weeks. Incubation is performed by both sexes and averages about 26 days, with a range of 23 to 28. Although these gulls are serious egg predators for other species, relatively few eggs are eaten or disappear within the nesting colony, and hatching success is often high. The chicks are relatively precocial, and though they are usually raised in the close vicinity of their nest they are also well able to run and elude danger from an early age. They fledge at ages of from 36 to 44 days, averaging 40 days.

Suggested Reading: Vermeer 1970; Baird 1976.

Ring-billed Gull
Larus delawarensis

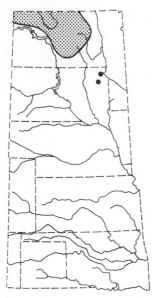

Breeding Status: Limited as a regular breeding species to North Dakota, where it breeds or has recently bred in Burleigh, Kidder, Stutsman, McClean, Ramsey, McHenry, Rolette, Burke, and Divide counties. It has also bred in Roberts County, South Dakota, and still breeds at Bitter Lake, Day County. Nesting in Minnesota occurs east of the limits of this book.

Breeding Habitat: Breeding occurs in colonies on isolated and sparsely vegetated islands of lakes and impoundments, the colonies varying in size from a few birds to more than 1,000 pairs.

Nest Location: Nests are usually on gravel or in matted vegetation on a flat substrate and are simple scrapes lined with readily available sticks, weeds, or grasses. In dense colonies nests may be less than a yard apart, and they are typically closer together than would be expected from random nest selection.

Clutch Size and Incubation Period: Normally 3 eggs, sometimes 2 or 4. The eggs are buffy to whitish with darker brown markings. The incubation period averages 25 days. Single-brooded.

Time of Breeding: Egg dates in North Dakota extend from May 17 to June 27, and flightless young have been seen from June 8 to July 28. Fledged young have been seen as early as mid-July.

Breeding Biology: Ring-billed gulls arrive on their nesting grounds well before the nesting season and establish nesting territories as early as possible, at times occupying exactly the same territory as in the previous year. Such behavior probably helps to maintain pair bonds and also results in birds returning to areas where successful breeding has previously occurred. As in other gulls, most pair-forming behavior consists of hostile postures and calls associated with territoriality. Eggs are laid at 2-day intervals and, as in the California gull, egg-laying is highly synchronized within colonies. Incubation is by both sexes, and apparently a major source of egg mortality comes from chilling as a result of disturbance to the colony. Once the eggs hatch, most chick mortality evidently comes from pecking by neighboring adults when a chick wanders too far from its parents. The young birds fledge in an average of 37 days.

Suggested Reading: Vermeer 1970; Tinbergen 1959.

Franklin Gull
Larus pipixcan

Breeding Status: Breeds in scattered colonies in central and eastern North Dakota, western Minnesota, and northeastern South Dakota. There is a single Iowa nesting record (3 nests) for Clay County and occasional records for Garden County, Nebraska (*Nebraska Bird Review* 34:63; 35:32).

Breeding Habitat: Large, relatively permanent prairie marshes with extensive stands of semiopen emergent cover are the primary breeding habitat in North Dakota, with more limited usage of shallow river impoundments.

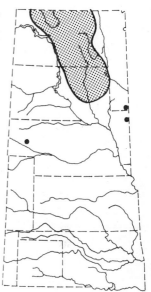

Nest Location: Nests are usually in emergent vegetation such as cattails, bulrushes, phragmites, and whitetop, in water as deep as 4-5 feet, and frequently among nesting black-crowned night herons. Emergent stands that are not extremely dense and that are close to open water are preferred. The nest is a floating mass of dead vegetation, anchored to live plants and with a well-formed cup.

Clutch Size and Incubation Period: From 2 to 4 eggs, usually 3. The eggs are mostly brown to greenish brown with darker brown blotching or scrawling. The incubation period has been estimated at 18-20 days. Single-brooded.

Time of Nesting: North Dakota egg records extend from May 23 to June 26, and young have been seen as early as June 11. Fledged young have been reported as early as July 11.

Breeding Biology: Franklin gulls nest in colonies, and after they arrive in their nesting grounds in spring they display on the previous year's nesting site, though they often shift to a new colony before the start of nest-building. Pairs apparently pick their nest sites on the basis of horizontal visibility and relative aggression. Where emergent vegetation is thick, reducing visibility, nests are closer together than where vegetation is less dense, and aggressive interaction between adjacent pairs is reduced. Both members of the pair assist in incubation, and they continue to add materials to the nest through the breeding period, presumably because of its floating nature. Unlike many gull species, Franklin gulls do not eat the eggs or young of their own species, but minks, great horned owls, and marsh hawks are major predators of young gulls as well as of adults. After hatching, the young remain on their nesting platform until they are 25-30 days old, and they do not learn to distinguish their own parents from other adult gulls until they are more than 2 weeks old. Likewise, parents will accept alien chicks less than about 2 weeks old, both of which suggest a slower development of parental and offspring recognition than is typical of ground-nesting gulls. Fledging occurs at 28–33 days.

Suggested Reading: Burger 1974.

Forster Tern
Sterna forsteri

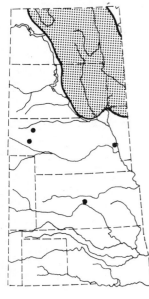

Breeding Status: Breeds uncommonly and locally in North Dakota east of the Missouri River, in western Minnesota, in eastern South Dakota and northwestern Iowa (Clay, Emmet, and Palo Alto counties). It is apparently only a local but regular breeder in Nebraska (Garden and Sheridan counties). Likewise, in Kansas breeding has only been reported from Barton County, in 1962.

Breeding Habitat: Large marshes with extensive areas of emergent vegetation or muskrat houses for nest sites are favored habitats of this species. It is also found around lakes, salt marshes, and coastlines but is more of a marshland species than the common tern. Small marshes seem to be avoided.

Nest Location: Nests are typically in or near water, usually on floating vegetational debris, but at times they are in a depression in sand or mud. The birds sometimes nest on small islands like common terns, but the two species rarely nest together. Studies in Iowa indicate that large muskrat houses are favored nest sites, especially when they are near the edges of open pools of water.

Clutch Size and Incubation Period: From 1 to 4 eggs (92 Iowa clutches averaged 2.5). The eggs are buffy to buffy olive with dark brown spotting. The incubation period averages 24 days. Single-brooded, but probable renesting has been reported.

Time of Breeding: The probable period of breeding in North Dakota is late May to late July, with newly hatched young recorded in late June and early July. Minnesota egg dates are from May 25 to June 24, with hatching reported as early as June 10. Iowa egg dates extend from the last week of May until the end of June.

Breeding Biology: Shortly after they arrive on their nesting marshes, Forster tern pairs begin to seek out nest sites. They are relatively colonial, and as many as five nests may be placed on a favorable site, such as a large muskrat house. The floating rootstalks of cattails may also serve as a nest site, but such locations are more often used by black terns. Nest-building is initiated almost simultaneously by all members of a colony, and both sexes incubate. Wind and wave action, house-building by muskrats,

176

and possibly intraspecific hostility are probably major causes of egg loss, which seems to be relatively high in this species. Little information is available on the growth of the young, but presumably they fledge in about a month, as is typical of the common tern.

Suggested Reading: Bergman, Swain, and Weller 1970; McNicholl 1971.

Common Tern
Sterna hirundo

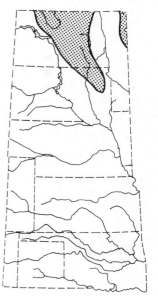

Breeding Status: Breeds locally in central and eastern North Dakota between the Missouri and Red River valleys, in Minnesota east of the Red River Valley, and in northeastern South Dakota.

Breeding Habitat: In our region breeding occurs on islands in large lakes or reservoirs. The species also breeds along coastlines on sandy beaches.

Nest Location: Sparsely vegetated islands are preferred nesting habitats in the Great Plains; grassy uplands and sandy beach areas are used in coastal regions. Nesting is in colonies, and nests are simple scrapes, often with little or no lining. Nest sites are usually near vegetation or other upright objects and are very often in previous nest hollows or natural depressions.

Clutch Size and Incubation Period: From 2 to 4 eggs, usually 3 (93 North Dakota nests averaged 2.8). The eggs are buffy to cinnamon, with dark brown spotting, especially at the larger end. The incubation period is usually 24–26 days, rarely as little as 21. Single-brooded, with renesting typical only when the first clutch is lost very early.

Time of Breeding: Egg dates in North Dakota range from June 8 to July 28, and dependent young have been seen from June 22 to July 31. The few available dates for Minnesota and South Dakota fall within these extremes.

Breeding Biology: As terns arrive on their nesting grounds, the first birds are those that have nested there formerly, and males soon begin to establish and occupy territories. In early aerial displays, or "fish flights," small fish are exchanged among the participants, but little or no sexual recognition is likely at this stage. After sexual recognition the true courtship is under way; aerial glides replace the typical fish flights, and terrestrial displays such as parading around the potential mate or incipient nest-building or scrape-digging are typical. Copulation begins at about

177

the time of scrape-making, and egg-laying soon follows. The egg-laying rate is rather variable, and incubation begins immediately, so that hatching is staggered. Both sexes incubate, but the females do so much more intensively and perform about three-fourths of the total incubation. The young are precocial and may leave the nest by the second day after hatching. Adults learn to recognize their young by the time they are 5 days old, but chickless adults sometimes adopt orphan young and care for them as their own. Young birds reach flight stage at an average age of 30 days but continue to beg food for several weeks thereafter. After the young birds leave the ternery they typically do not return for 3 years, until they are sexually mature.

Suggested Reading: Palmer 1941.

Least Tern (Little Tern)
Sterna albifrons

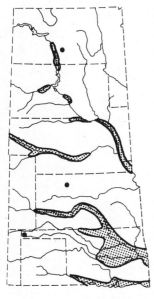

Breeding Status: Breeds locally and irregularly in the Missouri Valley from central North Dakota southward, in the Platte and Niobrara valleys of Nebraska, the Arkansas Valley of Kansas, and most of the larger river valleys of Oklahoma.

Nesting Habitat: In our region nearly all nesting occurs on river sandbars or islands, but in coastal areas nesting is also common on broad areas of sand or gravel beaches, on islands, and sometimes on newly cleared land. Gravel or pebble substrates are preferred over sandy ones. In Oklahoma nesting also occurs on salt plains, in habitats similar to those used by snowy plovers.

Nest Location: Nests are usually in colonies and consist of a simply scrape in sand or gravel, with little or no lining. Solitary nesting is frequent in the Great Plains.

Clutch Size and Incubation Period: From 2 to 4 eggs, but typically 2. The eggs are pale buffy with darker brown spotting. The incubation period is 20–21 days. Apparently single-brooded, but renesting has been reported in coastal areas.

Time of Breeding: A few North Dakota egg records are from July 2 to July 21. A larger series from Kansas are from May 21 to June 30, with the modal date of egg-laying June 5. Oklahoma egg dates are from May 31 to July 10, and chicks have been reported from June 23 to August 13.

Breeding Biology: In the Mississippi and Missouri valleys, least terns usually arrive in May, sometimes before sandbars suitable for nesting have been exposed by declining river levels. The exposure of these bars sets nesting in motion and thus synchronizes the breeding cycles of each nesting colony. During courtship a bird may make aerial glides while carrying fish, then alight and offer the fish to another bird. Sex recognition may be achieved in this way, since if a male is offered a fish it responds by attacking. Incipient nest-building by the male may stimulate the female to begin the actual nest, which is a simple scrape in the sand. Nest sites are usually widely spaced, lessening antagonism between nesting pairs. The eggs are laid on consecutive days or at 2-day intervals, and incubation probably begins with the first egg. At first the female incubates alone, but gradually the male assumes part of this duty. The eggs typically hatch on consecutive days, and the female does most of the brooding. Within a day the chick and parent have learned to recognize each other, and thus the parents feed no young other than their own. Within 2 days after hatching the young begin to wander away from the nest and usually do not return. They fledge on about the 20th day after hatching, and the colony is gradually deserted.

Suggested Reading: Hardy 1957; Tompkins 1959.

Caspian Tern
Sterna caspia (Hydroprogne caspia)

Breeding Status: Accidental. The only breeding record for the region is of a pair with two young, seen June 28, 1977 on Lake Williams, McLean County, North Dakota (*Prairie Naturalist* 10:23).

Breeding Habitat: Most breeding is near the coastline, usually on sandy or stony beaches, but breeding also occurs on offshore islands and sometimes on the shorelines of large inland lakes.

Nest Location: Nests are on the ground, on sand, shingle, or shell beaches. The nest is a shallow hollow, usually unlined but sometimes with a slight accumulation of plant debris. Nesting usually occurs in colonies, but at times single pairs are found in the vicinity of other tern species.

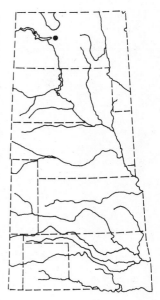

Clutch Size and Incubation Period: Usually 2 eggs, sometimes 3, rarely 1. The eggs are creamy to creamy buff with dark specks, spots, and blotches that tend to be small and rather evenly distributed. The incubation period is about 26 days and begins with the first egg; some references indicate an incubation period of 20-22 days. Single-brooded, but probable replacement clutches have been noted.

Time of Breeding: In the case of the North Dakota nesting, the young were still mostly downy when seen on June 28 but had

evidently fledged by July 12. The only definite Minnesota breeding is from Leech Lake, where two nests with eggs were found on
July 9, 1969 (*Loon* 41:83-84).

Breeding Biology: Nesting colonies of this species on islands in
northern Lake Michigan and Lake Huron tend to be rather large,
averaging about 150 pairs and sometimes going as high as 500
pairs. The colonies are also densely packed, with territories averaging less than 2 square yards, and the centers of adjacent nests
may be as little as 21 inches apart. Incubation is by both sexes,
and both parents tend the young. The fledging period has been
estimated to be as little as 4 weeks by some authorities and as long
as 6-8 weeks by others. After fledging, the juveniles rapidly
disperse and gradually work their way to coastal wintering
grounds. Immature birds are very sedentary and may spend a full
year on the wintering grounds. The birds become adult toward
the end of their third year of life.

Suggested Reading: Ludwig 1965; Bent 1921.

Black Tern
Chlidonias niger

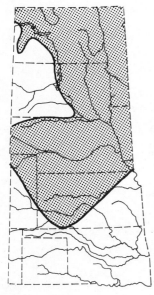

Breeding Status: Breeds locally over most of North Dakota ex
cept the extreme southwest, all of the included portions of Minne
sota and Iowa, eastern South Dakota, and Sandhills and other
wetlands of Nebraska and extreme northwestern Missouri. The
species apparently breeds locally in eastern Colorado, but there
are few published nesting records. There is only one definite
nesting record for Kansas (Barton County), although summering
birds and immatures have been seen in several other western
counties (Finney, Seward, and Meade).

Breeding Habitat: Favored breeding habitats are small to large
marsh areas containing both extensive stands of emergent vegeta
tion and areas of open water. Unlike common and Forster terns
this species feeds predominantly on insects and thus does not
compete strongly with fish-eating species.

Nest Location: Nesting is semicolonial in water from less than a
foot deep to about 3 feet, usually on floating emergent vegeta
tion, particularly cattail rootstalks. Muskrat houses are some
times used, but nest substrates tend to be smaller and lower than
those used by Forster terns.

Clutch Size and Incubation Period: From 2 to 4 eggs (151 Iowa
clutches averaged 2.6). The eggs are buffy to olive with extensive
blackish spotting. The incubation period is 21-24 days, usually

about 21 days. Single-brooded, at least in this region, but probable renesting has been reported.

Time of Breeding: North Dakota egg dates are from May 28 to July 24, and flightless young have been seen from June 21 to July 25. Iowa egg dates are from the last week of May to early July. In Kansas, complete sets of eggs have been seen between June 11 and July 12.

Breeding Biology: Prenesting behavior in black terns is marked by two types of display flights, including "fish flights" (the birds usually carry insects rather than fish), normally performed by two birds, and "flock flights" involving most or all of the birds of an entire nesting area. In the courtship phase one bird (probably the male) postures and calls while standing on a potential nest site. Also, the two birds make an aerial glide downward from several hundred feet while maintaining a fixed position relative to each other. Nesting sites of the previous year apparently are not reused. The nests seem to be built from materials gathered in the immediate vicinity of the nest rather than carried in. Both sexes assist in incubation, and both brood the young for at least 8 days after hatching. Little brooding is done after that time, though the chicks are unable to fly until they are more than 20 days old. Young birds are fed almost exclusively with insects and continue to feed on them for a time after they fledge.

Suggested Reading: Goodwin 1960; Bailey 1977.

FAMILY COLUMBIDAE
(PIGEONS AND DOVES)

Mourning Dove

Rock Dove (Domestic Pigeon)
Columba livia

Breeding Status: Introduced; now pandemic throughout the region.

Breeding Habitat: The original habitat consisted of cliffs, but in North America this species primarily frequents cities and farms, rarely straying far from human habitation. In North Dakota and Oklahoma feral birds have been found nesting in narrow, steep-walled canyons, cliffs, and similar natural habitats.

Nest Location: Buildings that provide narrow ledges similar to cliff ledges are the preferred nesting sites of this species. The nest is a flimsy platform of twigs, sticks, and grasses.

Clutch Size and Incubation Period: Normally 2 eggs, sometimes 1. The eggs are white with a smooth, glossy surface. The incubation period is 17-19 days. Multiple-brooded.

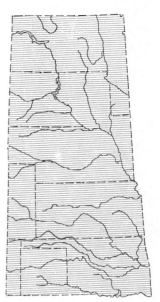

Time of Breeding: At least as far north as Kansas, eggs are laid every month of the year. In Kansas there is a spring peak of breeding, with early April being the period of maximum egg-laying. Probably much the same applies elsewhere in the region.

Breeding Biology: Because of the prolonged period of laying activity, pair-bonding is probably relatively permanent, but a period of sexual display precedes each nesting cycle. The most common display is bowing, which occurs in sexual, assertive, and defensive situations. When displaying sexually the male usually flies toward the female or leaps upward, clapping his wings sharply together, and begins cooing while inflating the neck. Tail-spreading is also a usual part of the sexual phase of this display. Males also perform an extended display flight, marked by loud wing-clapping and gliding with the wing held above the horizontal plane and the tail somewhat spread. Pigeons typically need from 8 to 14 days to complete their pairing and preincubation behavior. Both sexes help gather nesting materials, which is often scanty. Incubation requires an average of 17½ days, with the male usually roosting well apart from his sitting mate but helping to feed the nestlings as soon as they hatch. They are initially fed only on pigeon milk and are tended for an average of about 3½ weeks. They fledge when 35-37 days of age, but by the time the young are about 3 weeks old the pair begins a new nesting cycle. The first brood is tolerated in the territory until the second brood hatches, then is attacked and driven away by the parents. By overlapping successive breeding cycles, a pair might potentially rear as many as 10 or even 12 broods during a 12-month period.

Suggested Reading: Goodwin 1967; Murton and Clarke 1968.

Mourning Dove
Zenaida macroura (*Zenaidura macroura*)

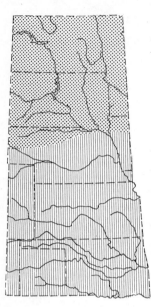

Breeding Status: Pandemic, breeding throughout the entire region. Occurs abundantly in virtually all areas.

Breeding Habitat: The most typical habitat is probably open woods and edge areas between woods and prairie or cropland, but the birds also breed in cities, on grasslands far from trees, and in cultivated fields.

Nest Location: Nests are most often in trees, at heights from about 4 to 20 feet and averaging 5 and 7 feet in two North Dakota samples. Of 55 North Dakota nests, 25 were on the ground; ground nests are more common in relatively treeless areas. A wide variety of trees are used, but when stiff-branched conifers are present they seem to be preferred over deciduous species. The nest is a frail platform of twigs with little or no lining.

Clutch Size and Incubation Period: Normally 2 eggs, sometimes 1 or 3. The eggs are white with a glossy surface. The incubation period is 13–15 days. Multiple-brooded, with up to four broods being raised in the southern parts of this region.

Time of Breeding: North Dakota egg dates range from April 16 to September 6, with a peak in May and June, and dependent young have been seen from May 8 to October 2. Kansas egg records are from April 1 to September 10, with a peak in mid-May. Egg records in Texas extend from January through December, but there is a breeding peak from March to September.

Breeding Biology: Mourning doves begin to form pairs at the onset of the breeding season, when males that are dominant in winter flocks mate with high-ranking females; such pairs are the first to establish territories and appear to be the most successful in their reproductive efforts. The availability of choice nesting materials (twigs) is important in determining territorial boundaries in captive birds, whereas food and water sites are not defended. The two eggs are usually laid at about 24-hour intervals, and incubation is by both sexes, the male normally incubating during the day and the female at night. Typically the eggs hatch on successive days, so that one chick tends to be larger and more aggressive in food-begging than the other. By the time the young are 12 days old they are ready to leave the nest, and they normally fledge when they are 13–15 days old. By that time the adults have generally begun a second clutch in a new nest. In subsequent nesting the two nests may be used alternately; in Texas as many as six nesting cycles have been reported by a single pair, using three different nest sites.

Suggested Reading: Hanson and Kossack 1963; Goforth and Baskett 1971.

FAMILY CUCULIDAE (CUCKOOS, ROADRUN-NERS, AND ANIS)

Yellow-billed Cuckoo

Yellow-billed Cuckoo
Coccyzus americanus

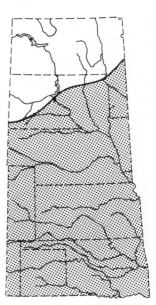

Breeding Status: This species is a rather common breeder through the wooded areas of the southern half of this region, extending northward through eastern Nebraska and Iowa into southeastern South Dakota and southwestern Minnesota. There are no definite breeding records for North Dakota, but a few records of birds in summer have accumulated.

Breeding Habitat: Moderately dense thickets near watercourses, second-growth woodlands, deserted farmlands overgrown with shrubs and brush, and brushy orchards are favored habitats for this species. It avoids extremely dense woods.

Nest Lcoation: Nests are frail, shallow platforms placed 2–20 feet above the ground, usually 4–8 feet high on horizontal limbs of hardwood trees, well concealed by surrounding foliage. There is usually a thin lining of soft vegetation.

Clutch Size and Incubation Period: From 2 to 5 eggs (54 Kansas clutches averaged 3.1). The eggs are pale greenish blue, gradually fading to yellowish. The incubation period has been estimated as 14 days, probably starting before the clutch is complete. Possibly double-brooded, at least in Oklahoma and other southern areas.

Time of Breeding: Kansas egg records are from May 11 to September 10, with a peak in early June. Oklahoma egg dates are from May 14 to August 11, and dates in Texas are from March 22 to September 5.

Breeding Biology: Cuckoos are relatively late spring migrants, arriving in Minnesota in May or even early June and inconspicuously taking up breeding territories. Their distinctive clucking and repeated hollow notes of *kaw* or *kowp* are frequently uttered, especially on cloudy days or at night. The birds gather nesting materials from trees by breaking off small branches and carrying them back one at a time to the nest. The eggs are laid at irregular intervals, and incubation apparently begins during the egg-laying period, since hatching is staggered. Apparently both sexes assist equally in incubation and also feed and brood the young. It has been suggested that when second broods are produced one parent may remain to tend the first brood while the other looks after the second clutch and brood. The young birds remain in the nest about 9 days but are still flightless when they leave the nest. At that time they are very agile in climbing about branches. When they leave the nest the young birds are somewhat more than half the adult weight; the most recently hatched and smallest young may be left alone in the nest, often to be neglected and even to starve.

Suggested Reading: Preble 1957; Bent 1940.

Black-billed Cuckoo
Coccyzus erythropthalmus

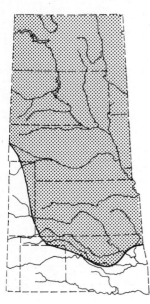

Breeding Status: Breeds in wooded parts of nearly the entire region except for higher elevations of the Black Hills, extending southward to southern Colorado and northern Oklahoma. There are no breeding records for the Texas panhandle or New Mexico.

Breeding Habitat: This species occupies somewhat more densely wooded habitats than does the yellow-billed cuckoo and is more generally northern in its distribution. It seems to prefer extensive areas of upland woods that provide a variety of trees, bushes, and vines for nesting cover.

Nest Location: Nests are more substantial than those of the yellow-billed cuckoo and are on horizontal limbs situated from a few inches above the ground to about 20 feet high, but usually less than 6 feet above the ground. The nest is made of loosely interwoven twigs and has a substantial lining of softer materials.

Clutch Size and Incubation Period: From 2 to 4 eggs (13 Kansas nests averaged 2.5). The eggs are a darker greenish blue and slightly smaller than those of the yellow-billed cuckoo; the two species sometimes lay in one another's nests and occasionally lay in the nests of other birds as well. The incubation period is only 10–11 days, beginning with the laying of the first egg, and hatching is staggered. Probably single-brooded.

Time of Breeding: Egg dates in North Dakota range from June 4 to June 20, with nestlings seen from June 20 to July 27. Kansas egg records are from May 21 to August 10, with a peak in early June, and Oklahoma records of eggs (or incubating females) extend from May 26 to June 20.

Breeding Biology: Black-billed cuckoos seek out nest sites that are well concealed by overhanging branches and leaf clusters, and they soon begin a protracted period of nest construction that may continue into the incubation period. Materials are gathered in the immediate vicinity, and eggs are laid at irregular intervals until the clutch is complete. Both sexes assist in incubation, and they usually make several changeovers each day. Hatching apparently is fairly rapid, and the newly hatched young emerge from the egg essentially dry, with a coal black, downless skin. Feeding begins almost immediately, and the young grow extremely rapidly, so that by 6–7 days after hatching they are able to leave their nests. At that age they are still relatively unfeathered and unable to fly, but they can run and climb with remarkable speed, thus eluding most predators. Initial flight occurs at 21–24 days.

Suggested Reading: Bent 1940; Spencer 1943.

190

Roadrunner
Geococcyx californianus

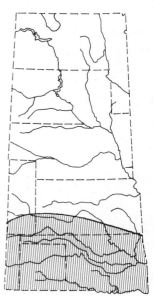

Breeding Status: Breeds throughout eastern New Mexico, the Texas panhandle, and all of Oklahoma but is more common westerly. It is locally uncommon breeder in southern Kansas and southeastern Colorado (Las Animas, Bent, and Baca counties).

Breeding Habitat: Roadrunners primarily frequent edge habitats provided by mixed open land and brush or forest and occasionally extend into either generally sparse or dense vegetation. They are found from near sea level to about 6,800 feet in Texas.

Nest Location: Nests are usually in low trees, thickets, or clumps of cactus, usually 3–15 feet above the ground, and rarely on the ground. The nest is built of sticks, with a lining of softer materials, and averages about a foot in diameter and 6–8 inches high. At least in desert regions it is constructed so that a band of shade crosses the nest during the middle of the day.

Clutch Size and Incubation Period: Normally 2–6 eggs (4 Kansas clutches averaged 4.5), but apparently double clutches of up to 12 eggs have been reported. The eggs are white with a chalky surface. Incubation begins soon after the first egg is laid and requires 17–20 days per egg. Since the eggs are laid at intervals of 1–3 days, the ages of the young in a nest vary greatly. Double-brooding is common, and three broods per year have occasionally been reported.

Time of Breeding: In Kansas, eggs are laid from at least early April to Mid-July. Oklahoma egg records extend from April 12 to September 4, and young are seen from early May to late September. Texas egg records extend from March 5 to October 10.

Breeding Biology: Roadrunners exhibit no obvious sexual differences in their plumage, but males have white skin in the unfeathered area immediately behind the eyes, whereas in females this area is pale blue. Courting males can also be recognized by various behavior patterns, including their distinctive cooing calls, often given from a perch, which attract and stimulate unmated females. Several other calls are uttered by males or by both sexes, and many visual displays are associated with territoriality and pair-formation. Before copulation the male performs a tail-wagging display and presents food. Nest-building lasts 3–6 days, and incubation begins with the first egg. Both sexes incubate, with the male incubating the entire night and part of the day. Fledglings remain in the nest about 12–13 days but do not fledge until they are 18–21 days of age. They become independent of their parents 30–40 days after leaving the nest.

Suggested Reading: Whitson 1975; Ohmart 1973.

FAMILY TYTONIDAE
(BARN OWLS)

Barn Owl

Barn Owl
Tyto alba

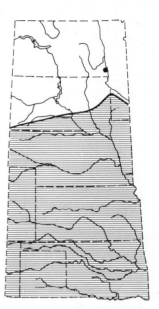

Breeding Status: Breeds locally and commonly throughout nearly the entire area except the northernmost portions. There is a single North Dakota breeding record (Cass County), and only one Minnesota record (Nobles County) since 1965. The species is considered rare in Iowa, breeds only locally in South Dakota, and in Nebraska apparently occurs uncommonly throughout the state, though there are few actual nest records. It likewise is a "low density" resident throughout Kansas, is a rare breeder in extreme northwestern Missouri, and is uncommon in Colorado. It is probably a casual breeder in eastern New Mexico (no definite records), is infrequent in the Texas panhandle and Red River Valley (breeding records for Hemphill, Wilbarger, and Cooke counties), and apparently is uncommon throughout Oklahoma.

Breeding Habitat: Extremely widely distributed, this species favors warm climates and open to semiopen habitats where small rodents are abundant and where hollow trees, old buildings, or caves provide roosting and nesting cover.

Nest Location: Natural cavities, crevices, nesting boxes, or deserted buildings are frequent nest sites; old crow nests or rooftops may also be used. Often no nest is built; the eggs are simply deposited on the substrate, with disgorged pellets or other rubbish often forming a nest lining.

Clutch Size and Incubation Period: Usually 4–7 eggs (4 Kansas nests averaged 4.7), but up to 11 have been reported. The eggs are white and elliptical. They are laid at intervals of 2–3 days and incubated from the first egg. The incubation period is 30–31 days. Double-brooded in favorable years, but not nesting at all in unfavorable years.

Time of Breeding: Eggs in Kansas are laid from at least April to July. Oklahoma egg records are from March 4 to October 1. In Texas, breeding occurs almost throughout the year, but most nesting is during winter, with eggs reported from November 13 to May 16.

Breeding Biology: Barn owls become mature in their first year of life, and once paired they probably remain mated indefinitely. The first indication of breeding activity is the increased screeching of males, reflecting territoriality. Males often screech while flying through their territories in a kind of "song flight," and they also greet their mates with a squeaking call. Females greet their mates by "snoring," which stimulates prey-presentation and copulation. Only the female incubates, but the male often visits the nest with food, and after hatching he brings food to the nest, which the female dismembers and feeds to the young. Later the female also gathers food, and the young are fed for a time even after they

195

fledge at about 8 weeks of age. About a month after fledging they leave the territory, and the pair may begin a second brood.

Suggested Reading: Bunn and Warburton 1977; Reese 1972.

FAMILY STRIGIDAE
(TYPICAL OWLS)

Barred Owl

Screech Owl
Otus asio

Breeding Status: Breeds fairly commonly throughout the wooded portions of the entire area, becoming progressively less frequent westerly, and is rare in the western Dakotas, including the Black Hills. It is local and uncommon in the wooded river bottoms of northeastern and southeastern Colorado and is not known to breed in northeastern New Mexico or in the Texas panhandle.

Breeding Habitat: Breeding occurs in a variety of wooded habitats, including farmyards, cities, orchards, and other man-related environments.

Nest Location: Nests are typically in natural tree cavities, particularly old woodpecker holes. Artificial cavities such as birdhouses and small kegs are also used, especially if sawdust is placed in the bottom. Nests are usually 5–30 feet above the ground but have also been found in stump cavities practically at ground level.

Clutch Size and Incubation Period: From 2 to 7 eggs, but usually 4 or 5. The eggs are pure white and often nearly spherical, with a glossy surface. The incubation period averages about 26 days but reportedly ranges from 21–30 days and begins before the clutch is completed. Single-brooded.

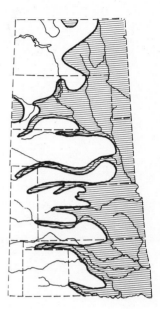

Time of Breeding: North Dakota egg records extend from April 10 to June 1. Kansas records extend from March 20 to May 10, and those from Oklahoma are from March 20 to May 4.

Breeding Biology: Screech owls are so small and inconspicuous that they may well nest in an urban backyard without the owner's ever being aware of their presence. Most often they can be detected by their distinctive wailing call, or a series of short whistled notes that often speed up and become a trill, similar to the noise of a ball bouncing to a standstill. From 8 to 9 days are needed to complete a clutch of 4 eggs. From the time incubation begins the male probably hunts for both members of the pair, but he does not incubate the eggs. When the young have hatched, over a period of about 3 days, they are fed about equally by both parents. The adults forage and return separately with their food items, and they present the prey either intact or after it has been partially dismembered. Early studies by A. A. Allen indicated that a surprising variety of prey is brought to the nestlings, including numerous adult songbirds such as sparrows, warblers, phoebes, and tanagers. In a 45-day period, 77 birds of 18 species were brought to the young, as well as numerous insects, mammals, salamanders, crayfish, and other prey. The young begin to fly when about 28–30 days of age but continue to be fed for some time.

Suggested Reading: Allen 1924; Van Camp and Henny 1975.

Great Horned Owl
Bubo virginianus

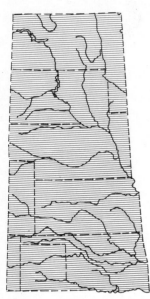

Breeding Status: Pandemic throughout the entire area, but most abundant in wooded or deeply eroded areas that provide both food and nest sites.

Breeding Habitat: This widely ranging species occurs in dense forests, in large city parks or farm woodlots, and in rocky canyons or gulleys well away from forest cover.

Nest Location: Nest sites are highly variable, and these owls often use an abandoned nest of a large bird such as a hawk, heron, or crow, or even a squirrel nest. They also use large tree cavities, crotches, stumps, caves, and ledges, and some nests have been found on the ground amid rocks, in logs, or under vegetation. The sites are generally used as they are found, with at most a few feathers added for lining.

Clutch Size and Incubation Period: From 1 to 4 eggs, usually 2 (22 North Dakota clutches averaged 2.6). The eggs are white and rough surfaced, sometimes nearly spherical. The incubation period is probably about 30 days, but estimates range from 26–35 days, starting with the first egg laid. Single-brooded.

Time of Breeding: Egg records in North Dakota extend from March 8 to May 5, and nestlings have been observed from April 20 to July 5. Kansas egg dates are from January 11 to March 20, with a peak of egg-laying about February 10. Oklahoma egg records are from February 6 to May 30, and nestlings have been reported as early as March. Texas egg records are from late December to June 12, with nestlings seen as late as July 8.

Breeding Biology: Horned owls are strongly monogamous, and pairs keep in contact by using their familiar hooting calls, *who-whoowhoo-whooo*. The male's call is appreciably lower in pitch than the female's. They begin their nesting season amazingly early, usually nesting in the same area and sometimes in the same nest they used the previous year. Incubation begins as soon as the first egg is laid, perhaps partly to keep the eggs from freezing, but also to ensure staggered hatching of the young. Both sexes reportedly incubate, but the female probably does most of it while the smaller male hunts for the pair. The young are hatched in a scanty down coating and do not open their eyes for a week or more. They are brooded by their parents for nearly a month and cannot fly until they are about 9–10 weeks old. Even after they fledge they continue to beg for food until they are driven away from the area by their parents.

Suggested Reading: Bent 1938; Errington, Hamerstrom, and Hamerstrom 1940.

Burrowing Owl
Athene cunicularia (Speotyto cunicularia)

Breeding Status: Breeds over the western portions of the Great Plains, extending locally eastward to the Red River Valley of North Dakota, extreme southwestern Minnesota (Stevens and Traverse counties, probably also Grant, Big Stone, and Swift), western Iowa (recent breeding in Plymouth and Woodbury counties), southeastern Nebraska (Lancaster County), east-central Kansas, and central Oklahoma. Originally probably ranged farther eastward, but reductions in prairie dogs and ground squirrels have caused range contraction and decreased abundance throughout the Great Plains. There is a single nesting record for Missouri *(Bluebird* 41:11).

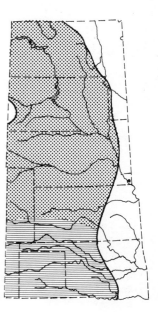

Breeding Habitat: This species is associated with heavily grazed grasslands, particularly where there are colonies of large rodents, such as prairie dogs and Richardson ground squirrels.

Nest Location: Rodent burrows are favored nest locations, but the cavities of badgers and tortoises or similar excavations may also be used. Sometimes the owls dig their own nesting burrows, typically making a 4- to 9-foot tunnel that terminates in a circular nest cavity, usually lined with vegetation or dried manure. Manure is often placed near the burrow entrance, apparently to provide scent camouflage.

Clutch Size and Incubation Period: From 3 to 10 eggs (18 North Dakota nests averaged 5.2). The eggs are white and relatively glossy but soon become stained. The incubation period is estimated at about 3 weeks. Single-brooded.

Time of Breeding: North Dakota egg records are from May 15 to August 23, with dependent young reported from June 19 to September 7. Kansas egg records are from April 11 to July 10, with a peak of egg-laying in mid-May. Colorado egg records are from May 9 to June 19. Texas egg dates are from April 13 to May 18, and dependent young have been seen as late as July 21.

Breeding Biology: Based on studies in New Mexico, burrowing owls arrive on their nesting areas either singly or paired, with males returning to the same burrows they occupied previously. Unpaired males display from their burrow locations by bowing and uttering their double-noted *coo-coooo* "song" through the night. Pair-formation may occur in a single evening, and copulation occurs near the nest entrance. Evidently only females incubate, and males feed their mates during pair-formation, incubation, and brooding. When the young are 3–4 weeks old the female begins to forage for herself and her brood, and at about this time the young birds are capable of flight.

Suggested Reading: Martin 1973; Grant 1965.

Barred Owl
Strix varia

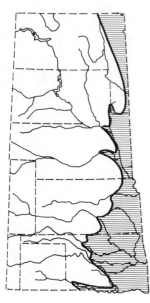

Breeding Status: Breeds locally in woodlands through the eastern portions of the region, extending west to extreme western Minnesota (nesting in eastern North Dakota is probable but unproved), southeastern South Dakota (extending west to the Black Hills but not known to breed there), the Missouri Valley of Nebraska, and perhaps locally in the Platte Valley westward toward Colorado, for which a single old nesting record exists (Phillips County). It is a local breeder in eastern Kansas, with no nesting records west of Morris County. In Oklahoma it is a resident breeder in the eastern half of the state, and in northern Texas it breeds along the Red River Valley.

Breeding Habitat: Dense river-bottom woods are favored habitats, and the bird is never far from forest environments. When both are present, coniferous woods seem to be preferred to hardwood habitats.

Nest Location: Tree cavities are the most typical nest sites, but old nests of other large birds such as hawks and crows are also used. No lining is added to the nest. Nests are typically in tall, old trees with cavities at least 25 feet above the ground, near the middle of large expanses of woods.

Clutch Size and Incubation Period: Usually 2–3 eggs, rarely 4. The eggs are white, elliptical, and have a dull surface. The incubation period is 28–33 days, starting with the first egg laid. Single-brooded, but renesting is known to occur.

Time of Breeding: There are few actual egg records for the area. Minnesota records are for March and early April. Three Kansas records are for the first half of March; egg-laying in Oklahoma is said to begin in late February, and eggs have been seen as late as April 23. Texas egg records are from February 2 to June 4.

Breeding Biology: Barred owls typically return to the same nesting place year after year, and their distinctive *Who cooks for you?* call is perhaps the best known and most distinctive of all North American owl calls. Calling is most evident before the egg-laying period. It is believed that the female does most if not all the incubation, and the young hatch down-covered but helpless, with their eyes closed. Within a week they begin to open their eyes, but they are brooded most of the time during their first 3 weeks of life. They are fed mainly rodents during the nestling period and fledge when they are 7–9 weeks old. However, they move out of their nest when about a month old and climb about in the trees with surprising agility. Should they fall out of the tree before they fledge, they can readily climb back up, using their beaks and feet simultaneously to "walk" up nearly vertical trunks of rough-barked trees. About 8 weeks after fledging the young can capture mice and crayfish on their own, but family bonds seem to persist well into the fall.

Suggested Reading: Dunstan and Sample 1972; Bent 1938.

Great Gray Owl
Strix nebulosa

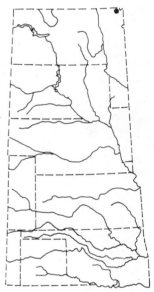

Breeding Status: Accidental in our area, with two nestings in Roseau County, Minnesota, in 1935 and 1970.

Breeding Habitat: This northern species is associated with dense coniferous forests of Canada and with montane coniferous forests of the western states.

Nest Location: Nests are usually in old nests of goshawks, red-tailed hawks, or other large hawks, 10–80 feet above the ground, in such trees as tamaracks, balsam poplars, aspens, and spruces.

Clutch Size and Incubation period: From 2 to 5 eggs, usually 3. The eggs are white, relatively small, and not glossy. The incubation period is probably 28–30 days. Single-brooded.

Time of Breeding: The few records from Minnesota extend from April 4 to May 22 (representing the last, unhatched egg of a clutch of 5).

Breeding Biology: Studies in Alberta indicate that these owls usually nest in poplar woodlands, preferably near muskeg areas and well secluded from human activities. During the breeding season the male utters a long, drawn-out four-noted call that lacks the depth and throatiness of the call of the great horned owl, while the female's response is shorter and more screechy. Nesting in Alberta begins in late March or early April; the birds usually take over an old raptor nest with little or no attempt to recondition it. Evidently the female does all the incubating, and the male provides food for his mate and later the brood as well, chiefly small rodents such as meadow voles and red-backed mice. Very few larger mammals such as squirrels are taken, and almost no birds have been reported among the prey. The young are helpless and covered with white down when first hatched, and the male must hunt all day and presumably during the night to keep the brood and the female supplied with food. The young leave the nest when about 24 days old and by then are able to climb trees effectively, though they are unable to fly well until they are nearly 6 weeks old. They continue to follow their parents for at least another month and probably are still fed to some degree.

Suggested Reading: Nero 1970; Oeming 1955.

203

Long-eared Owl
Asio otus

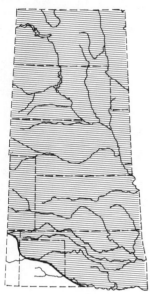

Breeding Status: Breeds in wooded areas throughout most of the region, especially in eastern portions. It is probably locally common in woodland areas of Minnesota, is rare to uncommon in North Dakota, South Dakota, and Nebraska, is considered a rare resident in Iowa and uncommon in Kansas, and is apparently uncommon to rare in Oklahoma. It is regarded as fairly common on the wooded streams and reservoir shorelines of eastern Colorado, it presumably is a very local and occasional breeder in eastern New Mexico, and there is a single breeding record for the Texas panhandle (Deaf Smith County).

Breeding Habitat: Breeding occurs in either coniferous or deciduous forests, with the former preferred, in open as well as dense woodlands, and in parks, orchards, or woodlots.

Nest Location: Old nests of large birds such as crows or hawks are most often used; squirrel nests and tree cavities are also sometimes utilized. Most often the nest is 15–30 feet above the ground, but ground nests have been reported. Unlike most owls, some nest rebuilding is typical, and rarely a bird will construct its own nest.

Clutch Size and Incubation Period: From 3 to 8 eggs, usually 4–5. The eggs are white, relatively round, and very glossy. The incubation period is probably normally 24–28 days and starts with the first egg. Single-brooded.

Time of Breeding: North Dakota egg dates range from April 28 to June 16, and nestlings have been seen from May 20 to July 15. Kansas egg records are from March 11 to April 10, and Oklahoma records of eggs or females ready to lay are from March 9 to June 14. Texas records of eggs range from March 9 to April 9.

Breeding Biology: A few weeks before egg-laying, courtship calling begins, marked by a series of short three-noted calls similar to mourning dove calls, uttered at intervals of about 3 seconds. Aerial display flights include wing-clapping noises as well as acrobatic flying maneuvers. The eggs are laid at irregular intervals of 1–5 days, and a clutch of 7 eggs may be completed in 10 or 11 days. Only the female incubates, and because of the early onset of incubation the young are hatched over a period of about 7 to 12 days. For the first 15 days of brooding the female does not leave the nest area and is fed by the male. By the time they are 25–26 days old the young are sufficiently developed to leave the nest and float to the ground, but they are not capable of full flight until they are about 30–32 days of age.

Suggested Reading: Armstrong 1958; Glue, 1977.

Short-eared Owl
Asio flammeus

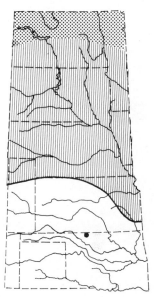

Breeding Status: Breeds in nonwooded habitats of the northern portions of the region, including virtually all of Minnesota, North Dakota, South Dakota, and Nebraska. It is a rare breeder in western Iowa and northwestern Missouri, is uncommon in eastern Colorado, and is not known to breed in New Mexico. Its southern breeding limits probably are in Kansas, where it is a local resident in eastern Kansas, but it may no longer nest in western portions of the state. There is a single old nesting record for Oklahoma (Woods County).

Breeding Habitat: Breeding occurs in such open habitats as grasslands, marshes, tundra, forest clearings, and brushy areas.

Nest Location: Nests are sometimes in rather loose colonies and are placed in slight depressions in the ground, either in rather exposed situations or in grassy cover. Rarely an excavated burrow will be used. The nest is sparsely lined with feathers and vegetation.

Clutch Size and Incubation Period: From 4 to 10 eggs (19 North Dakota clutches averaged 6.4). The eggs are white, gradually becoming nest-stained, and are not glossy. The incubation period is 21–28 days, starting before the clutch is completed. Single-brooded, but renesting occurs after clutch loss.

Time of Breeding: North Dakota egg dates range from April 4 to August 1, and nestlings have been seen from June 7 to July 30. Few egg dates are available from more southerly areas, but records from Nebraska, Kansas, and Illinois extend from April 8 to May 17.

Breeding Biology: The short-eared owl is one of the most diurnal of the Great Plains owls, and during spring it can sometimes be seen performing acrobatic courtship flights high above the prairies, marked by strong wing-clapping, swooping, diving, and somersaulting maneuvers and by a quavering, chattering cry as the bird plummets toward the ground. Copulation sometimes follows such aerial displays or may occur in their absence. Eggs are laid over a considerable period, at intervals of 2–7 days. The female incubates alone, but her mate brings food to her during this period. The eggs usually hatch at intervals of about 3 days, and about 2 weeks after hatching the young begin to move some distance away from the nest. When they are about 6 weeks old they begin to catch some of their own food, such as insects and amphibians, but even after they are flying well at the age of 2 months the adults continue to care for them. About 90 percent of this owl's food consists of rodents, which makes the species extremely valuable from the human standpoint.

Suggested Reading: Eckert 1974; Clark 1975.

Saw-whet Owl
Aegolius acadicus

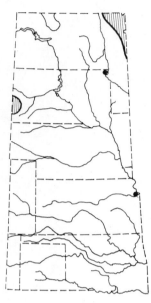

Breeding Status: The breeding status of this tiny and inconspicuous owl is hard to determine. It is probably most common in Minnesota, where it is thought to breed in wooded areas of the state south to the Twin Cities, but in the area covered by this book nesting records exist only for Kittson County. It is a hypothetical breeder in North Dakota (three breeding season records in the northeast) is a probable uncommon resident of the Black Hills of South Dakota, and nested in Roberts County in 1978. There are no definite nesting records for Nebraska (territorial birds regularly heard at Fort Robinson, Dawes County), Iowa or eastern Colorado, and only a single nesting record (Wyandotte County, 1951) for Kansas, which seems to be the southernmost breeding record for the region.

Breeding Habitat: Dense woods, especially swampy areas of coniferous or hardwood forests, are favored by saw-whet owls. They are likely to be found around tamarack bogs, alder thickets, or cedar groves. Cedar groves are also favored roosting sites, as are vine clusters.

Nest Location: The favored nest site is a flicker hole or a hole made by a woodpecker of similar or larger size, usually one from 18 to 50 feet above the ground. Sometimes birdhouses are also used. A few breast feathers are the only lining.

Clutch Size and Incubation Period: From 4 to 7 eggs, but usually 5-6. The eggs are pure white and lack gloss. The probable normal incubation period is 26–28 days and begins with the laying of the first egg. Reportedly double-brooded at times.

Time of Breeding: Minnesota egg records are from April 12 to May 9, and young in the nest have been reported as early as May 9.

Breeding Biology: The weak voice and relatively quiet nature of this species make its nesting easily overlooked; the courtship call consists of a note resembling the filing of a saw and is primarily heard during the early parts of the nesting period in March and April. Males court females by flying around them and landing nearby, often presenting a small prey. As soon as egg-laying begins the female becomes very reluctant to leave the nest, and the combination of a large clutch size and an egg-laying interval of 24–74 hours results in a highly staggered period of hatching. During the incubation and early brooding period the male is occupied with getting food which often consists of small mice, frogs, and occasionally birds. The young remain in the nest about 4 weeks and by the end of this period are able to fly moderately well. However, parental care continues until late summer, when the distinctive juvenile plumage is lost and the first adultlike plumage is assumed.

Suggested Reading: Eckert 1974; Santee and Granfield 1939.

FAMILY CAPRIMULGIDAE (GOATSUCKERS)

Common Nighthawk

Chuck-will's-widow
Caprimulgus carolinensis

Breeding Status: Breeding is limited to the southern half of the region, including the wooded portions of eastern Oklahoma and eastern Kansas (south of Wyandotte County and east of Shawnee, Greenwood, Stafford, and Sedgwick counties). The northern limit of probable breeding is northwestern Missouri, where summering birds occasionally occur at Squaw Creek National Wildlife Refuge. Birds have repeatedly been heard calling in extreme southeastern Nebraska, but nesting has not yet been proved (*Nebraska Bird Review* 35:50).

Breeding Habitat: Mixed oak and pine forests are the favored habitat, and evergreen oak groves are also important.

Nest Location: The eggs are usually laid at the edges of forests, near roads or other clearings, and are placed on the ground amid dead leaves, with no actual nest constructed. Usually there is little or no undergrowth about the eggs.

Clutch Size and Incubation Period: There are 2 eggs, creamy white with blotching of brown, lavender, and gray tones. The incubation period is 20 days, starting with the first egg, and the eggs hatch a day apart. Single-brooded, but a persistent renester.

Time of Breeding: Kansas egg records are from April 21 to May 31, with a possible peak in the third week of May. Oklahoma egg dates extend from May 4 to June 23, and unfledged or recently fledged young have been seen from May 31 to July 1.

Breeding Biology: Shortly after the birds arrive on their nesting grounds, males begin their distinctive nocturnal calling, the notes sounding like the species' vernacular name. Calling is strongest at this time but continues until after the eggs are hatched, and at the peak of calling it may also occur during daylight hours. When displaying to a female, the male sidles up to her with wings drooping and tail spread, while calling and maximally "inflating" himself. An aerial display involving wing-clapping several times in rapid succession is also known and probably serves in territorial defense or maintenance. The typical "song" is believed to function both in territoriality and in courtship, and several other calls are known to occur in this species as well. The two eggs are laid about 24 hours apart, and both sexes incubate. The folklore about these birds picking up their eggs or young in their beaks and moving them some distance after the clutch has been disturbed is probably untrue, but the birds are known to roll the eggs limited distances. The young hatch in a downy and relatively precocial state and crawl about actively shortly after hatching. They are apparently cared for only by the female, and they can fly fairly well by the 16th or 17th day of life.

Suggested Reading: Mengel and Jenkinson 1971; Ganier 1964.

Whip-poor-will
Caprimulgus vociferous

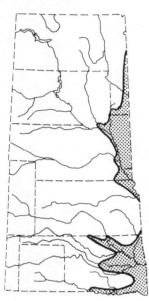

Breeding Status: Breeds in wooded areas along the eastern edge of the region, including central and southern Minnesota, southeastern South Dakota, eastern Nebraska in the Missouri River Valley (and Pawnee County) and presumably also adjacent Iowa, northwestern Missouri, northeastern Kansas (breeding records for Doniphan, Leavenworth, and Douglas counties), and probably the eastern third of Oklahoma, although specific records are still lacking for that state. There is a single breeding record for the Texas side of the Red River Valley, in Grayson County.

Breeding Habitat: Open hardwood or mixed woodlands, particularly younger stands in fairly dry habitats, seem to be the preferred habitat of this species. Stands with scattered clearings also seem to be favored.

Nest Location: The eggs are deposited on a carpet of dead leaves on the ground, usually in an area of dappled shade where there is no undergrowth. Sometimes the eggs are under a small bush, but they are placed in its shadow rather than close to its stems.

Clutch Size and Incubation Period: There are 2 eggs, creamy to grayish with darker spotting. The eggs are laid on successive days, and incubation lasts 19 days, starting with the second egg. Single-brooded.

Time of Breeding: Two Kansas records of eggs range from May 21 to June 20. Texas egg records are from mid-April to at least mid-June.

Breeding Biology: Evidently these birds begin to sing their distinctive songs immediately after arriving on their nesting areas, since the widespread initiation of their vocalizations occurs suddenly. The courtship displays have been witnessed only a few times, but the male seems to "waddle" or "dance" about the female while producing a variety of strange sounds as well as sidling up to the female and touching bills with her. An aerial display involving tail-spreading is also present. When incubation begins the male continues to sing and gradually moves closer to the nest. In one nest studied in Iowa, the female incubated during the daytime and the male incubated at night. Males have been seen at the nest on only a few occasions, and it is still uncertain as to whether they regularly help feed the young. When hatched, the young are covered with a thin down that is soon replaced with juvenile feathers. They begin to leave the area of the nest when 4–5 days old and can fly by the time they are 3 weeks of age.

Suggested Reading: Raynor 1941; Kent and Vane 1958.

Poor-will
Phalaenoptilus nuttallii

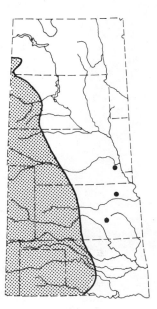

Breeding Status: Breeds in arid habitats from extreme southwestern North Dakota southward through western South Dakota, western Nebraska (rarely east to Lancaster County), western Kansas (rarely east to Leavenworth County), and western Oklahoma and the Texas panhandle, as well as adjacent New Mexico and Colorado.

Breeding Habitat: This species prefers rocky habitats with scrubby cover or xeric woodlands, but it also extends into prairie grasslands in some areas.

Nest Location: The eggs are placed on the ground, often in bare areas under dwarf scrub oaks, where dead leaves provide concealment for both adults and young.

Clutch Size and Incubation Period: There are 2 eggs, white to pinkish or creamy. The incubation period is about 18 days. Single-brooded.

Time of Breeding: Records of eggs in Colorado extend from May 20 to June 14, and dependent young have been recorded from June 6 to July 9. Kansas egg records are from May 1 to June 20, and those from Texas are from March 21 to July 26.

Breeding Biology: Like other species in this family, poor-wills are late spring migrants, and soon after arrival they begin to utter their distinctive *poor-will* or *poor-will-low* notes during the evening. Virtually nothing is known of their courtship displays, which are presumably similar to those of the whip-poor-will. Nests are extremely difficult to locate, and the adult usually remains motionless on the nest until very closely approached. Although females perhaps do most of the incubating, males have been seen incubating at night and also have been observed brooding the young. When incubating or brooding, both adults and young keep their eyes almost completely shut, which adds to the effective camouflage of these inconspicuous birds. When disturbed the adults often utter a loud hissing sound and maximally inflate themselves, which has a frightening effect on those unfamiliar with the birds. The young are hatched in a downy coat, soon replaced with a juvenile plumage similar to that of adults. The fledging period has not been well established, but is less than a month. Most poor-wills migrate south by early fall, but in a few locations such as in California individuals have been found torpid among rocks or vegetation during subfreezing temperatures. This is the first known example of semihibernation among birds, although some other birds such as hummingbirds also enter a torpid state when exposed to cold overnight temperatures.

Suggested Reading: Bent 1940; Bailey and Niedrach 1965.

Common Nighthawk
Chordeiles minor

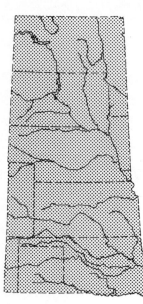

Breeding Status: Breeds throughout the entire region and varies from abundant to common in most regions, especially around cities.

Breeding Habitat: Open habitats such as grasslands, sparse woods, or cities are probably preferred by this aerial forager, but it seems highly adaptable to utilizing varied habitats.

Nest Location: No nest is built; the eggs are placed on flat substrates such as gravelly ground, burned-over areas, and gravel and asphalt rooftops.

Clutch Size and Incubation Period: There are 2 eggs, gray with darker spotting over most of the egg. The incubation period is 19 days, beginning with the second egg. Single-brooded.

Time of Breeding: North Dakota egg dates extend from June 11 to July 23, and nestings have been seen from June 29 to August 7. Kansas egg dates are from May 11 to June 30, and those from Oklahoma are from May 24 to June 22.

Breeding Biology: Nighthawks are fairly late arrivals on northern nesting areas and the males soon announce their presence by aerial displays. The most conspicuous of these is the *peent* call, uttered during a series of 4–5 wingbeats and serving to announce territorial ownership. Males also perform steep dives with down-flexed wings, each dive ending with a rush of air that produces a booming noise. Such dives are often almost directly over the nest site. Several other vocalizations are produced by males or by males and females. The females deposit their two eggs on almost any flat surface and often move them about in the course of incubation, sometimes as far as 5–6 feet from their original position. The eggs are rolled in front of the bird as the female settles on them for incubation. Some investigators report that only the female incubates, while others state that one sex incubates at night and the other by day, which seems most likely given the foraging behavior of the species. In any case, both sexes are known to help care for the young. As the adults bring food to their offspring they apparently place their bills inside the gaping mouths of the chicks and regurgitate food with a strong pumping of the head. Feeding is usually done at dusk, after sunset, and just before dawn, but not at night. About 3 weeks are required for the fledging of the young, and they are independent at about 30 days.

Suggested Reading: Sutherland 1963; Weller 1958.

FAMILY APODIDAE
(SWIFTS)

Chimney Swift

Chimney Swift
Chaetura pelagica

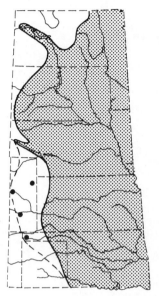

Breeding Status: Breeds fairly commonly in the eastern portions of the entire region, becoming progressively less common westerly and breeding locally in the western portions of the Dakotas, Nebraska, Kansas, eastern Colorado, most of Oklahoma (recorded west to Cimarron County), and northern Texas (Wilbarger County). During the twentieth century there has been a marked southward and westward range expansion.

Breeding Habitat: The chimney swift is not confined to any single habitat; its breeding range is largely dependent on suitable nesting sites. In historical times the birds have thus shifted from breeding in stands of mature trees to nesting around humans.

Nest Location: Originally adapted to nesting in caves and tree hollows; most nesting is now in manmade structures such as chimneys. The nest is a small, weak structure of twigs, glued together with saliva and attached to a vertical wall. It may be placed anywhere from near the top of the chimney to more than 20 feet below the top. No lining is present.

Clutch Size and Incubation Period: From 3 to 6 eggs, white and glossy. The incubation period is 18–21 days, probably starting with the next-to-last egg. Single-brooded.

Time of Breeding: Records of birds on apparent breeding territories in North Dakota extend from May 12 to August 2. Kansas egg records are from May 11 to June 30, and Texas egg records are from May 5 to July 5.

Breeding Biology: Chimney swifts arrive on their nesting grounds in rather large flocks. Little is known of courtship behavior, which occurs in the air. Even copulation is performed while in flight. Birds gather nesting materials by striking dead branches while in full flight, breaking off twigs, and carrying them back toward the nest site in their claws. Before reaching the nest, they transfer the twigs to the bill. The twigs are coated with glutinous saliva and pressed against the wall and one another until a small cuplike structure is built. Both sexes help build the nest, which requires from 3 to 6 days. Sometimes a previous year's nest is repaired and used again. Both sexes incubate eggs, and the tiny young are hatched blind and naked. For the first few days both parents feed them by regurgitation. Later the parents bring them tiny insects. Their eyes open by the time they are 2 weeks old, and soon thereafter they begin to venture away from the nest by clinging with their long claws to the vertical surface of the nest cavity. When nearly 4 weeks old they spend considerable time standing in the nest and exercising their wings, and they fledge at about 30 days.

Suggested Reading: Bent 1940; Fisher 1958.

White-throated Swift
Aeronautes saxatalis

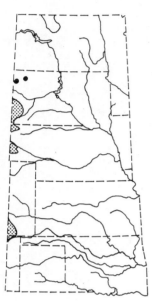

Breeding Status: Breeds locally in the extreme western portions of the region, including the Black Hills of South Dakota, the Pine Ridge, Scottsbluff, and Wildcat Hills areas of Nebraska, and locally in eastern Colorado (Baca County, possibly elsewhere). Listed as a summer resident of Capulin Mountain National Monument, New Mexico.

Breeding Habitat: Steep cliffs and deep canyons provide nesting habitat for this species, at elevations from near sea level to about 13,000 feet.

Nest Location: Nests are placed in cracks and crevices of cliffs or canyons and usually are completely inaccessible. The nest is made mostly of feathers and grasses, thoroughly glued together.

Clutch Size and Incubation Period: From 3 to 6 white eggs, usually 4. The incubation period is not known. Probably single-brooded.

Time of Breeding: Breeding in South Dakota probably occurs in June and early July, since copulations have been seen in late May and adults carrying food to a nest have been noted in late July. In Colorado, eggs have been collected during the latter half of June.

Breeding Biology: Most white-throated swifts arrive at their nesting areas in Colorado by late April and in South Dakota by mid-May. They soon begin to construct their unique nests, carrying individual feathers in their bills, sometimes apparently for miles. They also begin their aerial courtship, which may even include copulation while in flight. To initiate copulation, the birds fly toward each other from opposite directions, meet, and begin to tumble downward while clinging together, sometimes falling several hundred feet. But copulation apparently also takes place in the nesting areas, judging from some observations of egg-collectors. It may be presumed that both sexes incubate, but nothing specific is known about incubation behavior or the rearing of the young. During the nonbreeding period the birds often roost in communal quarters, much like chimney swifts. Observations on one such roosting site in California indicated that the birds, numbering 100–200, returned to their roosting crevice shortly after sunset. Within 5 minutes the entire flock had entered the crevice, passing through an entry only about 2–3 inches wide and about 2½ feet long! In very cold weather roosting birds may become torpid, although this is not known to be a regular adaptation of the species for coping with cold periods.

Suggested Reading: Bent 1940.

FAMILY TROCHILIDAE (HUMMINGBIRDS)

Ruby-throated Hummingbird

Ruby-throated Hummingbird
Archilochus colubris

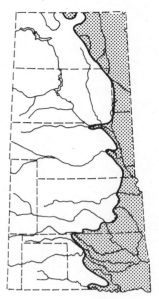

Breeding Status: Breeds generally throughout the eastern half of the region, west to central North Dakota (Turtle Mountains), eastern South Dakota, eastern Nebraska, central Kansas, west-central Oklahoma (probably to Woods County), and northeastern Texas (Cooke County).

Breeding Habitat: A variety of wooded habitats are used, from rather dense to open coniferous and hardwood woodlands and manmade environments (orchards, shade trees). Herbs or shrubs that provide tubular nectar-bearing flowers (honeysuckle, lantana, gilia, trumpet vine, etc.) are an important part of the habitat.

Nest Location: Nests are 6-50 feet above the ground, on fairly level or downward-slanting twigs or branches that are protected from above by larger branches or a leafy canopy. The nest is frequently near water, probably because favored flowers often grow there, and is more often in hardwood trees than in conifers.

Clutch Size and Incubation Period: There are 2 eggs, white and nonglossy. The incubation period is about 16 days. Reportedly but not definitely double-brooded in some areas and definitely known to renest after loss of a clutch.

Time of Breeding: A few nest or egg records for North Dakota are for late June and early July. Kansas breeding records are from May 21 to July 10, with a probable peak of laying during the last third of May. Oklahoma records of eggs or apparently incubating females are from May 16 to August 8, and nestlings have been seen as late as August 25.

Breeding Biology: Hummingbirds return to their Great Plains breeding grounds in April to late May, and territorial males advertise by flying back and forth along an arc of a wide circle, frequently passing within a few inches of a perched female at the lowest part of the arc. A male may spend as much as 2 months before attracting and mating with a female, and copulation occurs on the ground. It is apparently preceded by a period of aerial display by both birds, which hover in the air facing each other and ascend and descend vertically. The female spends several days constructing the nest, particularly in attaching lichens to the outside and later adding plant down for lining. Lining is added during incubation and even after the young are hatched. By the time the young are 10 days old they are nearly as large as their mother and are fed a combination of nectar and insects by regurgitation. The young fledged when 19-20 days old in one observed nest, but fledging records range from 14 to 28 days, suggesting considerable variability in this regard.

Suggested Reading: Bent 1940; Pickens 1936.

221

Broad-tailed Hummingbird
Selasphorus platycercus

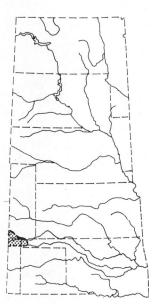

Breeding Status: Accidental or very rare breeder in the Great Plains, with early nesting reports for Cimarron County, Oklahoma, now considered doubtful (Sutton 1974). The species has been seen as far east as Baca County, Colorado, and in Union County, New Mexico, but has been proved to nest only in Las Animas County, Colorado. The black-chinned hummingbird (*Archilochus alexandri*) also occurs in summer east to Union County, New Mexico, and a possible nest of this species was found in Cimarron County, Oklahoma, in 1971. It might also nest in Baca County, Colorado. Hummingbird nests of uncertain species have also been reported from Woods, Caddo, and Beckham counties in Oklahoma (Sutton 1974).

Breeding Habitat: The species is generally associated with ponderosa pine forests in Colorado but extends from the plains all the way to the timberline.

Nest Location: Nests are often in shrubs near moist canyon walls but also are placed in aspens, Douglas firs, or other conifers. Nests are usually 4–15 feet above the ground and are typically on small horizontal branches. They are covered with spider webs, to which is attached vegetation that matches the surrounding nest support.

Clutch Size and Incubation Period: There are 2 white eggs, which are incubated 16–17 days. Probably single-brooded.

Time of Breeding: There is a single egg record for June 20 in Oklahoma. Colorado egg records are from June 20 to July 24, and nestlings have been seen from July 8 to July 24.

Breeding Biology: In central Colorado these hummingbirds arrive in early May, and the males soon become highly territorial, chasing other males from the vicinity. Their display consists of hovering in front of a female, orienting the brilliant red gorget toward her, then quickly climbing 30–40 feet and making a vertical dive downward, swooping directly past the female. The female spends several days gathering cottonwood or willow down and spider webs to construct her nest, which is usually on a horizontal tree branch with another branch or crook directly overhead. Of 10 Colorado nests studied, 5 were in aspens, 4 in spruces, and 1 in a subalpine fir. They were from 3 to more than 30 feet above the ground, and the nest core was coated with moss, lichens, and fragments of aspen bark. All incubation is by the female; the promiscuous males play no role in parental care. The young are initially fed on regurgitated food, but increasingly they are provided with tiny insects. The female feeds them by thrusting her bill into their throats and regurgitating with rapid pumping movements of the head. The young soon nearly outgrow their nest, which is well trampled down by the end of the 18-day nestling period. Females have been known to consume almost twice their own weight in sugar syrup during the day, which provides some measure of the metabolic rate of these tiny birds.

Suggested Reading: Calder 1973; Bailey and Niedrach 1965.

(FAMILY ALCEDINIDAE (KINGFISHERS)

Belted Kingfisher

Belted Kingfisher
Megaceryle alcyon

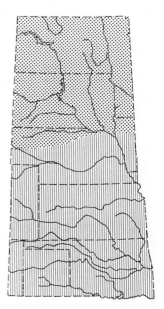

Breeding Status: A pandemic breeder in suitable habitats throughout the region.

Breeding Habitat: This kingfisher breeds near water supporting aquatic animal populations, where bluffs, road cuts, gravel pits, sandbanks, or similar nearly vertical earth exposures provide suitable nest locations.

Nest Location: The nest consists of a rather long burrow, often about 5 feet long but up to 15 feet, with an entrance about 3-4 inches in diameter (slightly wider than high) and usually within 3 feet of the top of the bank and at least 5 feet above ground level. The nest cavity is an enlarged area at the end of the burrow and is often lined with disgorged food pellets. Sandy clay soil seems to be the preferred substrate, and nests are usually near a dead or dying tree.

Clutch Size and Incubation Period: From 5 to 8 eggs, white and glossy. The incubation period is 23-24 days. Single-brooded, but a persistent renester.

Time of Breeding: The nesting season in North Dakota probably extends from mid-April to late July. In Kansas the egg-laying period is at least from April 21 to May 20, and Oklahoma egg dates are from April 26 to July 1, with nestlings reported from June 4 to July 15.

Breeding Biology: Belted kingfishers take up residence in suitable habitats that allow for large home ranges. At times they may forage up to 5 miles from the nest site, and a population density of about 1 pair per 1.8 square miles of habitat has been estimated at Lake Itasca in Minnesota. Small fish averaging about 3-4 inches long compose more than half their diet. Both sexes participate in nest excavation, which may require up to 3 weeks. Both sexes incubate, apparently beginning after the last egg is laid. After hatching the male also assists in getting food. At night the male usually roosts away from the nest, sometimes in a separate burrow or in a forested area. The young are relatively helpless and spend much time clinging to one another, apparently to maintain body warmth. They remain in the nest for at least a month, when they are first able to fly, then stay near it for the next few days while their parents teach them to catch fish. The adult captures a fish, beats it until it is nearly senseless, and drops it back into the water. The young are encouraged to capture these easy prey and gradually learn to catch normal fish. Within 10 days they are relatively independent and soon leave the vicinity of the adult pair.

Suggested Reading: Cornwell 1963; White 1953.

FAMILY PICIDAE
(WOODPECKERS)

Common Flicker

Common Flicker (Yellow-shafted and Red-shafted Flickers)
Colaptes auratus (including *C. a. cafer*)

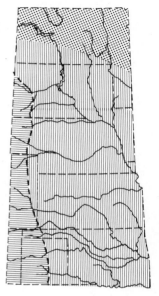

Breeding Status: As a species, pandemic throughout the region. Most of the area is represented by predominantly yellow-shafted forms, with the center of the hybrid zone extending through the western portions of South Dakota and Nebraska, eastern Colorado, and the panhandles of Oklahoma and Texas. West of this line the red-shafted (*cafer*) subspecies predominates, but hybrids are numerous throughout much of the region covered by this book (Short 1965, 1971).

Breeding Habitat: Habitats are diverse and include relatively open woodlands, orchards, woodlots, and urban environments. Generally, open country or lightly wooded areas are favored over dense forests.

Nest Location: Nests are 2–60 feet above ground, in cavities of trees that are either dead or have decaying interiors. Dead trees or stubs are preferred to live ones, and the nests are usually near the top of tree stubs. Hardwood species are greatly preferred to living coniferous trees, and utility poles or wooden buildings are sometimes used. The entrance hole averages slightly less than 3 inches in diameter, and the interior of the nest cavity is usually about 8 inches in diameter. Tree diameters at the nest hole average about 11 inches. There is no lining other than wood chips.

Clutch Size and Incubation Period: From 3 to 11 eggs (4 North Dakota nests averaged 7.2). The eggs are white and glossy. The incubation period is 11–12 days. Normally single-brooded, but double-brooding has been reported in Oklahoma, and persistent renesting follows egg removal.

Time of Breeding: Egg dates in North Dakota are from May 24 to July 3, and nestlings have been reported from May 27 to July 15. Kansas egg records are from April 11 to June 10, with a peak of egg-laying about May 10. Oklahoma egg dates extend from April 24 to June 28.

Breeding Biology: This species is relatively migratory over most of the region concerned, and when returning to the nesting area both sexes seek out their old territories and nest sites. Males tend to arrive a few days before females and soon begin uttering location calls and drumming as a territorial advertisement. Recognition of previous mates is apparently site-induced, and sex recognition is based on the "moustache" markings of the male. Courtship displays include exposing the undersides of wing and tail, bobbing, and billing ceremonies. Males apparently select the nest site, often using a previous year's nest or starting to excavate a new one. Most of the excavation is done by the male, and copulation typically occurs just before the nest is finished. The eggs are laid at daily intervals, and both sexes share incubation, with the male assuming most of the nocturnal responsibilities. Both sexes care for and brood the young, feeding them by regurgitation, and the males again take most of the responsibility. The nestling period is about 26 days, but the parents continue to feed their offspring for some time after they leave the nest.

Suggested Reading: Kilham 1959; Lawrence 1967.

Pileated Woodpecker
Dryocopus pileatus

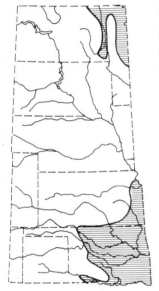

Breeding Status: Breeds locally in wooded bottomlands of the Red River Valley of North Dakota (only one definite record) and in wooded areas of Minnesota. Not currently known to breed in South Dakota, Nebraska, or western Iowa (before 1900 it bred in the Missouri River Valley). Breeds locally in eastern and southeastern Kansas (records for Cowley, Linn, and Cherokee counties) and eastern Oklahoma (definite records for McCurtain, Cleveland, Okmulgee, Delaware, Washington, Okfuskee, Murray, Marshall, and Love counties). Also breeds in northeastern Texas (record for Cooke County).

Breeding Habitat: The species is generally limited to mature forests—coniferous, deciduous, or mixed. Preferred habitats are near water and include mature lowland forests containing tall living trees with dead stubs.

Nest Location: Nests are usually excavated in the dead stubs of living trees and less frequently are in living trees, usually where the trunk is 15–20 inches in diameter. Hardwood species such as beeches, poplars, birches, oaks, and hickories are used. The opening is generally about 3½ inches in diameter and often faces south or east. Entrances range from 15 to 70 feet high, but average 45 feet. Typically a new cavity is excavated for each brood.

Clutch Size and Incubation Period: From 3 to 4 eggs, glossy white. The incubation period is 18 days, possibly starting before the clutch is complete. Single-brooded, but known to sometimes renest.

Time of Breeding: Egg records in Minnesota are from May 9 to May 23, and in Kansas eggs are laid at least during April. Eggs (or females bearing eggs) have been reported in Oklahoma from March 25 to April 28, and nestling young are reported from May 13 to June 4. Texas egg records are from March 27 to May 16.

Breeding Biology: Apparently pileated woodpecker pairs are very sedentary and maintain the same territory year after year. Terri-

232

tories are advertised by bouts of loud drumming that last about 3 seconds, performed throughout the year by both sexes, but especially the male. Paired as well as unpaired males drum; one unpaired male in Maryland was found to have a drumming territory 700 yards long. In Maryland, nest excavations begin in March, and the male does most of the work. Copulation apparently occurs about the time the nest excavation is completed. Both sexes incubate, with the male and female sharing duties during the daylight hours and the male remaining on the nest during the night. There is a recent observation of a female removing her eggs from a damaged nesting tree and carrying them in her bill to some undetermined location. The young are fed by regurgitation and remain in the nest for 26 days. They are also fed and cared for by the parents for some time after leaving the nest.

Suggested Reading: Hoyt 1957; Kilham 1959.

Red-bellied Woodpecker
Melanerpes carolinus (*Centurus carolinus*)

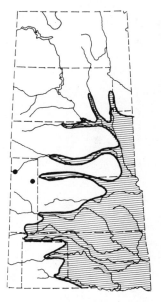

Breeding Status: Breeds in Missouri Valley woodlands from central South Dakota southward (one breeding record for Morton County, North Dakota), with probable western breeding limits in Cherry and Lincoln counties of Nebraska, Rawlins and Hamilton counties of Kansas, Baca County of Colorado, Cimarron County of Oklahoma, and Potter County of Texas. Apparently still extending its range westward along wooded river systems.

Breeding Habitat: The species is generally associated with slightly open stands of coniferous or deciduous forest, especially along rivers and near forest edges. It also frequents orchards, gardens, and similar areas near humans.

Nest Location: Nests are excavated in a variety of sites, including trees, stumps, poles, buildings, and so forth, and are usually less than 40 feet above the ground. Relatively soft deciduous tree species are preferred, as are dead trees or those with decayed stubs. The diameter of the nest limb averages about 9 inches. The entrance hole is usually less than 2 inches in diameter, and the cavities are about a foot deep. Wood chips are added to the cavity during incubation. Nesting trees are usually in more wooded areas than those of the red-headed woodpecker, and cavities are typically on the underside of a leaning branch.

Clutch Size and Incubation Period: From 3 to 8 eggs (7 Kansas clutches averaged 5.0), white and somewhat glossy. The incubation period is 12 days, beginning after the laying of the last egg, with hatching often spread over a 2-day period. Single-brooded in northern part of range, but reportedly double-brooded in Oklahoma.

Time of Breeding: Egg records in Kansas extend from the middle third of April through the first third of June, and nestlings have been noted from early May to mid-July. Breeding records in Oklahoma are from April 27 (eggs) to late July (newly fledged young). Texas egg records are from April 5 to July 9.

Breeding Biology: At least in most of this area, red-bellied woodpeckers are nonmigratory and remain on individual territories throughout the year, although typical territorial behavior is evident only during the nesting season. The major vocalizations of both sexes are a breeding call, *kwirr*, used at the onset of the nesting season, and a more general territorial call, *cha-aa-ah*, uttered throughout the year. Courtship includes three major components: mutual bill-tapping at a nest site, reverse mounting (the female mounting the male), and actual copulation. All of these occur from the start of nest excavation, and both sexes share incubation, with the male apparently incubating at night. The sexes also share about equally in the feeding of the young birds, which remain in the nest about 20 days. When a second nesting is performed, the young birds leave their parents within a few weeks but otherwise remain with them until fall, when family groups dissolve.

Suggested Reading: Kilham 1961; Jackson 1976.

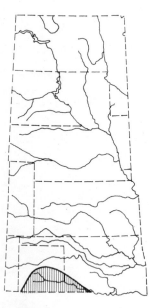

Golden-fronted Woodpecker
Melanerpes aurifrons (Centurus aurifrons)

Breeding Status: Breeds locally in southwestern Oklahoma (Harmon County) and adjacent parts of Texas (breeding records for Armstrong, Hardeman, and Wilbarger counties).

Breeding Habitat: The species is primarily associated with mesquite pastures and also is common in pecan groves on open or semiopen river bottoms. Large timber stands near mesquite, especially oaks on gravelly uplands, are also favored.

Nest Location: Living or dead trees, as well as utility poles or fence posts, are used for nests. The same cavity is typically used year after year. The cavities are usually between 6 and 25 feet above the ground and generally are in the live trunks of large trees, especially mesquite.

Clutch Size and Incubation Period: From 4 to 7 eggs, but usually 4 or 5; the eggs are white and have little gloss. The incubation period is reportedly 14 days, and possibly two broods may be raised in a season.

Time of Breeding: Texas egg records are from March 30 to July 6. A single Oklahoma breeding record is for small young on May 18.

Breeding Biology: This close relative of the red-bellied woodpecker is probably very similar to that species in behavior and breeding biology, but little has been written about it. Its vocalizations are very similar but are louder and harsher, which is believed to be adaptively useful, since it allows for better communication in the species' arid and open habitat. Besides drilling in such extremely hard woods as mesquite, they also drill in softer trees and do considerable damage to pine utility poles, for which they have been greatly persecuted. This, together with the elimination of mesquite from ranchland areas of Texas, has seriously reduced this species in some areas. During the breeding season both sexes work at nest excavation, which requires some 6-10 days. Both sexes also assist in incubation and in brooding responsibilities. By late summer, groups of adults and juveniles may be seen traveling together, but such family units separate before winter.

Suggested Reading: Bent 1939; Selander and Giller 1959.

Red-headed Woodpecker
Melanerpes erythrocephalus

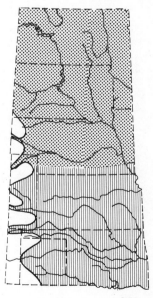

Breeding Status: A breeding species in nearly the entire region except for local treeless areas of westernmost Nebraska, eastern Colorado, eastern New Mexico (breeds in Union County), and the western panhandle of Texas.

Breeding Habitat: This species uses relatively open forests or woodlots, as well as urban parks and wooded housing areas. It occupies somewhat more open areas than does the red-bellied woodpecker and thus extends farther west.

Nest Location: Nests are typically in rather isolated, generally dead trees or dead tree limbs, especially those with no bark. Cavities are often excavated where there is an existing crack, and they range from about 10 to 50 feet above the ground. The entrance averages about 2 inches in diameter, and its shape is often affected by preexisting cracks.

Clutch Size and Incubation Period: From 4 to 7 eggs (8 Kansas nests averaged 4.0). The eggs are white and rather glossy. The incubation period is 12 days, starting somewhat before the last egg is laid, which results in an extended hatching period. Possibly double-brooded in southern parts of range, and a persistent re-nester.

Time of Breeding: Egg dates in North Dakota are from June 5 to July 23, and in Kansas they extend from the middle of May to the middle of June, with young recorded in the nest as late as early August. Egg dates for Oklahoma are from May 5 to June 2, and dependent young have been seen as late as early August.

Breeding Biology: The breeding behavior of this species is closely similar to that of the red-bellied woodpecker, although they differ in degree of migratory behavior. After males of this species return to their nesting areas, they call from their roosting and prospective nesting holes, apparently to attract mates to their excavations, and they also drum. When a female approaches, the male begins tapping from within the cavity, then typically flys away to allow the female to inspect the hole. He may also solicit copulation by inviting reverse mounting by the female while he is perched near the nest cavity. Mutual tapping at the nest hole seems to indicate that a pair bond is formed and that the female accepts the nesting cavity. Both sexes assist in incubation and brooding, with the male performing these activities at night. In one Illinois study, 3 of 15 pairs nested a second time in one season, sometimes while still feeding their first fledglings. The young birds tend to follow their parents for some time after leaving the nest, until they are chased away after about 25 days.

Suggested Reading: Kilham 1977*a*; Reller 1972.

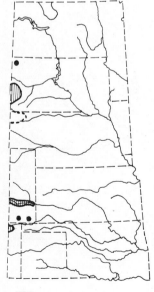

Lewis Woodpecker
Melanerpes lewis

Breeding Status: Breeds uncommonly in the Black Hills of South Dakota and rarely (probably no longer) in the Pine Ridge area of Nebraska (one possible nesting for central Nebraska, in Logan County). Occurs in eastern Colorado cottonwoods nearly to Kansas, and in canyons south of the Arkansas River. Also breeds locally in northeastern New Mexico (Capulin Mountain, Union County) and is probably a previous resident of Cimarron County, Oklahoma.

Breeding Habitat: In the Black Hills, this species prefers the edges of pine forests and streamside cottonwood groves with considerable dead growth, as well as burned-over areas with abundant tall stumps. Elsewhere in their range they also occupy oak woodlands and to a small extent orchards and piñon-juniper woodlands.

Nest Location: Nests average about 25 feet above the ground and are usually in dead trees, less often in dead portions of live trees. Conifers and deciduous species are used about equally, particularly cottonwoods, sycamores, and oaks. The entrance diameter is about 2½ inches, and the cavity is about 12 inches deep. Old nest sites or previously existing cavities tend to be used, probably since the species is poorly adapted to excavating.

Clutch Size and Incubation Period: From 5 to 9 white eggs, usually 6–7. The incubation period has been variously estimated from 12 to 14 or even 16 days, with 13–14 days probably normal. Single-brooded.

Time of Breeding: Colorado egg records are from May 15 to June 20, and nestlings have been seen from June 22 to late July.

Breeding Biology: Unlike other North American woodpeckers, this species is adapted to feeding on free-living insects and is remarkably adept at aerial flycatching. As the breeding season approaches the male begins to utter his harsh *churr* breeding call, which serves to attract mates and defend or announce nest sites. Males also drum, but mutual tapping and female drumming have not been reported. Copulation is typically preceded by reverse mounting, as in other woodpeckers. Males take the predominant role in selecting the nest site and defending the nest. Since old nest cavities are usually used, little excavation is needed. Males brood and incubate at night, and both sexes share these responsibilities during the day. The fledging time has not been definitely established but is probably between 28 and 34 days. A few days before the young can fly, they move out of the nest cavity and begin to climb about, which exposes them to hawk predation. As the pair leaves the nest vicinity each member takes part of the brood and continues to feed them occasionally until they are able to catch insects on their own.

Suggested Reading: Bock 1970; Bent 1939.

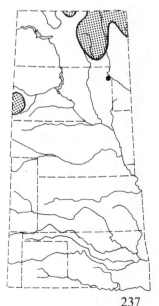

Yellow-bellied Sapsucker
Sphyrapicus varius

Breeding Status: Breeds in northeastern and north-central North Dakota, primarily in the Turtle Mountains, in northwestern Minnesota, and in the Black Hills of South Dakota.

Breeding Habitat: In North Dakota breeding is done in extensive tracts of deciduous forests in upland or lowland areas. Coniferous forests are also used in Minnesota and South Dakota; in the Black Hills, pine forests and aspen groves are preferred nesting habitats.

Nest Location: Nests are usually near water, often facing it, and typically are in either dead trees or live trees with decaying hearts. The entrance hole is very small, averaging about 1½ inches, and varies from 8 to 40 feet above the ground. Rarely is the same nest hole used in subsequent years, but a new cavity may be excavated in the same tree. Aspens and poplars are preferred nesting trees, perhaps because they often have rotted interiors.

Clutch Size and Incubation Period: From 4 to 7 eggs, usually 5–6. The incubation period is 12–13 days. Single-brooded.

Time of Breeding: Dates of active nests in North Dakota range from June 19 to July 17, and dependent flying young have been recorded from July 17 to August 22. June and July are likewise the probable nesting months in South Dakota.

Breeding Biology: Males of this migratory species are the first to arrive in breeding areas in spring, and they soon establish territories that they advertise by drumming and territorial conflicts. Most males begin new nest-hole excavations each spring, and unlike other North American woodpeckers they advertise them by a distinctive courtship flight, performed below the level of the partner, emphasizing the distinctive back patterning. Both sexes may perform this flight, which seems to stimulate the pair bond and to build a site attachment to the nest area. Males also perform tapping at a potential nest site, to which the female may respond similarly if she accepts the location. Both sexes help excavate new cavities, and males roost inside such excavations. Both sexes incubate, with the females gradually taking on a greater share. The period of brooding lasts for 8–10 days after hatching, and adults bring both sap and insects to nestlings. Fledging occurs 28 days after hatching.

Suggested Reading: Kilham 1962, 1977*b*.

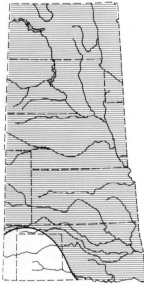

Hairy Woodpecker
Picoides villosus (*Dendrocopos villosus*)

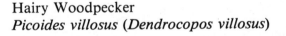

Breeding Status: Breeds in suitable habitats nearly throughout the area, excepting perhaps eastern New Mexico (breeds in the Cimarron Valley) and the Texas panhandle (no breeding records).

Breeding Habitat: Fairly extensive areas of coniferous or deciduous forest provide optimum breeding habitat, but streamside groves of trees are also used. Deciduous forests seem preferred over coniferous forests.

Nest Location: Cavities are usually in deciduous trees, frequently aspens, ashes, elms, or cottonwoods, but no special preference is evident. Both live and dead trees are used, but live trees with decaying centers are often selected, and the entrance may be from 5 to 30 feet or more above the ground. The entrance is usually vertically elongated, averaging about 2 by 2½ inches.

Clutch Size and Incubation Period: From 3 to 6 white eggs, usually 4. The incubation period is 11–12 days. Single-brooded, but known to be a regular renester.

Time of Breeding: In North Dakota the probable nesting season is from late April to mid-July. In Kansas the egg records extend from March 21 to May 30, with a peak of egg-laying in early May. Oklahoma breeding dates are from March 6 (nest ready for eggs) to May 24 (young still in nest).

Breeding Biology: Hairy woodpeckers are largely nonmigratory and begin to form pairs in midwinter, about three months before the start of nesting. Typically, males are attracted at this time to territories that the females establish the previous fall. During this time the pair performs drumming duets, and both sexes use drumming to locate a mate when they are visually separated. The male also uses drumming as a territorial display. When searching for a suitable nest site, the birds perform a slow tapping, which tends to attract the mate. At about this time females begin to solicit copulation, but copulation reaches a peak during actual excavation. The males do most of the excavation, working thoughout the day and sometimes sleeping in the cavity at night. During incubation the male continues to be most attentive to the eggs, even during daylight. Both sexes brood the nestlings for more than 2 weeks. By about the 17th day the adults begin to feed the young from outside the nest rather than going inside with food. By the time the young are 28–30 days old they emerge from the nesting hole and are able to fly strongly.

Suggested Reading: Lawrence 1967; Kilham 1966.

Downy Woodpecker
Picoides pubescens (Dendrocopos pubescens)

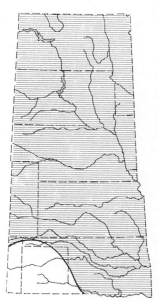

Breeding Status: Breeds in suitable habitats nearly throughout the area, except perhaps eastern New Mexico and the Texas panhandle. Rare breeder in the Oklahoma panhandle (one nesting record for Cimarron County).

Breeding Habitat: Basically the same breeding habitats are used as those of the hairy woodpecker, but this species forages in shrubs and tall herbaceous plants as well as in trees.

Nest Location: Nests are usually 8–50 feet above the ground in dead or dying wood but are sometimes in live tree branches. Dead

239

stubs of aspens are a favored nest site. Nest sites and branches used average very slightly lower and smaller than those of hairy woodpeckers (about 30 feet high and 10 inches in diameter in one study). The entrance is circular, about 1¼ inches in diameter, and is usually on the underside of a branch, or at least is protected from above in some way.

Clutch Size and Incubation Period: From 3 to 6 white eggs, usually 4–5. The incubation period is 12 days. Single-brooded in this area, possibly double-brooded farther south.

Time of Breeding: The breeding season in North Dakota is from mid-May to mid-July, with egg dates from May 30 to June 7 and nestlings reported as late as July 10. Kansas egg dates are from April 11 to June 10, and Oklahoma egg records extend from March 17 to June 7.

Breeding Biology: Like hairy woodpeckers, this species is nonmigratory, and birds spend the entire year near their breeding areas. Toward late winter territorial drumming and associated conflicts begin, and mates often drum at dawn to locate each other when their roosting holes are widely separated. In spring both sexes begin to seek out suitable nest sites, and when one has located a potential site it begins drumming and tapping to attract the other. Short courtship flights occur near the nest and may strengthen site attachment or stimulate copulation, which takes place near the nest. Both sexes help excavate the nest, but usually the female is most active. The male also assists with incubation and spends the night in the nest. For the first week after hatching one or the other adult remains at the nest at all times, but as the young birds develop both adults spend much time foraging. By the time the young are 2 weeks old they can climb to the nest entrance to be fed, and they are ready to fly less than 4 weeks after hatching. Pair bonds break down after the breeding season, and each sex excavates fresh roosting holes for use in winter, when each forages independently of its mate.

Suggested Reading: Kilham 1974; Lawrence 1967.

Ladder-backed Woodpecker
Picoides scalaris (*Dendrocopos scalaris*)

Breeding Status: Breeds from extreme southeastern Colorado (breeding records for Baca and El Paso counties) and southwestern Kansas (Morton County, possibly Hamilton County) southward through the Oklahoma and Texas panhandles as well as adjacent southwestern Oklahoma. Also breeds in eastern New Mexico (Union County, probably elsewhere).

240

Breeding Habitat: Generally coextensive with mesquite in Oklahoma; in Texas it occurs in mesquite- and cactus-covered areas, along brush-lined streams, in fencerows, and in cultivated areas near thickets or woods.

Nest Location: Nests are in trees, yucca stems, agave stalks, or utility poles, usually only a few feet above the ground. When in trees, either living or dead wood is used, and the trees may be mesquites, cottonwoods, willows, or a variety of other species. The entrance hole is only about 1½ inches across, and saplings as small as 3½ inches in diameter have been utilized.

Clutch Size and Incubation Period: From 2 to 6 white eggs, usually 4-5. The incubation period is about 13 days. Single-brooded.

Time of Breeding: Eggs or broods have been seen in Oklahoma between May 19 and June 29. Texas egg records extend from April 1 to July 9.

Breeding Biology: This is basically a desert-adapted species, with almost no close relatives in areas where it is abundant, at least in the Great Plains. The two sexes tend to forage in different locations, with females using smaller shrubs or cacti or smaller branches and twigs than males, presumably to reduce competition. Females also tend to specialize in their foraging, while males are more general in their behavior. Pair-formation has been studied rather little, but drumming seems to be infrequently used and not associated with fixed drumming areas. Vocalizations include several call types, and many of the vocal or postural displays are based on aggressive tendencies. They resemble closely the calls and postures of hairy and downy woodpeckers, and the mode of pair formation is probably very similar. Apparently both sexes incubate, and both sexes attend the young. Little is known of the nestlings' life, and the fledging period is still unreported.

Suggested Reading: Short 1971; Austin 1976.

Red-cockaded Woodpecker
Picoides borealis (Dendrocopos borealis)

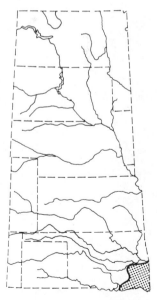

Breeding Status: Breeding is limited to a small area of southeastern Oklahoma, with recent populations known only from McCurtain, Latimer, and Bryan counties.

Breeding Habitat: This woodpecker frequents open pine woodland, and in Oklahoma specifically shortleaf pines.

Nest Location: Nests are exclusively in live pines that are diseased with a fungus (*Fomes pini*) that causes rotting of the heartwood.

241

The nest is placed from 18 to 100 feet up, in a roosting cavity of the male, which is bored through the hardwood area into the rotted core area. The birds chip away the bark above and below the nest, producing a sap flow near the entrance that makes the nest hole conspicuous.

Clutch Size and Incubation Period: From 3 to 5 white, glossy eggs. The incubation period is slightly over 10 days, among the shortest known. Single-brooded.

Time of Breeding: Few Oklahoma records are available, but nesting apparently starts early, with hatching possible in early April or even earlier. Young may be fed for as long as 5 months after hatching.

Breeding Biology: These woodpeckers remain territorial throughout the entire year and seem to recognize territorial boundaries. Pairs feed mainly on pines, with the males foraging on the limbs, branches, and higher trunk while the female uses the lower trunk. The center of the territory is the roosting tree, and pairs may use the same tree for many years, or several generations may use the same tree. Typically each member of the pair or family has its own excavated roost. Generally the eggs are deposited on the floor of the male's roosting cavity, but at times the female's roosting cavity may be used. Incubation is performed by both members of the pair, and probably begins before the clutch is complete, so that the eggs hatch over a period of several hours. The young are brooded almost continuously for their first 4 days of life, and both brooding and feeding may be shared with a "nest helper," which at least in one case was known to be a male offspring of the previous year. The young fledge 26–29 days after hatching but remain dependent on their parents or the nest helpers for a remarkably long period thereafter. They apparently remain with their parents until at least the following spring and thus are unlikely to breed in their first year.

Suggested Reading: Ligon 1970; Steirly 1957.

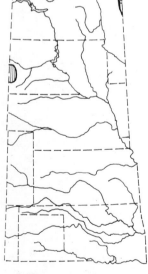

Black-backed Three-toed Woodpecker
Picoides arcticus

Breeding Status: Breeds locally and infrequently in northern Minnesota (nesting records for Clearwater and Becker counties) and uncommonly in the Black Hills of South Dakota.

Breeding Habitat: Associated everywhere with coniferous forests, particularly those having an abundance of dead trees, as in areas that have recently been burned or logged.

Nest Location: Nests may be in live trees (primarily those with dead interiors), dead trees, stubs, or utility poles. Usually nests are no more than 15 feet above the ground and have entrances less than 2 inches in diameter, usually strongly beveled at the lower side. The bark around the entrance may also be removed, making the cavity conspicuous.

Clutch Size and Incubation Period: From 2 to 6 white eggs, often 4. The incubation period is probably about 14 days. Single-brooded, but renesting has been reported.

Time of Nesting: Active nests in Minnesota have been reported from May 9 to June 22, and nearly grown young accompanied by their parents have been seen in mid-July. The estimated nesting season in the Black Hills is also May through July, with young birds reported from June 20 to July 25.

Breeding Biology: This is a relatively eruptive species, coming into areas shortly after logging or forest fires, breeding for a few seasons, then disappearing again. The birds seem to feed exclusively on conifers, and usually nest in them, but have at times been found nesting in aspens. Dead tamarack and spruce swamp areas are favored foraging areas for this species in northern Minnesota, and the birds are usually found in pairs or probably family groups of three or four individuals. Little is known of pair-forming behavior, but both sexes help in excavating the nesting hole and, as might be expected, both sexes incubate. Probably the male does most of the feeding of the young, and the species mainly eats the larvae of wood-boring beetles. Thus the bird is extremely beneficial in controlling this serious enemy of coniferous forests. There is no specific information available on the length of time to fledging.

Suggested Reading: Roberts 1932; Bent 1939.

Northern Three-toed Woodpecker
Picoides tridactylus

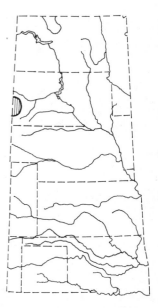

Breeding Status: Breeds very rarely in northern Minnesota (one breeding record for Clearwater County) and the Black Hills of South Dakota (one nesting record).

Breeding Habitat: Coniferous forests, especially spruces and tamaracks, provide nesting habitat for this species.

Nest Location: Nests are usually in live or dead coniferous trees, especially those in burned areas. The entrance is usually less than 40 feet above the ground and is about 2 inches in diameter, with the lower side strongly beveled.

Clutch Size and Incubation Period: Typically 4 white eggs. The incubation period is estimated to be 14 days. Single-brooded.

Time of Breeding: The only definite Minnesota breeding record is of a young bird seen with its parents in early July. A nest of 3 fresh eggs was found in the Black Hills in mid-June.

Breeding Biology: This relatively little-studied species, like the other three-toed woodpecker, seems to be a fire-adapted form that rapidly colonizes stands of recently burned trees that are being attacked by bark-boring beetles. It also occupies undisturbed stands of virgin forest where there are old trees with diseased or decayed hearts. About three-fourths of the food of both three-toed species consists of wood-boring beetle larvae, with caterpillars being of secondary significance. The birds often strip bark from large areas of the trees as they search for woodborers. They also are relatively tame so their presence is usually easy to detect. Almost nothing is known of their social behavior, but it is probably little different from that of the other American *Picoides* species. Both sexes are known to assist in incubation and in rearing the young, with the male often taking on most of these responsibilities. They are rather sedentary birds and probably maintain foraging territories throughout the year.

Suggested Reading: Bent 1939; Gibbon 1966.

FAMILY TYRANNIDAE
(TYRANT FLYCATCHERS)

Scissor-tailed Flycatcher

Eastern Kingbird
Tyrannus tyrannus

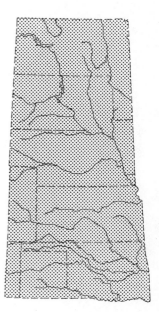

Breeding Status: A pandemic breeder throughout the region, generally more common in the eastern portion than in the west.

Breeding Habitat: The species frequents open areas having scattered trees or tall shrubs, and with forest edges or hedgerows. It is usually found in forests where the canopy level is uneven, allowing vantage points for foraging.

Nest Location: Nests are on tree branches well away from the main trunk or at times on shrubs or even manmade structures. They are usually less than 20 feet above the ground and are relatively bulky cups of herbaceous plant materials, lined with finer materials such as grasses.

Clutch Size and Incubation Period: From 3 to 5 eggs (21 North Dakota nests averaged 3.9), white with large dark brownish spots. Probably single-brooded.

Time of Breeding: North Dakota egg records are from June 6 to July 11, and nestlings have been reported from July 6 to early September. Kansas egg dates are from May 11 to July 20, with about 70 percent of the eggs laid in June. Oklahoma nest-building or egg dates are from May 10 to July 5, and unfledged young have been seen as late as July 21.

Breeding Biology: Eastern kingbirds arrive on their breeding grounds when insect populations begin to become noticeable and soon become extremely conspicuous as the males begin territorial behavior and associated courtship. Aerial displays are common then, with the bird flying erratically in a series of swoops and dives not far above the ground and uttering harsh screams. Chases and fights between birds on adjacent territories are also prevalent at this time. Kingbirds have a strong tendency to return to the same nesting territory in subsequent years, although the specific nest site varies. Males help build the nest, but the female typically does the incubating. After the eggs hatch both sexes are kept constantly busy bringing food to the young, the female generally being more active in feeding and brooding than her mate. The young remain in the nest for approximately 2 weeks. Thereafter they remain as a group, often on a wire or an exposed tree branch, waiting for their parents to come and feed them. The young begin to catch flies after they are about 8 days out of the nest, and they continue to improve in flying ability for the next month. They are not fed by the adults beyond about 35 days after fledging.

Suggested Reading: Bent 1942; Morehouse and Brewer 1968.

Western Kingbird
Tyrannus verticalis

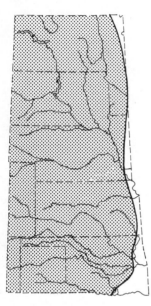

Breeding Status: Breeds throughout nearly the entire region, with the possible exceptions of extreme eastern Oklahoma and Kansas, northwestern Missouri, and west-central portions of Iowa and Minnesota.

Breeding Habitat: This species is strongly associated with edge habitats, such as shelterbelts, hedgerows, orchards, woodland margins, tree-lined residential districts, and the like.

Nest Location: Nests are in a variety of trees or tall shrubs and on artificial structures. When available, trees are chosen over small plant species, but nests usually are less than 50 feet above the ground. They often are on the lower dead branches of tall cotton-woods or elms, where good vantage points are available. In size and materials they are much like those of the eastern kingbird.

Clutch Size and Incubation Period: From 3 to 6 eggs (averaging about 4.5 in North Dakota and Kansas). The eggs closely resemble those of the eastern kingbird. The incubation period has been variously estimated as 12–14 days. Normally single-brooded, but records of second broods exist.

Time of Breeding: North Dakota egg dates are from June 12 to July 16, while those from Kansas extend from May 11 to July 31, with a peak in mid-June, more than 70 percent of all clutches being laid in June. Dates of eggs or newly hatched young in Oklahoma are from June 1 to July 5, and Texas egg dates range from May 16 to July 12.

Breeding Biology: Like the eastern kingbird, this species is highly territorial and generally is extremely intolerant of larger birds such as crows and hawks in the vicinity of its nest. The birds are also at least as noisy as the eastern kingbird, and during the early stages of territorial establishment and pair-formation they are particularly conspicuous for their calling and singing. Yet, in spite of their overt aggressiveness, there are reported cases of several pairs occupying the same tree or sharing a small grove for nesting. Little is known of their nest-building behavior, but a traditional return to a previous year's nesting territory is evident, as in eastern kingbirds. The female does most of the incubating, but males have been seen on the nest as well. Likewise, both parents actively feed the young, which remain in the nest for about 2 weeks. In one observed case in Oklahoma, a pair began a new nest only 4 days after its first brood fledged, but presumably the young birds remain at least partially dependent on their parents for some weeks after fledging.

Suggested Reading: Bent 1942; Hespenheide 1964.

Cassin Kingbird
Tyrannus vociferans

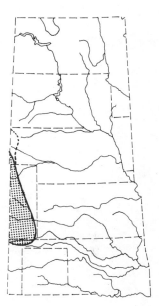

Breeding Status: Probably breeds from the western panhandle of Nebraska (no definite nesting records, but see *Nebraska Bird Review* 39:72) southward through eastern Colorado (records for the plains area are also lacking) to the Cimarron Valley (nesting records for Cimarron County, Oklahoma, and Union County, New Mexico, but not for Kansas, where sporadic breeding is probable).

Breeding Habitat: The species is associated with open country such as plains and semideserts with scattered trees and also with open woodlands. It is more western and montane-adapted than the western kingbird, but widely overlaps with it.

Nest Location: Trees, bushes, and posts are reportedly used for nest sites, but nearly all nests are in fairly tall trees at substantial heights (20–70 feet). In one Arizona study, 35 of 44 nests were in sycamores, and none were in shrubs or yucca, where western kingbirds were found to nest frequently. The nest is constructed much like those of the eastern and western kingbirds but is slightly larger and bulkier.

Clutch Size and Incubation Period: From 2 to 5 eggs, usually 3 or 4. The eggs are white, with a wreath of brown spots around the larger end. The incubation period is probably between 12 and 14 days. Reportedly double-brooded at southern end of range.

Time of Breeding: Few dates are available for this region, but in Oklahoma nest-building has been seen in early June and a clutch in late June. In western Colorado the breeding season is said to be from mid-May to the first of July.

Breeding Biology: In a study area in Arizona, where both western and Cassin kingbirds occur together, it was found that the greatest abundance of western kingbirds was in desertlike habitats, while this species was most prevalent in transitional areas between desert and pine or oak woodlands. Evidently the Cassin kingbird will nest in almost any habitat, provided tall trees are available for nesting sites. Apparently there is considerable intolerance between these species, and yet their nests are at times placed in close proximity. Little has been written on the role of the sexes in breeding, but apparently the female does most if not all of the incubating. Both sexes do care for and feed the young, which remain on the nest for approximately 2 weeks. The young follow their parents for some time thereafter, until they have perfected their own insect-catching abilities.

Suggested Reading: Bent 1942; Hespenheide 1964.

Scissor-tailed Flycatcher
Muscivora forficata

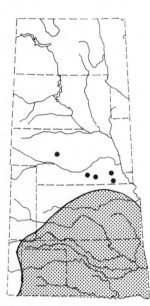

Breeding Status: Breeds sporadically in southeastern Nebraska (records for Adams, Lancaster, Logan, Gage, and Clay counties) and regularly from north-central Kansas southward and westward to extreme southeastern Colorado (Baca County), the panhandles of Oklahoma and Texas, and extreme northeastern New Mexico (Union County).

Breeding Habitat: The species breeds in open to semiopen habitats with a scattering of trees or other elevated sites such as buildings, or in wooded areas with openings.

Nest Location: Nests are typically in cottonwoods, elms, or other hardwood species, in exposed sites between 5 and 50 feet above the ground. Isolated trees are preferred to those growing in groves or heavier cover. Windmills, the crossbars of utility poles, and buildings are sometimes also used. The nests are relatively variable in size and poorly constructed, consisting of a shallow cup of twigs and weeds with a lining of softer materials.

Clutch Size and Incubation Period: From 2 to 6 eggs (17 Kansas nests averaged 3.2), white with variable brown spotting. The incubation period is 12–13 days. Frequently double-brooded, with late clutches distinctly smaller than earlier ones.

Time of Breeding: Kansas egg records are from May 21 to July 10, with a peak in late June. Oklahoma egg records extend from May 22 to July 28, and those from Texas are from March 31 to August 10.

Breeding Biology: This species is highly territorial, although for a time after spring arrival the birds remain in small flocks that roost together. Females apparently take the initiative in choosing nest territories, which center on a suitable nesting tree and include a radius of 30–40 yards around it. The female also apparently picks the nest site, since she does all the nest construction while the male watches from nearby. She gathers all the nest materials, though the male guards the nest site in her absence. The female likewise does all the incubating and is more active than the male in providing food for the young, and she apparently cares for the nestlings at night. Males congregate for roosting, but they appear at the nest shortly after sunrise to begin feeding the offspring. Besides males, unsuccessful females and unmated birds join in these roosting congregations, and shortly after fledging the young birds also begin to join these roosting groups, which may number as many as 250 birds. The nestling period is about 14 days.

Suggested Reading: Fitch 1950; Bent 1942.

250

Great Crested Flycatcher (Crested Flycatcher)
Myiarchus crinitus

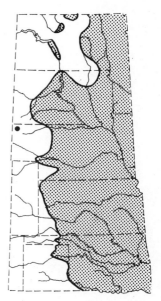

Breeding Status: Breeds locally from the eastern half of North Dakota and western Minnesota southward through the eastern portions of South Dakota and Nebraska (west to Sioux and Deuel counties), probably extreme eastern Colorado (Sedgwick and Yuma Counties), nearly all of Kansas except for Morton County, Oklahoma west at least to Beaver County, and the eastern panhandle of Texas. Also reported to breed in Union County, New Mexico.

Breeding Habitat: Breeds in fairly extensive hardwood forests, especially those with fairly open canopies. Limited to river valleys at the western edge of its range.

Nest Location: Nests are in woodpecker holes or natural cavities of trees, usually between 10 and 20 feet above the ground. The cavity is filled with a variety of materials, often including a castoff snakeskin or similar material such as plastic or cellophane. Artificial structures such as birdhouses, drain spouts, and other hollows may also be used, with little preference shown for the shape of the opening or the cavity expanse, which gradually becomes filled in.

Clutch Size and Incubation Period: From 4 to 8 eggs, usually 5 or 6. The eggs are yellowish with extensive streaks, blotches, or spots of brown or purple. The incubation period is 13–15 days. Single-brooded.

Time of Breeding: Minnesota egg records are from June 4 to June 26, with dependent young seen as late as August 3. Kansas records are from May 11 to July 10, with a peak in early June. Oklahoma records of eggs extend from May 18 to June 12, and Texas records are from April 30 to June 8.

Breeding Biology: Shortly after returning to the nesting area, males become highly territorial, and a good deal of calling and chasing by males is evident. At least in some instances pairs of the preceding year are reformed; one banded pair nested in the same box for three consecutive years. The pair spends a good deal of time in making their chosen cavity suitable for nesting, nearly filling it with whatever materials are readily available. There is no special reason for using snakeskins to line the nest beyond their being soft and pliable. The female does most or all of the incubating, with the male in close attendance. During feeding the pair often return to the nest together, one waiting outside the cavity until the other has passed on its supply of food, primarily insect larvae, to the young. Estimates of the nestling period vary from 12 to 21 days, with two estimates of 18 days. The family remains together for some time after leaving the nest, while the young birds gradually learn to forage for themselves.

Suggested Reading: Mousley 1934; Bent 1942.

Ash-throated Flycatcher
Myiarchus cinerascens

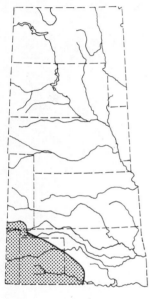

Breeding Status: Breeds in eastern New Mexico (records for De Baca and Quay counties) eastward across the Texas panhandle to southwestern Oklahoma (Harmon, Tillman, and Comanche counties, also Cimarron and possibly Beaver counties in the panhandle). Reportedly also breeds along the Cimarron Valley of southeastern Colorado.

Breeding Habitat: The species is generally associated with mesquite and cactus deserts, and open piñon-juniper woodlands, but in northern Texas and Oklahoma it is more typical of open stands of cottonwoods, willows, or mesquites, or areas near partially wooded stream courses or dry gulches.

Nest Location: Nests are in natural cavities of trees or stumps, in woodpecker holes, or sometimes in old nests of the cactus wren. The cavity is filled and lined with a variety of materials, often including hair and sometimes including snakeskins. It is usually less than 20 feet above the ground.

Clutch Size and Incubation Period: From 3 to 7 eggs, usually 4. The eggs are buff to creamy with brownish to lavender spots, lines, and streaks. The incubation period is probably 15 days. Apparently single-brooded.

Time of Breeding: Breeding records in Oklahoma range from June 4 (eggs) to August 7 (fledged but dependent young). In Texas, egg records are from April 12 to June 17, and young in the nest have been reported as late as July 14.

Breeding Biology: This desert-adapted species has been studied rather little, but in many respects it resembles the great crested flycatcher in its ecology and behavior. Its call notes are closely similar to those of that species, and two of its most diagnostic calls are *ha-whip* and *ha-wheer*, given with a vertical tail-flick. Birds develop these characteristic calls by the time they fledge at about 16 days of age, even when they are reared in isolation, indicating their innate basis. During nest-building the female gathers the materials while the male sings and guards the nesting area. Both parents feed the young, mostly on the soft parts of insects, which at first are regurgitated but later are fed to the young birds directly. For some time after the young birds leave

252

the nest they remain close to their parents, and the adults help train them to capture live prey by releasing slightly injured insects directly in front of them.

Suggested Reading: Bent 1942; Lanyon 1961.

Eastern Phoebe
Sayornis phoebe

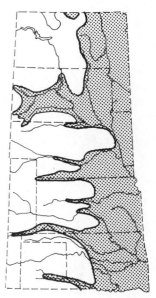

Breeding Status: Breeds from the eastern half of North Dakota and western Minnesota southward through most of South Dakota as well as the Black Hills, eastern Nebraska (west locally at least to Sioux and Lincoln counties), nearly all of Kansas (to Cheyenne and at least Comanche counties), southeastern Colorado (Baca County), northeastern New Mexico (Union County), the Black Mesa area of Oklahoma and most of eastern Oklahoma extending locally along the Red River into the Texas panhandle (Brisco County).

Breeding Habitat: The species frequents woodland edges or wooded ravines near water, cliffs, and habitats providing natural or artificial ledges for nesting, such as bridges and farm buildings. It is usually found near lakes or streams, in partially wooded areas.

Nest Location: Nests are most often associated with bridges over rivers or with other manmade structures near lakes, but ledges on rock bluffs, ravines, and similar natural sites are also used. The nest is constructed of mud and vegetation with a lining of hair and fine grasses.

Clutch Size and Incubation Period: From 3 to 6 white or sparsely spotted white eggs (58 Kansas clutches averaged 4.2). The incubation period is 15–16 days. Usually double-brooded, with second clutches averaging smaller than initial clutches. Frequently parasitized by cowbirds.

Time of Breeding: North Dakota nest-building or egg records are from May 7 to June 21. Kansas egg dates extend from March 21 to July 20, and Oklahoma egg dates range from March 28 to June 28.

Breeding Biology: This is one of the earliest of the flycatchers to arrive in spring. Males typically precede the females, immediately establishing and patrolling territories. Females soon follow, and pairs are formed when the female enters a male's territory and is accepted. Males soon begin to follow their mates closely and often attempt to mount them, but they are frequently rebuffed. The female builds the nest alone, while the male perches nearby and repeatedly utters the familiar *phoebe* notes. These actually consist of two different song types that have rather different communication functions. The female does all the incubating, although the male helps feed the young. They fledge at about 15–16 days, after which a second clutch may be started. At this time they may use the same nest, but they often build a new one or may superimpose the new nest over the initial one.

Suggested Reading: Smith 1969; Bent 1942.

Say Phoebe
Sayornis saya

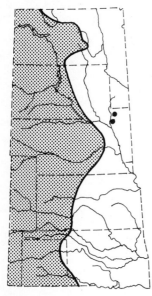

Breeding Status: Breeds from the east-central portions of North Dakota south through eastern South Dakota, possibly including Rock County, Minnesota *(Loon* 47:13), to extreme northwestern Iowa (Sioux and Plymouth counties), central Nebraska (east to Cuming, York, and Clay counties), western Kansas (east to at least Cloud County), eastern Colorado and New Mexico, and extreme western Oklahoma (Cimarron County). There are no breeding records for the Texas panhandle (the nearest being from Wilbarger County), where they apparently are local breeders.

Breeding Habitat: The Say phoebe is associated with open and arid regions, especially rocky habitats that provide nesting sites. Sunny canyons, mountain meadows, and open areas near buildings are all used. Unlike the eastern phoebe, this species is independent of surface water.

Nest Location: Nests are on rocky ledges, on horizontal ledges under bridges or in other manmade structures, or even built on old nests of barn or cliff swallows. Caves or abandoned mine shafts are also frequently used. The nest is much like that of the eastern phoebe.

Clutch Size and Incubation Period: From 2 to 7 eggs, usually 4 or 5, pure white or with a few small brownish spots. The incubation period is about 14 days. Double-brooded over most of range; sometimes three broods are raised.

Time of Breeding: Kansas egg records are from May 1 to July 20, with a peak of laying in late May. Texas egg records are from April 7 to June 27, and dependent young have been seen as late as August 25.

Breeding Biology: Male phoebes arrive on the nesting grounds before females, and when the females return pairs form or reform rather rapidly. Nest-building or the repair of a previous nest may begin only a week or two after the birds arrive, and is presumably done by the females. Often the same nest is used in subsequent years or for successive clutches, and at favored nest sites the loss of one or both members of a pair brings a rapid replacement. Males do not participate in incubation but remain nearby on a convenient lookout post. However, the male guards the nest and feeds the female, as well as feeding the young virtually alone during the first week or so of their lives. When they are ready to leave the nest at about 2 weeks of age the male takes over and teaches them to capture insects, while the female prepares to produce a second clutch of eggs. Apparently she assumes the entire job of feeding herself and her second clutch, although reportedly the male may again appear to take care of the brood when it fledges, freeing the female for a possible third brood.

Suggested Reading: Schukman 1974; Ohlendorf 1976.

Acadian Flycatcher
Empidonax virescens

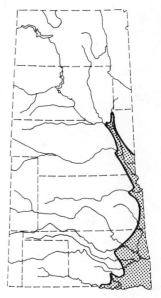

Breeding Status: Breeds locally in the Missouri Valley from extreme southeastern South Dakota (no record since 1921) southward through Nebraska, Iowa, northwestern Missouri, eastern Kansas, and eastern Oklahoma, west to Payne and Murray counties.

Breeding Habitat: The species breeds in shady and humid river-bottom forests, forested swamps, and wooded uplands.

Nest Location: Nests are on forks of horizontal branches well away from the main trunk of a tree; lower branches of beeches and dogwoods are favored sites. They are often situated over water and usually are only 10–20 feet above the ground or water. The nest is a frail and rather shallow cup of plant stems and fibers, loosely woven and airy in appearance. Open space is characteristic below the nest, so that the birds can approach it easily.

Clutch Size and Incubation Period: From 2 to 4 eggs, usually 3; the eggs are white to buffy with sparse brown spotting. The incubation period is 14–15 days. Frequently double-brooded, and a persistent renester.

Time of Breeding: Kansas breeding records are sparse but suggest that most eggs are laid in late May or early June. Oklahoma egg dates are from May 23 to June 18, and brooding or incubating adults have been seen as late as July 16.

Breeding Biology: Males take up territories immediately upon arriving at their nesting grounds, and most territories are occu-

255

pied within a week. Territories average about 2 acres and are advertised by a characteristic *tee-chup* song repeated several times a minute. Evidently pair bonds are quickly formed after the females arrive, since they often begin depositing eggs within 10 days of their arrival. The female searches for suitable nest sites and may test a number of forks for "fit" while the male watches nearby. Only the female constructs the nest, which may take from 6 to 9 days to make ready for eggs, but at least at times the birds simply repair a nest of the past year. Only the females incubate, but apparently they are not fed by the males, which periodically visit the nest. Cowbird parasitism is prevalent in some areas and affects nesting success. The young have a nestling period of 13–14 days, and both parents feed them about equally. However, a female may alternately feed her first brood and work on the construction of a second nest. Likewise, some males have been known to father two broods and to feed both simultaneously, an unusual divergence from monogamy in this group of birds. Besides second clutches, renesting is frequent, and in one study five nestings were attempted in a single season.

Suggested Reading: Mumford 1964; Bent 1942.

Willow Flycatcher (Traill Flycatcher)
Empidonax traillii

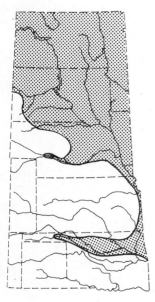

Breeding Status: Breeds in suitable habitats throughout North Dakota and western Minnesota, southward through most of South Dakota and western Iowa, the major river valleys of Nebraska (west at least to Cherry, Thomas, and Keith counties), northeastern Kansas (Doniphan, Douglas, Jefferson, and Wyandotte counties, and probably others), and eastern to central Oklahoma (west locally to Alfalfa County and perhaps Beaver County).

Breeding Habitat: Edge habitats that include thickets or groves of small trees and shrubs surrounded by grasslands are optimum, as are the edges of gallery forests along rivers or streams.

Nest Location: Nests are usually in horizontal forks or upright crotches of shrubs or small trees, usually between 3 and 15 feet above the ground and averaging about 4–6 feet. The nest is a small, compact cup of weeds and fibrous materials lined with cottony or silky fibers. Nests are placed at the outer edge of a shrub or thicket so they can be easily approached.

Clutch Size and Incubation Period: From 2 to 5 creamy eggs (22 Kansas clutches averaged 3.4), with variable-sized spots and blotches of brown. The incubation period is 12–15 days. Single-brooded, but known to renest at least twice after clutch loss.

Time of Breeding: North Dakota egg dates are from July 2 to July 17, and dependent young have been seen as late as August 8. Kansas egg dates are from May 21 to July 10, with a peak in early June. In Oklahoma, eggs have been seen from May 26 to June 25.

Breeding Biology: Recent studies have suggested that this "species" really includes two species, although in this circumscribed region the breeding birds are all likely to be of the willow or *fitz-bew* song type, with the *fee-bee-o* type possibly encountered only in the extreme northeastern areas of boreal forest such as Itasca Park. This account is based on the assumption of this identification. In southern Michigan males arrive at the nesting areas somewhat before females and begin to establish territories that average about 2 acres, always including shrubs and small trees as well as clearings. Water is present either on the territory or very close to it. Birds usually sing from the highest point on the territory, up to 30 songs a minute. Nests are built by females, usually in upright crotches of shrubs that the returning bird can fly to directly. Probably the male's role in nesting is the same as that indicated for the Acadian flycatcher, since only the female is known to incubate. The young are fledged in 12–16 days, and they remain in their parents' territory until fall. They continue to beg for food until they are about 24–25 days old, when they have become fairly adept at catching insects.

Suggested Reading: Walkinshaw 1966; Holcomb 1972.

Least Flycatcher
Empidonax minimus

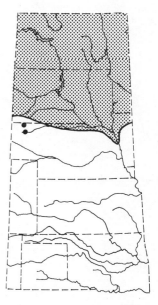

Breeding Status: Breeds locally throughout North Dakota and western Minnesota, southward into South Dakota and rarely northwestern Iowa (Emmet County). There are no recent breeding records for Nebraska, but there are early reports of breeding for Omaha and Dakota City, and pairs have been seen on territories in Brown County (*Nebraska Bird Review* 33:2). Farther south, summering birds have been seen in Kansas, though no breeding records exist, and there is one dubious breeding record for Oklahoma.

Breeding Habitat: Favored habitats include floodplain forests in prairie areas, scattered grovelands on the prairies, wooded margins of lakes, shelterbelts, and urban parks or gardens.

Nest Location: Nests are in upright crotches or on horizontal forks of deciduous or coniferous trees, usually saplings or small trees. They range in height from 2 to 60 feet but usually are only about 5–20 feet above the ground, at the edge of a clearing. The

257

nest is a compact structure with a deep cup, lined with plant down, hair, or other soft materials.

Clutch Size and Incubation Period: From 3 to 6 white eggs, usually 4. The incubation period is about 14 days. Single-brooded, but possibly two broods in southern parts of the range.

Time of Breeding: North Dakota egg dates are from June 13 to June 29. In Minnesota, nests with eggs have been seen from late May to July 12.

Breeding Biology: Shortly after returning to their nesting grounds, male least flycatchers establish breeding territories that are surprisingly small (averaging .18 acres in one study) and that usually but not always include exclusive foraging areas as well as nest sites; sometimes neutral or communal foraging sites are shared. Territories are advertised by the males' songs and are defended primarily by males. Females defend only a small area around the nest, and nest-building is done by the female. The female also does all the incubating and brooding, though the male remains near the nest throughout the entire period, occasionally feeding the female. He also begins to feed the young when they hatch and at least initially provides most of the food. The young birds leave the nest at an age of 13-15 days and may leave the territory in a few days or remain within it for up to 13 days, but they usually do not become independent of their parents until about 3 weeks of age. There seems to be no good evidence that second nestings are typical of this species, even as far south as Virginia.

Suggested Reading: Davis 1959; Nice and Collias 1961.

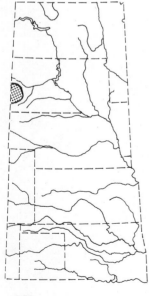

Dusky Flycatcher
Empidonax oberholseri

Breeding Status: Apparently confined as a breeding species to the Black Hills of South Dakota, where it is at least locally common (Spearfish Canyon).

Breeding Habitat: In the Black Hills the species is associated with deciduous shrubbery, aspen groves, and open deciduous woods. Brushy, logged-over slopes are favored in Montana.

Nest Location: Nests in the Black Hills have been found in deciduous shrubs or trees, 3-8 feet above the ground, and often in small birches. In California, studies suggest an average nest height of only about 3 feet, in shrubs or trees 3-40 feet high. Nests are usually built on upright or pendant twigs or in crotches. Black

258

Hills nests resemble those of the least flycatcher but are bulkier and lack the usual downlike lining common to that species.

Clutch Size and Incubation Period: From 2 to 4 white eggs, infrequently with brown spotting. The incubation period is about 15 days. Not reported to be double-brooded.

Time of Breeding: The breeding season in the Black Hills is probably June and July. Egg records are from June 13 to July 1, and fledged but dependent young have been seen as late as July 22.

Breeding Biology: Dusky flycatchers closely resemble other small flycatchers such as willow flycatchers and western flycatchers, and in the Black Hills they occupy drier habitats than the western flycatcher. The dusky flycatcher has a three-syllable territorial advertisement sounding like *prillit, prrddrt, pseet*, with the second syllable low and burred. Territories are marked by this song, by pursuit flights, and by trill calls, the last given as the male perches above the female. Trill-calling plays an important role in pair-formation. In Montana the nests are typically built in crotches of small bushes, and the eggs are laid at the rate of one a day. Only the female incubates, starting incubation after the second egg is laid. The eggs hatch over a period of 2–3 days, and both sexes feed the young, although only the female broods. The nestling period is 15–17 days.

Suggested Reading: Bent 1942; Sedgwick 1975.

Western Flycatcher
Empidonax difficilis

Breeding Status: Limited as a regular breeder to the Black Hills of South Dakota, where it is a common nester. Probably also occurs regularly in Nebraska's Pine Ridge area, but there is only one state breeding record for Sioux County (*Nebraska Bird Review* 43:18).

Breeding Habitat: In the Black Hills this species occurs in hollows, canyons, and sometimes also on mountain slopes where coniferous or mixed forest provides shade and usually where there are streams or other moist habitats.

Nest Location: Most Black Hills nests have been found on ledges or crevices of canyon walls, often concealed by ferns or clumps of mosses. Tree nests are typically supported from below and from the rear, either in a crotch or on a limb projecting far from a main trunk. They are constructed of a variety of materials, but

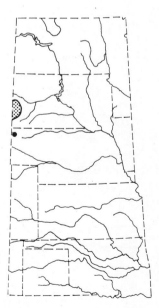

they often contain moss and are almost invariably lined with fine, dry grasses.

Clutch Size and Incubation Period: From 3 to 5 white eggs, spotted or blotched with browns and purples. The incubation period is 14–15 days. Sometimes double-brooded.

Time of Breeding: Breeding in the Black Hills occurs during June and July. Nests with eggs have been seen from June 17 to early July and nestlings reported as late as July 24.

Breeding Biology: Western flycatchers are highly aggressive, and their territorial aggression is directed not only toward their own species but also toward other species of similar size. Unmated males sing their advertising *ps-seet'-ptsick seet* notes throughout most of the day, whereas mated males sing only at dawn. Nests are constructed by one of the pair, presumably the female, over a period of 4 or 5 days, and the first egg is laid within a day or two of the completion of the nest. Incubation is performed by a single bird, also presumably the female, while the mate occasionally feeds her on the nest. During the first few days only one of the pair does all the brooding and most of the feeding, again most probably the female, but soon both parents are kept busy feeding the growing brood. The nestling period lasts from about 14–18 days, and for a period after the young depart from the nest the adults continue to feed them at a rate even greater than when they were in the nest. After about 4 days of this the female may stop feeding the brood and begin her second nest. As the young birds grow they become more independent and gradually drift away from the original territory.

Suggested Reading: Davis, Fisher, and Davis 1963; Bent 1942.

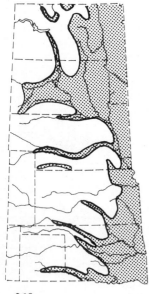

Eastern Wood Pewee
Contopus virens

Breeding Status: Breeds from the eastern half of North Dakota and western Minnesota southward through eastern South Dakota (west to Meade County), eastern Nebraska (west locally to Dawes, Dundy, and Deuel counties), possibly extreme eastern Colorado (no breeding records), eastern Kansas (west to the Cimarron River, where it may hybridize with *C. sordidulus*), Oklahoma except for the panhandle, and extending locally into the Texas panhandle, where it may be an occasional breeder.

Breeding Habitat: The species is generally associated with deciduous forest, including floodplain and river-bluff forests at the western edge of its range. It is also found in woodlots, orchards, and suburban areas planted to trees.

Nest Location: Nests are on horizontal tree limbs, often dead, usually well out from the trunk, and 15–65 feet above the ground. They are placed on the tree bark, often but not necessarily in a crotch, and are well camouflaged by spiderwebs and lichens, so that it may be easily overlooked. The nest is a surprisingly small, shallow cuplike structure lined with fine grass or hairs.

Clutch Size and Incubation Period: From 2 to 4 eggs, usually 3. Eggs are white, with brownish spots around the larger end. The incubation period is 12–13 days. Single-brooded.

Time of Nesting: In North Dakota the probable breeding season is from late May to mid-September, with singing males seen from May 27 to September 13. Kansas egg dates are from June 1 to July 20, with most clutches laid in mid-June. Oklahoma breeding records are from May 30 (nestlings) to August 5 (dependent young).

Breeding Biology: A rather late arrival among the flycatchers, pewees usually reach the northern states in late May, near the end of the spring migration period. In Minnesota the species favors oak woodlands for nesting and continues to sing its distinctive three-noted song through nearly the entire summer. As is typical of the Tyrannidae, only the female constructs the nest, which is usually on the same branch year after year. In one case a fork of an elm tree was used as a nest location by this species every year for 35 years. Incubation is done by the female, but the male occasionally feeds her and remains near the nest to help feed the young when they hatch. Like the nest, the juveniles closely resemble the surrounding bark and lichens, and by the 15th to 18th day after hatching they are ready to leave the nest. They are probably dependent on the parents for food for some time after fledging, until they have become skilled in catching flies.

Suggested Reading: Bent 1942; Craig 1943.

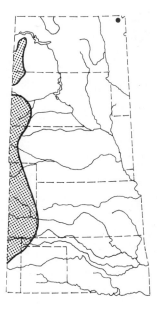

Western Wood Pewee
Contopus sordidulus

Breeding Status: Breeds from southwestern North Dakota (Little Missouri Valley), southward through western South Dakota (Black Hills), western Nebraska (Sioux, Dawes, and Scottsbluff counties, possibly hybridizing with *C. virens* in the Niobrara Valley west of Valentine), eastern Colorado (east to Logan County), Kansas (Cimarron Valley), the Oklahoma panhandle (Cimarron County), and adjacent New Mexico (Quay County). There are no breeding records for the Texas panhandle, but one extralimital record exists for Minnesota (*Loon* 49:169).

Breeding Habitat: The species uses diverse western habitats, including pine-oak woodlands, floodplain forests, and wooded canyons. Open, mature pine forests are used in the Black Hills, but not dense spruce woods. The birds are generally adapted to drier environments than the eastern wood pewee and use areas dominated by conifers.

Nest Location: Horizontal branches of trees, especially dead branches, are used for nesting; as with the eastern species they may be placed in a fork or on the top of the branch. Nests are somewhat larger and more deeply hollowed than those of the eastern wood pewee, and usually the lichen covering is either lacking or replaced by other materials. Spider webs are always present, and sometimes the nest is lined with feathers rather than hair.

Clutch Size and Incubation Period: From 2 to 4 eggs, usually 3, identical to those of the eastern species. The incubation period is probably 12 days, and single brooding is likely.

Time of Breeding: South Dakota egg or nest dates are from June 1 to June 21, and dependent young have been seen as late as July 26. Colorado egg dates are from June 25 to July 9, although nests have been found as early as late May. There is an Oklahoma egg date for June 3.

Breeding Biology: There are few significant biological differences between the eastern and western wood pewees, and where their ranges come into contact they are suspected of hybridizing. One such area where this might occur is in the western Niobrara Valley of northern Nebraska (*Nebraska Bird Review* 29:15) and another is the Cimarron Valley of southwestern Kansas (Rising 1974). In such areas of contact the western species can usually be identified by its two-syllable *pee-a* or *pee-we* song, as opposed to the *pee-a-wee* song of the eastern species. However, some birds in the area of overlap have been known to sing the eastern song type and have plumage characteristics of the western species, so song identification may not be definitive proof of species where hybridization is possible.

Suggested Reading: Barlow and Rising 1965; Eckhardt 1976.

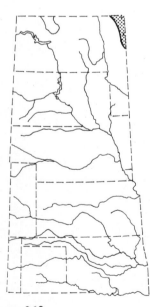

Olive-sided Flycatcher
Nuttallornis borealis

Breeding Status: Limited as a breeder to north-central and northwestern Minnesota (Kittson to Becker counties), although specific nesting records for the area seem to be lacking. It has been seen in summer but has not yet been proved to breed in the Black Hills.

Breeding Habitat: Coniferous forests, mixed forests, boggy areas, and burned-over forest areas provide nesting habitat in Minnesota.

Nest Location: Nests are usually well hidden in a cluster of needles and twigs on the horizontal branch of a conifer, well away from the trunk. They are usually between 15 and 50 feet above the ground. The nest is a loosely constructed cup of twigs, lichens, mosses, and needles, about 5–6 inches in diameter.

Clutch Size and Incubation Period: Usually 3 eggs, rarely 4. The eggs are white to creamy with a wreath of brownish spots and blotches. Incubation lasts 16–17 days. Single-brooded.

Time of Nesting: In Minnesota, nests or eggs have been found in mid-June, and fledglings a few days out of the nest have been reported on July 20.

Breeding Biology: Males of this species establish territories along the edges of tall coniferous forests, where they can sally out to obtain insect prey and can also sit on some high, exposed perch such as a dead tree or branch, uttering their loud and distinctive three-syllable song, *whip-whee'-peooo,* variously interpreted as "look, three deer," or "quick, three bears." Courtship and pair-formation have not been extensively studied in this species but probably closely resemble those of the other tyrant flycatchers. The nestling period is usually 15–19 days, but in one instance the young remained attached to the nest for 23 days, flying to and from it with their parents during the last few days.

Suggested Reading: Bent 1942; Roberts 1932.

Vermilion Flycatcher
Pyrocephalus rubinus

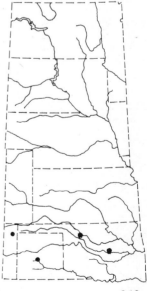

Breeding Status: Very rare or accidental breeder in Oklahoma (nesting records for Major and Lincoln counties). The species is scarce and local even in the Texas panhandle, but there is a breeding record for Amarillo (*Audubon Field Notes* 13:440), and one for Clayton, New Mexico (Hubbard, 1978).

Breeding Habitat: In Texas, this species is found in largest numbers in widely spaced junipers and oaks, but it also occurs near cottonwoods, willows, oaks, mesquites, and sycamore-lined water areas. It is usually found rather near water, but it does not extend into canyons.

Nest Location: Nests are in bushes or trees, typically on a small horizontal, forked branch 8–20 feet above the ground. The nest is usually sunk down on the fork so that it is inconspicuous and scarcely projects above the branch level. As with pewee nests, spider webs are incorporated, and the nest is decorated with lichens. It is also lined with soft materials, such as plant down, fur, or small feathers.

Clutch Size and Incubation Period: From 2 to 4 eggs, usually 3. The eggs are white to creamy, with heavy spotting on the larger end. The incubation period is 14–15 days. Frequently double-brooded.

Time of Breeding: In Texas, egg records are from March 25 to June 23. The Oklahoma nesting records are for May and June.

Breeding Biology: The males of this brilliantly colored species are highly territorial and not only chase other flycatchers but have been seen expelling swallows, warblers, and finches from their territories. One advertisement display is a song flight, during which the bird repeatedly utters a *pur-reet* note. When a pair bond has been formed, the male performs a nest-showing display, consisting of crouching down in a potential nest site, performing slight nest-building movements, and fluttering his wings while calling. The female may join the mate and perform much the same display but nevertheless may begin to build her nest at some other location in the same tree. Only the female builds the nest, which requires at least 4 days. Likewise, only the female incubates, while the male provides food for her. Both parents care for the nestlings, although only the female broods the young birds. The fledging period has not yet been reported, but in one case the pair began constructing a second nest only 4 days after their first day-old brood was found dead. New nests are constructed for second nestings, but materials from the first nest may be used in their construction.

Suggested Reading: Taylor and Hanson 1970; Smith 1970.

FAMILY ALAUDIDAE
(LARKS)

Horned Lark

Horned Lark
Eremophila alpestris

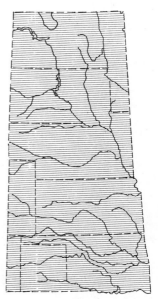

Breeding Status: Pandemic, breeding in suitable habitats throughout the region.

Breeding Habitat: In the Great Plains the horned lark is associated with natural or planted low-stature grasslands and cultivated fields; it is also found in deserts and alpine regions.

Nest Location: Nests are typically placed in a depression so that the upper edge of the nest is level with the ground surface. There is often little or no vegetational cover around the nest, or it may be next to and partly hidden by a clump of grass. The nest is a cup of coarse grass stems, lined with finer materials.

Clutch Size and Incubation Period: From 3 to 6 eggs (12 North Dakota nests averaged 4.0, and 16 Kansas nests averaged 3.6). The eggs are grayish white peppered with brown spots. The incubation period is 11 days. Frequently double-brooded.

Time of Breeding: North Dakota egg dates are from April 7 to July 10. Kansas egg dates are from March 11 to June 10, and those from Oklahoma are from March 16 (female with egg in oviduct) to August 4.

Breeding Biology: In the midwestern states, horned larks begin to establish and defend territories in January and February, while pairs are being formed. Territories are large (averaging about 4 acres in two studies) and are defended by males, but only against other males. Two advertisement songs are uttered, either on the ground or in the air. These songs seem related to courtship rather than to territorial defense and are most common after losing a mate or fledging a brood. Courtship feeding and other displays are also performed at this time. The female selects a nest site almost anywhere in the territory and constructs the nest alone. She digs a cavity with her bill and feet, often "paving" it on one side with various objects, for still uncertain reasons. The paving may cover and hide the fresh dirt that has been dug out or may help keep the nest lining from blowing away during early stages. The female may begin incubation slightly before the clutch is completed but more often begins when the last egg is laid. There is a relatively short nestling period of 9–12 days, averaging about 10, and the young birds leave the nest when their flight feathers are only about one-third to one-half grown. Until they are some 15 days old the young are able to fly only a few yards. The young birds begin to flock soon after leaving the nest, and in many areas the female shortly begins a second nesting.

Suggested Reading: Beason and Franks 1974; Verbeek 1967.

FAMILY HIRUNDINIDAE (SWALLOWS)

Tree Swallow

Violet-green Swallow
Tachycineta thalassina

Breeding Status: A common breeder in the Black Hills of South Dakota and also common in the Pine Ridge, Wildcat Hills, and Scottsbluff areas of northwestern Nebraska.

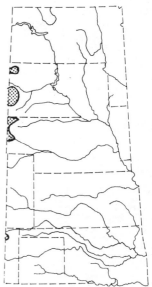

Breeding Habitat: Widespread in the western states, the species is generally associated with open forests such as ponderosa pines and extends into urban areas as well.

Nest Location: Nests are built in old woodpecker holes, in natural tree or cliff cavities, on building ledges, or sometimes in birdhouses. When nesting in birdhouses, the birds favor an entrance hole about 1¼ inch in diameter. The nest is a collection of weed stems and grasses with a feather lining.

Clutch Size and Incubation Period: From 4 to 6 white eggs. The incubation period is 14–15 days. Usually single-brooded.

Time of Breeding: In the Black Hills, nesting occurs during June and July. Nest construction has been seen as early as June 11, and nests with unfledged young have been seen as late as July 24.

Breeding Biology: The violet-green swallow is an unusually early migrant, usually arriving before other swallow species. It thus may not begin nesting for nearly a month after arrival. It spends some time seeking out a suitable cavity, and apparently the female makes the choice, with the male playing a minor role. But once the site is chosen both sexes begin to bring in nesting materials, the female doing most of the carrying. About 6 days are spent in building the nest, and the female roosts on it at night. Eggs are laid at daily intervals, and the female may begin incubating before the clutch is completed, though normally this does not happen. Thus the period of hatching is sometimes rather staggered and has been noted to require as long as 5 days. The female does most of the feeding and also broods during the first 10 days or so after hatching. The fledging period is somewhat variable but averages about 23–25 days. After leaving the nest, neither the adults nor the young return to it.

Suggested Reading: Edson 1943; Combellack 1954.

Tree Swallow
Iridoprocne bicolor

Breeding Status: Breeds in North Dakota from the Souris and James Rivers eastward through western Minnesota, south

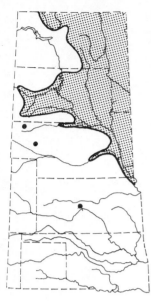

through eastern South Dakota, and in Nebraska west possibly to Cherry and Hall counties. It is uncommon in extreme north-western Missouri (Holt County) and a probable local breeder in extreme northeastern Kansas (Doniphan County). It has bred once in Barton County, Kansas (*American Birds* 31:1156).

Breeding Habitat: Open woodland, usually close to water, is the preferred habitat. The species is often found in the vicinity of dead trees, especially aspen and willows, which are favorite nesting trees.

Nest Location: Nesting is usually in old woodpecker holes in dead trees, dead limbs of live trees, old fence posts, or birdhouses. The nest entrance is normally between 3 and 15 feet above the substrate, often over water. The nest is a cup built of grass and straw and typically is lined with feathers, especially white ones.

Clutch Size and Incubation Period: From 4 to 7 white eggs (14 North Dakota nests averaged 5.5, and 6 Kansas nests averaged 4.8 eggs). The incubation period is from 13–16 days. Single-brooded.

Time of Breeding: North Dakota egg dates are from June 3 to June 20, with nestlings seen as late as July 23. Kansas egg records are from May 21 to June 20, with a peak of egg-laying in late May.

Breeding Biology: One of the earliest spring migrants of the eastern species of swallows, these birds reach Minnesota and the Dakotas in late April, at least a month before nesting gets under way. Much of the courtship apparently occurs in the air, and it includes synchronized flying by a pair. In one reported case a male grasped the female's breast in midair and the two birds tumbled downward until they almost reached the ground. The female then flew to the vicinity of the nest and perched, where-upon the male glided above her and landed on her back. The female constructs the nest with little or no assistance from the male, at times carrying in more than 100 feathers for nest lining. Evidently the male brings food to his incubating mate only rarely; instead, she leaves the nest a few times during daylight to forage for herself. The males often spend the evening perched near the nest, leaving it for their own roosting sites only after dark. The nestling period varies considerably, depending on brood size and thus on rate of feeding, so that the young may spend as few as 16 days or as many as 24 days in the nesting cavity. Once the young leave the nest, however, they rarely return to it.

Suggested Reading: Chapman 1955; Stocek 1970.

Bank Swallow
Riparia riparia

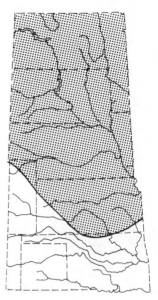

Breeding Status: Breeds in suitable habitats over the northern half of the region, being rather common southward through Nebraska. It is only locally common in eastern Colorado and western Kansas but is more common to the east. The southern breeding limits of this species are extremely ill-defined. It is reported to be a common nesting bird at Sequoyah National Wildlife Refuge in eastern Oklahoma and an abundant summer resident at Salt Plains National Wildlife Refuge, although Sutton (1974) stated that there are still few records of successful nesting in the state. In Texas the species is scarce and extremely local, with only one nesting record (Wilbarger County) for the area under consideration.

Breeding Habitat: A variety of open habitats, such as grasslands are used, but the species is usually found near water and is dependent on suitable nesting sites.

Nest Location: Nests are invariably in vertical banks of clay, sand, or gravel; the bird is characterized by colonial rather than solitary nesting. The openings are near the top of the bank and are about 1½ by 2¼ inches. The burrows average 2 feet in length and are turned slightly upward. The nest is a platform of vegetation, usually with feathers for lining.

Clutch Size and Incubation Period: From 4 to 6 white eggs, usually 5 (60 Kansas clutches averaged 4.8; 6 North Dakota clutches averaged 5.3). The incubation period is 15 days. Normally probably single-brooded, but two broods were reported in one study.

Time of Nesting: North Dakota egg dates range from June 5 to July 5, while Kansas records extend from May 11 to June 20, with most of the clutches laid between May 21 and June 10.

Breeding Biology: Shortly after bank swallows arrive in a nesting area, they begin to gather near the breeding site. Unpaired birds (a male in at least one determined instance) apparently select a burrow site, which may be the same burrow they used the previous year. Thereafter they defend the area from intrusion, although potential mates continue to return to a defended spot until one is eventually tolerated and accepted. Sexual chases of the female by the male are a common feature of pair-formation, accompanied by male song. Another vocalization, the mating song, is uttered by both members of a pair as they sit side by side or facing each other in the burrow opening. This behavior may be a preliminary to copulation, which probably occurs in the nest chamber. When a burrow needs to be dug or deepened, both sexes share equally in the task, then gather materials such as feathers and grass for nest lining. Incubation is by the female,

273

and may begin before the clutch is completed. Thus some eggs may hatch as early as 13 days after the clutch has been completed. Both parents alternate at brooding the young, and both feed the young and keep the nest clean. Birds as young as 20 days of age may be able to fly but often do not leave the nest for some time thereafter; or they may return to their burrows after initial flights.

Suggested Reading: Peterson 1955; Bent 1942.

Rough-winged Swallow
Stelgidopteryx ruficollis

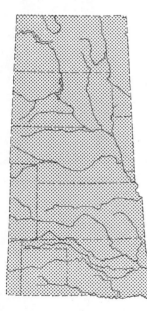

Breeding Status: Breeds virtually throughout the region, but variably abundant. Generally common in the eastern parts of the Dakotas and in western Minnesota and Iowa, but uncommon in southwestern North Dakota and western Nebraska. In eastern Colorado the species is locally common, and it is only moderately common in western Kansas. It is more common in eastern Oklahoma than in the west, it is infrequent in northern Texas (Wilbarger County is the only nesting record), and there are no nesting records for northeastern New Mexico.

Breeding Habitat: The species is associated with open-country habitats, including open woodlands, usually near water. Like the bank swallow, its local distribution is dependent on nesting sites.

Nest Location: Nests are excavated in banks of clay, sand, or gravel. They are much like those of the bank swallow but are solitary rather than in colonies. Natural rock crevices, fissures, and even drainpipes are sometimes also used. Unlike bank swallows, the birds do not use feathers for a nest lining, and their nest is much bulkier.

Clutch Size and Incubation Period: From 4 to 8 white eggs, usually 5–6. The incubation period is between 12 and 16 days, probably closer to the latter. Single-brooded, but renesting efforts are frequent.

Time of Breeding: Dates of active nests in North Dakota range from May 10 to July 15, with eggs seen as late as June 26. Kansas egg records are from May 11 to June 30, with most eggs laid between May 21 and June 10.

Breeding Biology: Almost as soon as they arrive on their nesting grounds, these swallows begin to show interest in suitable nesting sites, and they may seek out old kingfisher or bank swallow excavations that are still usable. Males establish a limited territory around a potential nest site, perching near it and pursuing females from it. Females carrying nesting materials are especially pursued, although this behavior may be associated more with copulation than with courtship. Copulation has not been described and presumably occurs in the nesting cavity. Evidently only the female gathers and carries nest-lining material; apparently neither bird does any excavating. An average of about 6 days is needed to construct the rather bulky nest, but it may take as long as 20 days. The female usually starts incubating with the laying of the next-to-the-last egg, and hatching may extend a few hours or as long as several days. Brooding is done primarily if not exclusively by the female, but both sexes feed the young. The young birds are able to fly some days before they leave the nest, which usually occurs at 18–21 days of age. Young birds rarely return to the nest after they leave it, and there is no evidence on how long the young remain dependent on their parents for food after fledging.

Suggested Reading: Lunk 1962; Bent 1942.

Barn Swallow
Hirundo rustica

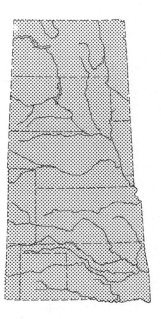

Breeding Status: Pandemic, breeding throughout the region wherever habitats permit.

Breeding Habitat: Favored habitats include open forests, farmlands, suburbs, and rural areas with buildings that provide nest sites.

Nest Location: Originally, nests were on cliffs or in caves or rock crevices, but now most nests are built on the horizontal beams or upright walls or beams of buildings. They are constructed of mud, with a lining of feathers, usually from domestic poultry. The nest is cone-shaped, with a semicircular opening. It may be supported only by the side, which is more common, or from the bottom. Usually colonial.

Clutch Size and Incubation Period: From 3 to 7 eggs (43 Kansas clutches averaged 4.7), white with brown spotting. The incubation period is 15 days. Often double-brooded, at least over most of the breeding range.

Time of Nesting: North Dakota egg dates are from June 13 to July 9, but active nests have been seen as late as September 6. Kansas egg records are from May 1 to August 10, and egg dates from Oklahoma span the period April 18 to August 2.

Breeding Biology: Within about 2 weeks after their arrival in nesting areas, most barn swallows have formed pair bonds. Pair-

275

formation takes place on fences and utility lines near nesting areas, with unpaired birds perching alone and singing, and perching or flying between paired ones. Both sexes gather mud for nests; when available, horsehair is added to the mud cup, and feathers are added later for lining. Many times an old nest is used, with new materials added as necessary. An average of about 6 days is needed to build a nest, and eggs are not laid until the nest is completed. Only the female incubates in most nests, but in some cases males also participate. An average of 21 days, with a range of 18–27 days, has been observed as the nestling period in this species, and courtship behavior soon begins again, such as "song flights" by flocks of swallows flying high and chasing each other. Partners are not changed between broods, and the same nest is usually used again, often with more mud and feathers added to it. There is a gap of about a month between nesting cycles, and only about a third of the swallows in one New York study raised second broods. Second clutches most often have 4 rather than 5 eggs, but egg and nesting mortality rates are similar in the two nesting cycles.

Suggested Reading: Samuel 1971; Bent 1942.

Cliff Swallow
Petrochelidon pyrrhonota

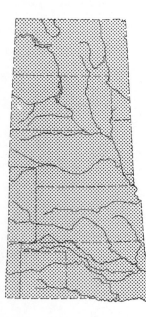

Breeding Status: Pandemic, breeding throughout the region in suitable habitats.

Breeding Habitat: Suitable habitats are widespread, occurring over open areas of farmlands, towns, near cliffs, around bridges, and other areas near mud supplies and potential nest sites.

Nest Location: Nests are gourdlike structures of dried mud, attached to the vertical and overhanging surfaces of cliffs, buildings, bridges, and other structures. The nest has a tubular rounded entrance on the lower side and is lined on the inside with vegetation, but rarely feathers. Nesting is strongly colonial, with up to several hundred nests occurring in favorable locations.

Clutch Size and Incubation Period: From 3 to 6 eggs, usually 4–5. The eggs are white with brown spotting. The incubation period is 15 days. Infrequently double-brooded.

Time of Nesting: North Dakota egg dates are from June 13 to July 9, although active nests have been reported from May 8 to September 6. Kansas egg dates are from May 21 to June 30, with most clutches laid between May 21 and June 10. Active nests or recently fledged birds have been reported in Oklahoma from April 4 to August 6.

Breeding Biology: At least in the northern states, cliff swallows begin to pair immediately upon arrival of their nesting grounds. This activity takes place at or near the nest, and the pair bond apparently consists primarily of mutual tolerance at the nesting site. Male "primary squatters" persistently return to specific perching places, and their singing attracts secondary visitors to that location, some of which are unpaired females. Both sexes defend the nest site, and both bring mud to construct the nest, which requires nearly 2 weeks of effort. When the nest is nearly completed, copulation occurs in the nest cup, and copulatory behavior continues until the middle of the laying period. Many cliff swallows occupy old nests if they are still usable; otherwise they construct entirely new ones. Incubation may begin before the clutch is complete, and males regularly participate. There is a relatively long nestling period in this species, averaging about 24 days, and a relatively low proportion of females (27 percent in one study) attempt a second clutch. In at least some cases females change mates for their second nesting, and a considerable amount of courtship activity is evident between broods.

Suggested Reading: Samuel 1971; Emlen 1954.

Purple Martin
Progne subis

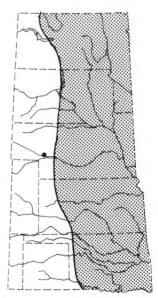

Breeding Status: Breeding occurs generally throughout the eastern half of the region, becoming progressively rarer to the west, and is increasingly limited to a few urban areas. In North Dakota it is rare and local west of the Missouri River, as is also true of South Dakota. In Nebraska it breeds west to Brown and Garden counties, in Kansas it breeds locally in Scott, Ford, and Stevens counties, in Oklahoma it does not currently breed in the panhandle, and in Texas the nearest breeding record is Wilbarger County. There are apparently no breeding records for eastern Colorado or northeastern New Mexico.

Breeding Habitat: The species is widespread in urban, suburban, or rural habitats, usually near water and always where there are suitable nesting cavities.

Nest Location: Nesting is colonial, usually in artificial birdhouses. Ideally these are 15-20 feet above the ground in open settings, preferably near suitable perches such as wires. The birds will use hollowed gourds hung in clusters, as well as old woodpecker holes and crevices in old buildings. The nest cavity is filled with a variety of vegetation and other materials and lined with fresh leaves.

Clutch Size and Incubation Period: From 3 to 8 white eggs, usually 4–5. The incubation period is 15–16 days. Evidently consistently single-brooded, but renesting is frequent and has been confused with double-brooding.

Time of Breeding: North Dakota records of active nests range from April 29 to August 24. Kansas egg dates are from May 11 to June 20, and active nests have been reported in Oklahoma from March 25 to June 25.

Breeding Biology: Purple martins arrive on nesting areas surprisingly far before egg-laying (almost 2 months in Kansas), perhaps because of competition for suitable nesting sites. Some birds arrive already paired—perhaps mates from the previous year—and these are mostly among the early arrivals. Pair-formation is achieved simply and is associated with nesting sites; females choose a male that has occupied a suitable nesting cavity. Copulation is on the house or on the ground and occurs from the time of nest-building to the laying of the first egg. Both sexes gather nesting materials, but only the female incubates, with the male guarding the nesting site in her occasional absences. Incubation lasts 15–16 days, and the young hatch in the same sequence as the eggs were laid. The young birds spend 27–35 days in the nest, averaging about 28 days. Both sexes feed the young, but only the female broods them for extended periods. There is a relatively strong tendency for young birds to return to the area where they hatched when they breed the next year.

Suggested Reading: Allen and Nice 1952; Johnston and Hardy 1962.

FAMILY CORVIDAE (JAYS, MAGPIES, AND CROWS)

Blue Jay

Gray Jay (Canada Jay)
Perisoreus canadensis

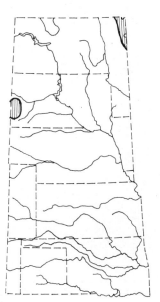

Breeding Status: A common resident of the Black Hills of South Dakota and the coniferous forest areas of north-central Minnesota.

Breeding Habitat: The species frequents coniferous forests, especially dense spruce and pine forests, occasionally extending into mixed woods.

Nest Location: Nests are typically in conifers, either in a crotch or on a horizontal branch near the trunk, often less than 10 feet above the ground. The nest is a fairly bulky accumulation of sticks, twigs, and bark strips, often decorated outside with plant down, cocoons, and spider webs or insect nests. It is usually 7–8 inches in diameter and 3–5 inches high. The lining is of feathers, fur, and plant down, or sometimes of pine needles.

Clutch Size and Incubation Period: From 2 to 5 eggs, usually 3–4. The eggs are grayish to greenish white with small dark spots. The incubation period is 16–18 days. Single-brooded.

Time of Breeding: Eggs have been reported in the Black Hills in early June, and fledged but dependent young have been seen in mid-June. Egg-laying in Minnesota has been reported as early as March 14, with fledging as early as April 18, and family groups are often seen from June to August.

Breeding Biology: Like many corvids, gray jays regularly cache food, but, in contrast to other species, these birds produce a special saliva that helps bind food particles together so that the mass can be firmly held in position among conifer foliage. Breeding in this boreal species begins very early, with nest-building sometimes beginning in February. Both sexes help build the nest, the female doing most of the actual construction in one observed case. As with other corvids, only the female incubates, and she typically sits on the nest from the time the first egg is laid, although incubation does not begin immediately. Both sexes feed the young, but the male brings most of the food during the first few days after hatching. Fledging occurs approximately 15 days after hatching, and it is likely that the young birds remain with their parents through the first winter of life.

Suggested Reading: Dow 1965; Goodwin 1976.

Blue Jay
Cyanocitta cristata

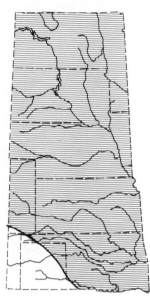

Breeding Status: Breeds nearly throughout the region, but increasingly infrequent to the southwest, with the general breeding limits probably in southeastern Colorado, the Oklahoma panhandle (Cimarron County), and the northeastern panhandle of Texas (Roberts, Hemphill, and Wheeler counties). The only New Mexico breeding record is for Roosevelt County (*Southwestern Naturalist* 17:432).

Breeding Habitat: The blue jay is widely distributed, in deciduous forests, parks, suburbs, cities, and almost anywhere trees are found in grassland areas.

Nest Location: Nests are in trees, 5–70 feet above the ground. They are fairly large (7–8 inches in diameter), constructed of twigs, bark, and leaves and lines with rootlets. They are normally well hidden, in the forks, crotches, or outer branches of trees, especially coniferous species.

Clutch Size and Incubation Period: From 3 to 6 eggs (15 Kansas clutches averaged 4.1). The eggs are buff or olive with small darker spots. The incubation period is 17–18 days. Single-brooded in the north, multiple-brooded in the south.

Time of Breeding: In North Dakota, nest-building and nests with eggs have been reported from May 7 to June 2, with dependent young seen as late as August 14. In Kansas, egg dates are from April 10 to July 10, with a peak in mid-May. Nest-building and nests with eggs have been seen in Oklahoma (where two broods are common) from March 19 to July 9, and dependent young have been seen as late as August 17.

Breeding Biology: Blue jays are generally found in pairs or family groups, with larger flocks sometimes occurring around feeding areas or during migration, when the young of the year typically leave their parents and move varying distances southward. Paired birds, however, usually winter on their breeding territories. At the onset of breeding, both sexes gather materials and begin nest construction. "False nests" may be initiated when the male brings twigs to the female while she crouches in a particular location, but the actual nest is always constructed elsewhere. Although some first-year birds do breed, this is infrequent, and most actual breeders are at least 2 years old. Normally only females incubate, and males bring food to them while they are on or off the nest. The fledging period is 17–21 days, and fledglings may obtain some food from their parents for as long as 2 months after leaving the nest. However, where two or even three broods are regularly raised it is unlikely that parental care extends much beyond the fledging period.

Suggested Reading: Hardy 1961; Goodwin 1976.

Steller Jay
Cyanocitta stelleri

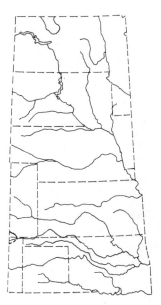

Breeding Status: Hypothetical. Possibly nests in northeastern New Mexico (considered a rare transient at Capulin Mountain National Monument), but there are no specific breeding records for our area. Also reported as a resident of the Black Hills (A.O.U. *Check-list*), but there is only one recent observation of the species there. Frequently reported in western Nebraska in the nonbreeding season, but no breeding-season observations.

Breeding Habitat: Associated with coniferous forests throughout its range, the species especially frequents the ponderosa pine zone but extends in limited numbers downward into the lower piñon zone and upward into the Engelmann spruce and Douglas fir zones.

Nest Location: Nests are 4–40 feet above the ground, usually on a horizontal branch of a young conifer. They are bulky, composed of twigs, weed stems, and sticks, and are lined with rootlets and grasses.

Cluch Size and Incubation Period: From 3 to 7 eggs, light blue with darker spotting that varies in color and intensity. The incubation period has not been specifically determined but presumably is 17–18 days. Single-brooded.

Time of Breeding: Colorado egg dates range from May 3 to May 30, and nestlings have been reported as early as June 1.

Breeding Biology: Steller jays are usually to be found in pairs, but they sometimes form small parties when aggregating at a food source or mobbing a predator. Paired birds tend to remain in or near their breeding territories throughout the year, but immature birds may wander about in the winter. Little has been written on nesting biology, but both sexes help build the nest and probably only the female incubates, although it has been suggested that both sexes incubate in Alaska. The fledging period has not been specifically determined, but the young birds are fed for a month or more after they fledge and presumably remain with their parents for most of their first year.

Suggested Reading: Brown 1964; Goodwin 1976.

Scrub Jay
Aphelocoma coerulescens

Breeding Status: Apparently limited mostly to northeastern New Mexico (Quay and Union counties; no breeding records, but an

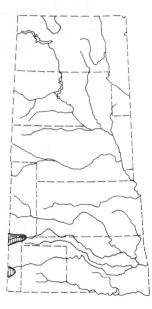

abundant permanent resident at Capulin Mountain National Monument). Also breeds in northwestern Cimarron County, Oklahoma, and in Baca County, Colorado.

Breeding Habitat: The species is associated with scrub oak, piñon-juniper, and less frequently with mixed oak and ponderosa pine habitats. Borders of brushy ravines and wooded creek bottoms are highly preferred.

Nest Location: Nests are usually in piñons, oaks, or tall shrubs, less than 10 feet above the ground, and are built of interlaced twigs forming a platform 6–9 inches in diameter. The cup is lined with rootlets or horsehair.

Clutch Size and Incubation Period: From 2 to 7 eggs, usually 4–6. The eggs vary in color from pale green to reddish or bluish, with darker spotting. The incubation period is about 16 days. Probably single-brooded.

Time of Breeding: In Oklahoma, nests with eggs as well as young about to fledge have been reported in late May. In southern Colorado the average date for fresh eggs is about the first week of May, but nestlings have been seen in Baca County as early as May 7.

Breeding Biology: Like most members of this family, scrub jays tend to remain rather permanently paired and live in such pairs or at most in family groups after the breeding season. They are highly territorial. When one member of a pair is on the ground or in low cover, the other member often stations itself as a "sentry" at some convenient vantage point. This often occurs during nest-building, when both pair members gather materials. At least in the Florida race, pairs often allow one or more immature birds (typically their own offspring) to share their territory. These birds not only help defend the territory but also may help rear the young. The female does all the incubation and brooding, but she is fed by the male. As many as three nonbreeding "helpers" have been observed at a single nest, but only older offspring of the pair have been observed actually feeding young. The young birds remain in the nest for about 18 days, but after becoming independent of their parents for food they may live with them for a year or more. By the time they are 2 years old they usually leave their parents' territory permanently and attempt to breed.

Suggested Reading: Woolfenden 1975; Hardy 1961.

Black-billed Magpie
Pica pica

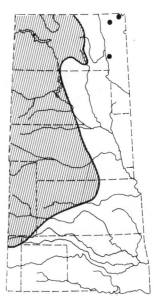

Breeding Status: Breeds generally over the western parts of the region, regularly east to the James River in North Dakota (rarely to Clay, Marshall, and Roseau counties in Minnesota), the Missouri Valley of South Dakota, east-central Nebraska (to Greeley, Howard, and Clay counties), north-central Kansas (east to Clay County), the Oklahoma panhandle (Cimarron County), and northeastern New Mexico (Union County).

Breeding Habitat: The species generally frequents river-bottom forests and forest edges but ranges out into more arid environments wherever there are thickets of shrubs or small trees.

Nest Location: Nests are in dense bushes or small trees, especially thorny ones. Nests are masses of sticks of varying sizes and may be up to several feet across. They have a lateral entrance leading to an egg chamber consisting of a mud cup lined with rootlets, grass, and hair.

Clutch Size and Incubation Period: From 4 to 7 eggs (7 North Dakota nests averaged 5.4). The eggs are greenish gray with heavy brown blotching. The incubation period is 16–20 days, averaging about 18. Single-brooded.

Time of Breeding: In North Dakota the breeding season is from late April to early August, with egg dates extending from April 29 to June 16. Kansas egg dates are from April 11 to June 20, with a peak in mid-May, and Oklahoma egg dates are from April 17 to May 10.

Breeding Biology: At least in some areas, pairs often remain in the general vicinity of their breeding areas through the winter period, and many use old nests for nighttime brooding during cold weather. Rarely, however, are old nests used again for nesting; new ones are typically built each year and for each breeding attempt. In eastern Wyoming, birds may begin carrying mud to anchor nest bases in late February, but intensive nest-building does not occur until mid-March or April. Both sexes gather materials, the male bringing more sticks than the female, and rarely each partner will begin a nest at a different location. A surprisingly long average period of 43 days is required to complete a nest, and during the latter part of this time intensive displaying also occurs, especially courtship feeding of the female. The female does all the incubating, but her mate feeds her throughout the incubation period. Both sexes feed the nestlings equally, and they remain in the nest for an average of nearly 4 weeks. After the young are able to fly well, the family gradually wanders out of its nesting area. Although it is known that birds

285

sometimes acquire territories and breed when a year old, it is likely that most initial breeding occurs in the second year.

Suggested Reading: Erpino 1968; Goodwin 1976.

Common Raven
Corvus corax

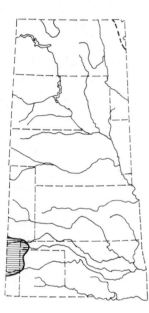

Breeding Status: Resident in coniferous forests of north-central Minnesota. Once bred in North Dakota but was extirpated; now also apparently absent from southwestern South Dakota, western Nebraska, and western Kansas. Ravens still breed in northeastern New Mexico (locally at Conchas Reservoir and Capulin Mountain National Monument, casually to Clayton, Union County), locally in Cimarron County, Oklahoma, and rarely in the western panhandle of Texas (Deaf Smith County).

Breeding Habitat: Common ravens are generally associated with wilderness areas of mountains and forests; in our area they are mostly limited to a few bluffs or cliffs providing inacessible nest sites, or to dense coniferous forests.

Nest Location: Nests are usually in tall coniferous trees or on cliff ledges. Large branches and sticks provide the bulk of the nest, which is deeply cupped and lined with bark shreds, grasses, and hair.

Clutch Size and Incubation Period: From 3 to 6 eggs, usually 4–5. The eggs are greenish with darker brown to olive markings. The incubation period is 18–20 days. Single-brooded.

Time of Breeding: In New Mexico, nests with eggs have been seen in early April, and nestlings have been noted from late May to early June. Nestlings in Colorado have been seen as early as mid-April.

Breeding Biology: The largest of all passerine birds, ravens also have the broadest worldwide distribution of any of the Corvidae. But apart from their size the birds are typical crows. Pairs form rather large territories and largely remain within them, whereas immature birds and adults lacking territories tend to roam about in flocks. In agonistic and sexual situations ravens perform a "self-assertive" display in which they raise the feathers above the eyes, and later all the head and throat feathers, producing a very shaggy-headed appearance, often followed by bowing and crowing while spreading the tail. The nest is built in an inaccessible location by both members of the pair, and the same site is sometimes used in successive years. From 1 to 4 weeks are spent in nest-building, and an additional day is required for each egg that is laid. Although the female spends most of her time on the nest as soon as the first egg is laid, incubation does not begin until the clutch is complete or nearly complete. Only the female incubates, but she is fed by the male, and both sexes feed the young. The female broods them for about 18 days, but they do not fledge until they are about 6 weeks old. Thereafter they remain in the care of their parents for nearly half a year.

Suggested Reading: Bent 1946; Goodwin 1976.

White-necked Raven
Corvus cryptoleucus

Breeding Status: Has bred rarely in south-central Nebraska (Adams and Kearney counties, one record for Dundy County); the northern edge of the normal breeding range is northwestern Kansas (Rawlins County). There are a few breeding records for eastern Colorado (Cheyenne, Kiowa, Kit Carson, and Lincoln counties), and nesting also occurs in the Oklahoma panhandle southward to the Red River, possibly as far east as Jefferson County. Breeding also occurs locally in the Texas panhandle and northeastern New Mexico (Union, Quay, and Roosevelt counties).

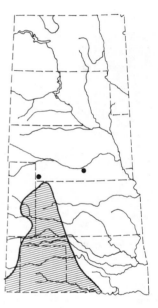

Breeding Habitat: Open and arid grassland habitats, with scattered trees, cactus, or yucca are favored; the species is generally not associated with river valleys or forested areas.

Nest Location: Nests are placed on windmill towers, isolated trees, and telephone poles. Typically the nests are made of sticks, but in some areas wire scraps are commonly incorporated. The nests are only slightly larger than those of crows (about 20 inches in diameter) and are usually lined with grass, bark, and hair or fur. They are often used year after year and gradually increase in size. Nests are 5–50 feet high, averaging about 20 feet.

Clutch Size and Incubation Period: From 3 to 8 eggs (averaging 4.7 in a New Mexico study). The eggs are pale green to grayish green with highly variable darker spotting or streaking. The incubation period is 20–22 days. Single-brooded.

Time of Breeding: In Kansas, the eggs are laid from late March to early May. In Texas, egg records extend from March 11 to July 3, but the latest date for nestlings in Oklahoma is June 16.

Breeding Biology: Outside of the breeding season these ravens often form rather large flocks and roost communally in canyons or gulches. Almost nothing is known of their social behavior, but they perform a self-assertive display in which they expose the hidden white bases of the neck feathers in hostile situations. Nesting sometimes is semicolonial, but this probably reflects a shortage of suitable nest sites rather than true colonial tendencies. Nesting occurs relatively late and is spread over a rather long period, perhaps because of the timing of late spring rains rather than temperature limitations. Other aspects of breeding biology are probably much like those of typical ravens.

Suggested Reading: Davis and Griffing 1972; Goodwin 1976.

American Crow (Common Crow)
Corvus brachyrhynchos

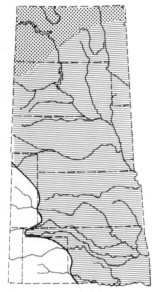

Breeding Status: Nearly pandemic, breeding in suitable habitats throughout the entire region except the nearly treeless areas of the Texas panhandle, adjacent New Mexico, and southeastern Colorado.

Breeding Habitat: Forests, wooded river bottoms, groves, orchards, suburban areas, parks, and woodlots are favored habitats.

Nest Location: Nests are in deciduous or coniferous trees, with conifers and oaks seemingly preferred. They are usually 20–60 feet above the ground, on horizontal branches near the trunk, and are about 2 feet in diameter. Where trees are lacking, as on prairies, nesting on the ground, on shrubs, or on telephone pole crossbars is fairly common. The large platform of sticks, twigs, bark, and similar materials is well cupped and lined with various softer materials such as rootlets, hair, and feathers.

Clutch Size and Incubation Period: From 3 to 8 eggs (10 North Dakota nests averaged 4.9, and 19 Kansas clutches averaged 4.2). The eggs are bluish to grayish green with darker spotting of gray or brown. The incubation period is 18 days. Probably single-brooded in our area.

Time of Breeding: North Dakota egg dates are from April 11 to June 10, and those from Kansas range from March 10 to May 31, with most eggs laid between March 21 and April 10. Oklahoma breeding dates are from March 5 (eggs) to May 31 (recently fledged young).

Breeding Biology: In our area crows begin to flock after the breeding season, and at least in northern areas they tend to migrate some distance southward, where they use massive roosting areas. It is likely, however, that pairs are maintained within these large flocks, and shortly after returning to the breeding areas the birds typically become well spaced and territorial. Crows utter a surprisingly broad range of notes, including more

288

than a dozen distinct calls, and in addition they commonly mimic other species. Both sexes help build the nest, and although it has been reported that both sexes incubate this seems unlikely in light of what is known of related species. The young birds fledge in about 36 days but remain with their parents for a protracted period.

Suggested Reading: Chamberlain and Cornwell 1971; Goodwin 1976.

Fish Crow
Corvus ossifragus

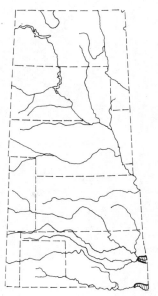

Breeding Status: Limited to eastern Oklahoma, where it is resident along the Arkansas and Red rivers, west to Muskogee and Idabel respectively.

Breeding Habitat: Fish crows are associated with forest-lined rivers and lakes, coastal marshes and beaches, and brackish bays. They are often found near heronries, from which they steal eggs.

Nest Location: Nests are solitary or in loose colonies near the tops of deciduous or coniferous trees, usually 10–90 feet from the ground, or, rarely, in tall shrubs. The nest is usually near water and is placed in a large fork or on a horizontal limb close to the trunk. In size and appearance it closely resembles that of the common crow.

Clutch Size and Incubation Period: From 4 to 5 eggs, rarely more. The eggs are slightly smaller than those of the common crow but otherwise identical. The incubation period is 17–18 days. Single-brooded.

Time of Breeding: One nest with eggs was found in Oklahoma in late April; the fish crow probably nests about the same time as the common crow in that region, during March and April.

Breeding Biology: Although rather large groups of these birds may be seen during the winter, the pair is the nuclear social unit, as in other crows. Territoriality is not highly developed, however, and colonial nesting is frequent, with nests typically situated in neighboring trees. Fish crows eat the eggs and young of other birds, fish found dead or stranded, crabs, shrimp, and other aquatic animals, and they often hover above water when seeking food. Little has been written on their breeding biology, but it presumably differs little from that of the common crow. Apart from this crow's rather different ecology and its hoarser and more nasal calls, the two species are very similar.

Suggested Reading: Bent 1946; Goodwin 1976.

Pinyon Jay
Gymnorhinus cyanocephalus

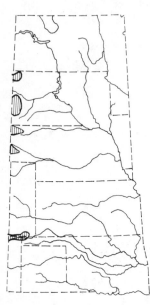

Breeding Status: A common permanent resident of the lower elevations of the Black Hills of South Dakota and also a resident of the Cimarron Valley of northeastern New Mexico, adjacent Colorado, and extreme northwestern Oklahoma.

Breeding Habitat: In the Black Hills, pinyon jays are found in pine forests where the soil is dry and the trees are scattered and small. They are associated with the piñon-juniper zone of southern Colorado and northeastern New Mexico.

Nest Location: Nests are 6–20 feet above the ground, usually in small piñon pines, and normally are rather exposed, at some distance from the center of the tree, often on the lowest horizontal limb. Nests are constructed of twigs with a lining of yucca, sagebrush, rootlets, hair, and sometimes feathers.

Clutch Size and Incubation Period: From 3 to 6 eggs, usually 4–5. The eggs are bluish white to grayish white with brown dots and spots. The incubation period is 18 days. Single-brooded, but known to renest.

Time of Breeding: Nesting in the Black Hills occurs in April and May, with dependent young seen as early as mid-April and as late as June 19. Colorado egg dates range from April 10 to May 5.

Breeding Biology: Pinyon jays are highly gregarious, usually gathering in flocks of up to 50 birds for much of the year. Pair bonds are probably rather permanent in these flocks, and as early as mid-November males begin feeding their mates by transferring pine seeds or other morsels. This behavior is also performed by first-year birds, although initial breeding may not occur until they are at least another year older. Later, females actively solicit feeding by courtship begging, a display that continues through nesting and stimulates the male to feed his incubating mate. In late stages of courtship, a male may pick up a bit of vegetation, present it to his mate, then fly up into a nearby tree, as if to lure her away from the flock. In this way, specific courtship crotches or branches are established, although the actual nest is often built in another location. Nests are usually placed on the south side of trees, probably for warmth. They are built by both members of a pair, usually over about a week, and the first egg is laid about 2 days later. Most birds in a colony begin and complete their nests at nearly the same time; the colony's location is dependent on the caches of pine seeds from the previous fall. Fledging occurs about 3 weeks after hatching, and the parents remain with their young for a prolonged period, continuing to feed them well after they are fledged.

Suggested Reading: Bent 1946; Balda and Bateman 1973.

FAMILY PARIDAE
(TITMICE, VERDINS, AND
BUSHTITS)

Black-capped Chickadee

Black-capped Chickadee
Parus atricapillus

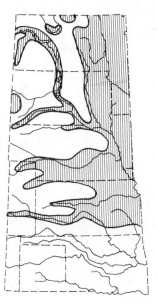

Breeding Status: Breeds in suitable habitats nearly throughout the region, becoming less common southward and reaching its southern limits in the southernmost tier of counties in Kansas. It is not known to breed in Oklahoma, northeastern New Mexico, or the Texas panhandle.

Breeding Habitat: This chickadee breeds both in deciduous and coniferous forests, as well as in orchards and woodlots, wherever suitable nesting cavities exist.

Nest Location: Nests are usually in edge situations or open areas of forest. Old woodpecker holes are most often used, but the birds also excavate cavities in rotted wood of dead stubs. Excavated holes are usually within 10 inches of the stub tip. Birdhouses are sometimes also used. Entrances of excavations are about 1 3/8 inches in diameter and 4-20 feet above ground. The cavity is usually lined with hair, feathers, or cottony materials.

Clutch Size and Incubation Period: From 4 to 7 eggs (10 Kansas clutches averaged 5.4). The eggs are white, rather evenly spotted with reddish brown. The incubation period is 12-13 days. Occasionally double-brooded.

Time of Breeding: Breeding activity in North Dakota has been reported from late April (nest excavation) to late July (flying dependent young). Kansas egg records are from March 21 to June 10, with a peak in mid-April and most eggs laid between April 11 and 30.

Breeding Biology: Chickadees are largely nonmigratory, but winter flocking does occur. Pair bonds are weak or absent during this time, although there is enough contact to allow frequent re-pairing with past mates. Courtship is apparently simple, consisting mainly of the loud *phoebe* song by males. Territories are not established until later, when the pair begins to excavate a nest site. Both sexes excavate, the female taking the lead, and both birds work intermittently during daylight. Eggs are laid daily and are covered with nesting material by the female, who also sleeps in the nest but does not begin incubating until the clutch is complete. Only the female incubates, but the male feeds her at intervals. During the first week after hatching the behavior of adults is similar to their behavior during incubation, but the male stops feeding the female and both parents feed the young. After brooding is terminated both sexes feed the young at about an equal rate, and the young birds leave the nest at 16-17 days of age. Fledglings are able to forage for themselves about 10 days after leaving the nest, but they remain with their parents for 3-4 weeks. A small proportion of adults attempt a second brood.

Suggested Reading: Odum 1941-42; Brewer 1963.

Carolina Chickadee
Parus carolinensis

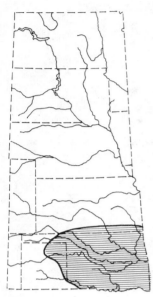

Breeding Status: Breeds from the southernmost tier of counties in Kansas southward through Oklahoma at least as far west as Harper County. Also breeds in the northeastern panhandle of Texas.

Breeding Habitat: Associated with deciduous and coniferous woodlands, the Carolina chickadee's habitat (forest and forest edge) is identical to that of the black-capped chickadee, though the two species normally do not overlap. Hybridization may occur in areas of contact between these species.

Nest Location: Generally the same as that of the black-capped chickadee; the nests and eggs of these species cannot be distinguished.

Clutch Size and Incubation Period: From 4 to 9 eggs, averaging about 7, significantly more than the black-capped chickadee. Like those of that species, the eggs are white with reddish brown spots. The incubation period is 13 days. Single-brooded.

Time of Breeding: Oklahoma egg dates are from April 3 to April 29, but nest-building has been observed as early as February 21 and recently fledged broods as late as May 26.

Breeding Biology: In nearly all major aspects of their biology, the Carolina and black-capped chickadees appear to be nearly identical. The black-capped chickadee is clearly adapted to breeding in cooler and perhaps drier habitats, whereas the Carolina chickadee can attain higher densities in more southerly and perhaps moister forests. Hybrids do occur where the species meet, but apparently they are less well adapted than either of the parental types, thus maintaining a genetic barrier between the species. Wintering birds form small flocks, consisting of at least partially paired birds and organized in a linear social hierarchy with paired birds at the top. Pairs are apparently formed when a female persistently associates with a male and endures his attacks on her until he accepts her presence. This can occur at any time during the year. Territories are established by singing, territorial skirmishes, and patrolling, and by restriction of a pair's activities to a particular area. About a month is spent in looking for suitable nest sites. Both sexes participate but the female evidently makes the final choice. Males often assist with excavation, but the female gathers nest-lining materials and does all the incubating. Males often feed females during the excavation period and continue to feed them while they are incubating. Fledging takes approximately 17 days, and for the first few days afterward the family remains in dense vegetation and the young birds beg vigorously. They continue to beg for at least 5 weeks after fledging but are fed only rarely during that time, and the parents may eventually make abortive attacks on them, causing dispersal.

Suggested Reading: Smith 1972; Brewer 1963.

Boreal Chickadee (Brown-capped Chickadee)
Parus hudsonicus

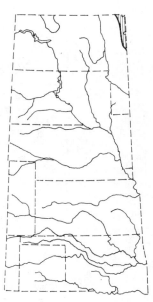

Breeding Status: Limited to northwestern Minnesota (nests in Clearwater County, possibly Roseau and Beltrami counties).

Breeding Habitat: The species is found in coniferous forests, boggy areas, and muskegs.

Nest Location: Nests are in trees or stumps with soft and decayed heartwood but hard outer layers. The entrance cavity frequently faces upward, rather than opening laterally like the nests of other chickadees, and is usually 1-12 feet above the ground. The cavity is lined with fur, cottony plant material, feathers, or similar soft materials.

Clutch Size and Incubation Period: From 4 to 9 eggs, usually 5-7. The eggs are white with small reddish brown spots. The incubation period is usually 15 days but ranges from 11 to 16. Single-brooded.

Time of Breeding: In Minnesota, nest excavation has been seen as early as May 13, and active nests have been seen as late as July 6.

Breeding Biology: The winter is spent in flocks, that tend to break up late in the season as aggressive activity increases. Several kinds of chases and attacks are common, including chases by either sex, short chases of a female by a male, and downward spiraling chases of a male by a female from a treetop, with the male uttering musical calls. Pair-formation may involve some of these chases, but hole-inspection is perhaps also a part of pair-forming behavior. Once pairing occurs it is probably for life. The birds soon establish a territory and the female begins food-begging. Both sexes search for suitable nesting sites, and they usually do not nest in the same cavity in succeeding years. The pair may spend anywhere from a day to at least 10 days excavating, and the nest lining is added by the female alone. Egg-laying begins before the nest is complete, and the male continues to feed his mate during egg-laying and incubation. The usual length of the nestling period is 18 days, and after fledging the family leaves the vicinity of the nest. However, for about 2 weeks the young remain within the nesting territory and are fed with decreasing frequency.

Suggested Reading: McLaren 1975; Bent 1946.

Tufted Titmouse
Parus bicolor

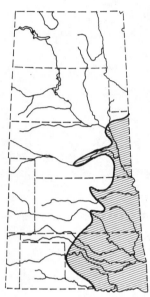

Breeding Status: Breeds in the Missouri Valley of South Dakota (north at least to Walworth County), in the Missouri and Platte valleys of Nebraska (west at least to Buffalo County), southward through the eastern half of Kansas (east of Cloud, Harvey, and Sumner counties) and most of Oklahoma except the panhandle. Occurs locally along the eastern border of the Texas panhandle, but there are no definite breeding records.

Breeding Habitat: The species breeds in coniferous and deciduous forests, orchards, woodlots, and suburban areas. At the western edge of the range it is limited to bottomland forests.

Nest Location: Nests are in natural tree cavities, old woodpecker holes, fenceposts, and sometimes pipes or birdhouses. Openings are usually 10-20 feet above the ground but have been reported from 2 to nearly 90 feet high. The nest is lined with fur, soft vegetation, or other soft materials including snakeskins.

Clutch Size and Incubation Period: From 4 to 8 eggs (6 Kansas clutches averaged 4.5). The eggs are white with small brownish spots. The incubation period is 13-14 days. Normally single-brooded, but a few instances of double-brooding are known.

Time of Breeding: Kansas egg records are from March 21 to June 10, with a peak in late April and most eggs laid between April 11 and 30. Oklahoma breeding dates range from March 18 (nest under construction) to June 10 (nestlings).

Breeding Biology: As in other members of this family, pair bonds seem to be permanent and may be formed as early as the first fall of life or well before territories are established in the spring. Females are subordinate to their mates and to other males, and pair-formation is at least partly related to chases of females by mates or potential mates. After pairs are formed the mates may spend considerable time searching for suitable nest cavities. Unlike chickadees, no excavation of cavities is typical, and both sexes help in locating a suitable site. From nest-site selection through the hatching period the male performs courtship feeding, feeding his mate either in the nesting cavity or away from it during the incubation period. Apparently only the female brings nesting material, and she may continue this through the egg-laying period. The female incubates alone, but when the eggs hatch she attracts the male by calling and wing-quivering, which stimulates him to begin bringing food to the brood. The nestling period is 17-18 days, but the young continue to beg for food from their parents until they are nearly 2 months old. A very few instances of second broods (in Pennsylvania and Tennessee) have been encountered.

Suggested Reading: Brackbill 1970; Offutt 1965.

296

Black-crested Titmouse
*Parus atricristatus**

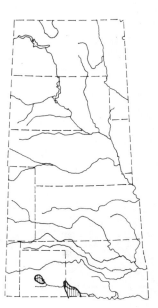

Breeding Status: Breeds in southwestern Oklahoma (Tillman, Jackson, and Harmon counties) and the central part of the Texas panhandle (Randall and Armstrong counties), especially in canyon areas. These two populations are apparently now not in contact.

Breeding Habitat: In Oklahoma this species is limited to bottomland woods along the Red River, while the subspecies breeding in the Texas panhandle is found in cottonwood groves of river canyons. In Texas it generally occurs widely in mesquite, open live oak groves, and oak-juniper woodlands.

Nest Location: Apparently the nesting needs of this species are the same as those of the tufted titmouse, but few nests have been described. Abandoned woodpecker holes are favored sites, and nests are usually in groves of open timber.

Clutch Size and Incubation Period: From 4 to 7 eggs, usually 6, resembling those of the tufted titmouse. The incubation period is also probably the same. Believed to be double-brooded.

Time of Breeding: In Texas, egg records extend from February 24 to June 11.

Breeding Biology: During the winter, these titmice may be found in flocks, pairs, or as single birds, but presumably pair bonds are relatively permanent. Territories are established by pairs, and territorial encounters seem to be settled by bluffing more frequently than by actual fighting. The calls and songs of black-crested and tufted titmice are very similar, and the similarities of their ecology and behavior are emphasized by the hybridization that occurs in areas of contact. Evidently the black-crested titmouse is more tolerant of foraging among open plant cover than is the tufted titmouse, and its calls are more nasal and extended. No obvious display differences have been noted between the two species, although male tufted titmice tend to be more aggressive toward their conspecifics as well as toward other species. Only the female incubates, but the males remain nearby during incubation and undoubtedly feed their mates. Nestling periods have not been established for this species but no doubt are the same as for the tufted titmouse.

Suggested Readings: Dixon 1955; Bent 1946.

*The A.O.U. has recently (*Auk* 93:878) recommended merging this form with *P. bicolor*.

297

Plain Titmouse
Parus inornata

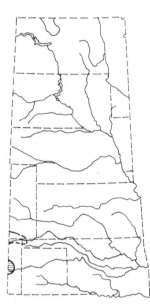

Breeding Status: Breeds in northeastern New Mexico (east to Colfax and probably Quay counties) in southeastern Colorado (Baca County), and in extreme northwestern Oklahoma (Cimarron County). Not reported from the Texas panhandle.

Breeding Habitat: In this region the plain titmouse is limited to upland habitats of piñons, junipers, and scrubby oaks. In other parts of its range it also occurs in river-bottom groves and suburban areas.

Nest Location: Nests are usually in natural cavities or woodpecker holes in trees between 3 and 30 feet above the ground. At least sometimes the birds excavate their nests in rotted heartwood, or even in clay banks, and birdhouses are sometimes also used. The nest is lined with fur, hair, feathers, or other soft materials.

Clutch Size and Incubation Period: From 6 to 8 white eggs, usually 7. The incubation period is 14–16 days, and there is probably only a single brood.

Time of Breeding: Active nests have been reported from Furnish Canyon, Baca County, Colorado, in late April and late May. A nest with young nearly ready to fledge was found in Oklahoma in early June.

Breeding Biology: At least in the interior populations of this species, a certain amount of winter flocking seems typical, although coastal populations remain on territories during this period. Apparently pairs are formed during the flocking period and before territories are established. Pair-formation is marked by singing and males' making "approach threats" toward females and chasing them in a sexual flight that represents attempted copulation. A submissive display by females, involving wing-quivering and a soft call, stimulates feeding by the male and helps to establish a pair bond. Apparently only the female searches for a nest site, and she also is the only one that gathers materials to fill the chosen cavity. But the male continues to feed her during this period and during incubation, and both sexes help feed the young, about equally, at least after the female terminates her brooding. The nestling period is approximately 20 days, and after leaving the nest the young birds continue to forage within a narrow radius of their nest for some time. By the time they are 5 weeks old they are foraging for themselves, and they gradually begin to disperse from their parental territory. In some cases they establish temporary territories of their own while still juveniles, but they may be nearly a year old before they successfully obtain suitable breeding territories.

Suggested Reading: Dixon 1949, 1956.

Verdin
Auriparus flaviceps

Breeding Status: Breeding is apparently restricted to an area of mesquite woodland along Sandy (= Lebos) Creek in Jackson County, Oklahoma (*Bulletin of the Oklahoma Ornithological Society* 5:32). A nest has also been found in Harmon County (G. M. Sutton, pers. comm.). This is some distance to the north of typical verdin habitat in southern New Mexico and the southern Staked Plain of Texas, and the only other possible breeding record for our region is from Wilbarger County, Texas.

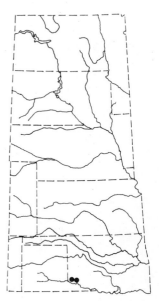

Breeding Habitat: The species breeds in brushy valleys, oak slopes, and other semiarid habitats where there are stiff-twigged and thorny bushes or trees, often far from surface water. Although arid-adapted, it prefers brushy areas over open deserts and also avoids dense timber areas.

Nest Location: Nests are built in a variety of shrubs or trees, including mesquites, hackberries, catclaws, palo verdes, live oaks, and many other thorny trees and shrubs. The nest is very firmly constructed of as many as 2,000 thorny twigs, forming an oval or globular structure up to 8 inches in diameter, with the thorns projecting upward. It is well anchored to a limb, from as low as 2 feet above the ground to nearly 20 feet up. The cavity is lined with leaves, grass fibers, and abundant feathers.

Clutch Size and Incubation Period: From 3 to 6 eggs, usually 4. The eggs are pale greenish to bluish with fine dark spots, either scattered or concentrated. The incubation period is from 14 to 17 days. Double-brooded.

Time of Breeding: Egg records in Texas are from March 25 to September 15, with a clustering between April 18 and May 6.

Breeding Biology: Based on studies in Arizona, verdins appear to occupy rather large home ranges of nearly 25 acres, but they defend only the area near the nest site. However, males do become more aggressive in spring; one researcher reported that unpaired birds construct display nests—incomplete nests that may serve to attract females. Unpaired males frequently utter *tseet* notes that likewise attract females and that later serve as contact notes between paired individuals. According to another observer, nests are not made until after pair bonds are formed, and nest-building apparently reinforces the pair bond. Two types of nests are constructed, breeding nests and roosting nests, and the latter are typically built from late summer through winter. Both sexes help build the breeding nests, and the male does much of the early work. Male roosting nests are usually some distance from the breeding nest, and males apparently never feed incubating females. The male does begin to feed the young about a week after they hatch, and from then until they fledge (about 17–19 days after hatching) he actively feeds them. After fledging, the young are cared for by one or both parents for about 18 more days. Males assume most of the postfledging care and roost with the fledglings while females often begin a second clutch.

Suggested Reading: Taylor 1971; Bent 1946.

Common Bushtit
Psaltriparus minimus

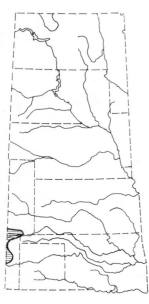

Breeding Status: Breeds locally in southeastern Colorado (to Baca County), extreme northwestern Oklahoma (Cimarron County), and northeastern New Mexico (east at least to Folsom, Union County).

Breeding Habitat: In this region bushtits are essentially confined to piñon-juniper habitats, but they occur elsewhere in tall sagebrush, mountain mahogany, brushy or tree-lined river bottoms, and in hillside aspen groves.

Nest Location: Nests hang on shrubs or trees up to 15 feet above the ground, being woven from mosses, spider webs, lichens, oak leaves, cottony plant fibers, and the like. Nests average nearly 10 inches in length and about 4 inches in width. The entrance is normally near the top and to one side and is about an inch in diameter. The bowl is lined with feathers, plant down, or spider webs.

Clutch Size and Incubation Period: From 4 to 6 white eggs, with a few clutches of 12–14 reported that presumably are laid by two females. The incubation period is 12–13 days. Single-brooded, but known to renest.

Time of Breeding: In Colorado, eggs have been found as early as April 26 and nestlings reported as late as May 18. In Oklahoma, eggs have been found as late as June 10 and young seen as early as May 4. Active nests in New Mexico have been seen from April 10 to June 6.

Breeding Biology: Bushtits are found in small flocks during the nonbreeding period, moving about in close-knit groups from about mid-September until the first of April. Courtship begins in the flocks and consists of sexual posturing, trills, and excited location notes. Territories are poorly defined and rather variably defended, probably depending on the abundance of nesting materials and food they contain. Nests are built by both members of a pair, and in one case a third bird of unknown sex was seen helping with nest-building and incubation. Nest-building is a long and intricate process, requiring from 13 to as many as 51 days in eight observed instances. Rarely, nests from the past year are usable, and even in renesting efforts a new nest might be built, often using materials from the first nest. In renesting efforts the pairs often dissolve and the members may take new mates. Incubation apparently is equally shared by the two sexes, and both birds sleep in the nest at night. Both sexes feed the young and remain in the nest at night during the entire nestling period, which lasts about 14 days. After leaving the nest the family forms a small flock, with the adults initially doing all the foraging for the young, but about 14 days after leaving the nest the young birds are independent.

Suggested Reading: Addicott 1938; Bent 1946.

FAMILY SITTIDAE
(NUTHATCHES)

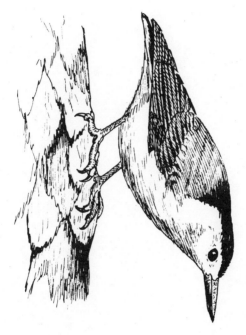

White-breasted Nuthatch

White-breasted Nuthatch
Sitta carolinensis

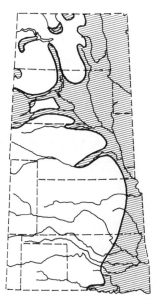

Breeding Status: Breeds nearly throughout the region in suitable habitats, but considerably more common in the eastern half. In western North Dakota it is largely limited to the Missouri and Little Missouri valleys, in western South Dakota it is common only in the Black Hills, in western Nebraska it breeds uncommonly in the Pine Ridge area, and in Kansas it is not known to breed west of a line from Douglas to Montgomery County. In Oklahoma it regularly breeds west to Osage, Lincoln, and Comanche counties, infrequently to Alfalfa County, and rarely to Texas County. There are no nesting records for the Texas panhandle.

Breeding Habitat: Deciduous or mixed forests, orchards woodlots, or trees in urban areas are the usual breeding sites. In the upper Missouri Valley the species is associated with late successional stages of floodplain forest.

Nest Location: Nests are in natural tree cavities ranging from 15 to 50 feet above the ground, or in old woodpecker holes, rarely in birdhouses. Knotholes often serve as cavity entrances, or natural crevices may be used. Strips of bark, hair, and other materials line the cavity.

Clutch Size and Incubation Period: From 5 to 10 eggs, frequently 8. The eggs are white with numerous brown to lavender spots. The incubation period is probably 12 days. Single-brooded.

Time of Breeding: North Dakota nest-building or egg dates range from April 13 to May 3, and dependent young have been seen as late as August 5. Egg dates in Kansas are March and April, and nest-building or eggs have been reported in Oklahoma from February 28 or March 27, with flying young seen as early as April 29.

Breeding Biology: Apparently white-breasted nuthatches maintain their pair bonds throughout most of the year and perhaps permanently, although during winter the birds roost in different areas and maintain little contact with each other. The male begins to sing in late winter, uttering early-morning "rendezvous songs" from tall trees to attract the female. Males also sing and display directly to their mates when they arrive and may keep in touch with them during foraging by uttering a series of *wurp* notes. The female takes the initiative in choosing a nest site and does all the nest-building, but both sexes participate in "bill-sweeping" in and around the nest. This behavior is of uncertain significance, but consists of arclike movements of the bill near the tree or cavity surface, sometimes while holding an insect or other object. It has been suggested that the odors thus spread may repel squir-

305

rels. The female does the incubating, but the male feeds her during egg-laying and incubation, and males later help feed the young. The fledging period is approximately 2 weeks.

Suggested Reading: Kilham 1968, 1972.

Red-breasted Nuthatch
Sitta canadensis

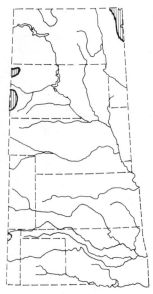

Breeding Status: Breeds locally in northwestern Minnesota (probably at least Clearwater County), commonly in the Black Hills of South Dakota, and probably also in the central Niobrara Valley and Pine Ridge areas of Nebraska (*Nebraska Bird Review* 35:30). Probably also breeds on Sierra Grande, New Mexico (Hubbard, 1978).

Breeding Habitat: In the Black Hills this species breeds in coniferous (pine and spruce) forests, and it is also closely associated with conifers in Minnesota.

Nest Location: Nests are made in abandoned woodpecker holes or excavated in the rotting wood of stubs or snags. The nest may be from 5 to 40 feet above the ground but is usually less than 20 feet up. The entrance is typically about 1½ inches in diameter and always is surrounded by sticky resin placed there by the adults. The nest is lined with shreds of bark and other soft materials.

Clutch Size and Incubation Period: From 4 to 7 eggs, usually 5–6. The eggs are white with variable amounts of brown spotting. The incubation period is about 12 days. Single-brooded.

Time of Nesting: Nest preparation has been observed in Minnesota in late May, and fledglings have been seen in early July. Likewise, the breeding season in the Black Hills is from May through July, with young birds observed in early July.

Breeding Biology: During the winter, red-breasted nuthatches may remain paired if the food supply is good or if the birds are close to a feeding station. At this season the birds maintain contact by uttering location calls, but by late winter unpaired males begin singing a series of plaintive *waa-aans* from tall trees, which probably serve both territorial and courtship functions. A major behavior during pair-formation is courtship feeding of the female by her mate, which continues through the incubation period. Pairs seek out nesting sites together, with the female making the final choice and also doing the initial excavating. Courtship chases of the female are frequent during nest-building and may end in copulation. When the nest excavating is nearly finished, both members of the pair bring resin in their bills, land in the nest entrance, and spread the resin above and below the hole. This sticky material probably deters other animals from entering the hole. The female does all the incubation but both sexes feed the young, which fledge in periods that have been estimated to range from 14 to 21 days.

Suggested Reading: Kilham 1973; Bent 1948.

Brown-headed Nuthatch
Sitta pusilla

Breeding Status: Limited to southeastern Oklahoma, where it has been reported breeding only in McCurtain County.

Breeding Habitat: Large short-leaf pines are the habitat of this species in Oklahoma, and it generally is found in open pine woods, particularly in burned-over areas or clearings where there are dead trees or old stumps.

Nest Location: Nests are usually in self-excavated holes, although birds occasionally use natural cavities or old woodpecker holes. They are often less than 5 feet above the ground, in dead trees or fire-blackened stumps. The cavity is excavated 6–8 inches and is partially filled with inner bark strips, strips of corn husks, and similar materials. The lining tends to be of pine seed wings.

Clutch Size and Incubation Period: From 3 to 9 eggs, often 5–6. The eggs are white with a covering of dark reddish brown spots. The incubation period is about 14 days. Single-brooded.

Time of Breeding: In Oklahoma, nest excavation has been observed in early March, and young recently out of the nest have been seen in late April.

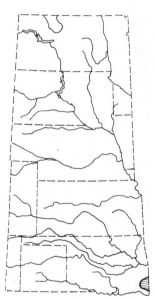

Breeding Biology: Male brown-headed nuthatches probably maintain rather permanent territories, which average 7 to 8 acres. The territory includes one or more potential nest sites, which are chosen by the male, as well as food sources. Cavity excavation begins early and may take the pair a month or more. Threesomes have been observed excavating nest sites, as also noted in the pygmy nuthatch. The female does all the incubating. Where threesomes have been observed, the extra male has not been seen roosting with the incubating female. The young are fed by both parents and by any nest helpers that may be present, and the fledging period averages 18–19 days. The period of parental dependence is about 45 days. Apparently families tend to remain together through fall and winter, and they have been found roosting together in nesting boxes close to their place of breeding.

Suggested Reading: Norris 1958; Bent 1948.

Pygmy Nuthatch
Sitta pygmaea

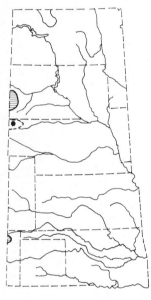

Breeding Status: Probably breeds uncommonly to rarely in the Black Hills of South Dakota (no specific nesting records) and periodically also in the Pine Ridge area of Nebraska (one state nesting record in Sioux County (*Nebraska Bird Review* 40:70). It is probably also a rare local breeder in northeastern New Mexico (Capulin Mountain National Monument) and is a possible resident in Cimarron County, Oklahoma (Sutton 1974).

Breeding Habitat: The species is generally associated with ponderosa pines, in the lower coniferous forest zone, especially in open, parklike forests.

Nest Location: Nests are in dead trees or in stubs, about 5-60 feet above the ground and often at least 25 feet up. They are often near the tops of larger snags, where the wood is well rotted, and have irregular openings 1¼-2 inches in diameter. The cavity is excavated to a depth of 8-9 inches and lined with bark shreds, feathers, moss, hair, and other soft materials.

Clutch Size and Incubation Period: From 4 to 9 eggs, usually 6-8. The eggs are white with variable reddish brown spotting. The incubation period is 15-16 days. Single-brooded.

Time of Breeding: There is no specific information on nesting in our region, except that adults carrying food to young were reported in northwestern Nebraska in late June. In Colorado, eggs have been found as early as May 5, and numerous clutches have been reported for May and June.

Breeding Biology: Pygmy nuthatches are more or less permanently territorial; males hold small territories and limit most defense to the nest site. This may be an existing cavity, or the pair may excavate a new one. Sometimes three or more birds have been seen excavating a single site, and at least in some cases the extra birds are males. Up to a month or more may be needed for excavation, and it takes one day to lay each egg. Only females incubate, but both sexes sleep in the cavity at night, and in the observed threesomes all the birds roosted there. Females are fed on or off the nest by the male or males. The eggs typically hatch within a 24-hour span, and the young are fed by both adults and, when present, additional helpers. The young fledge in 20–22 days but do not gain independence from the adults until they are approximately 45–50 days old.

Suggested Reading: Norris 1958; Bent 1948.

FAMILY CERTHIIDAE
(CREEPERS)

Brown Creeper

Brown Creeper
Certhia familiaris

Breeding Status: Positive records are few, but the bird is known to be an uncommon resident of the Black Hills area of South Dakota and a probable resident of the coniferous wooded areas of north-central Minnesota, with specific nesting records limited to Clearwater and Hubbard counties. There are several old (pre-1900) breeding records for eastern Nebraska and some recent ones for Sarpy County (*Nebraska Bird Review* 44:80, 46:14). Nesting in the Pine Ridge area is also suspected.

Breeding Habitat: In the Black Hills, creepers are associated with pine and spruce forests throughout the year.

Nest Location: The nest is behind the base of a loose piece of bark, on a live or dead tree, and is generally 5–15 feet above the ground. It is a crescent-shaped structure made of twigs, bark fibers, and similar materials.

Clutch Size and Incubation Period: From 4 to 8 eggs, white with reddish brown spotting. The incubation period is 15–16 days. Single-brooded.

Time of Breeding: Active nests in Minnesota have been seen from early May (eggs) to late June (ready to fledge). The estimated breeding season for the Black Hills is mid-May to mid-July.

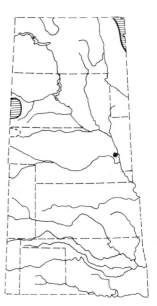

Breeding Biology: Although a few creepers may remain in the northern states through the winter, they are generally migratory, and in early spring small groups may be encountered foraging in loose flocks, maintaining contact by delicate *cree-cree-cree-ep* notes. At least in the European race of creeper it is known that during cold winter nights the birds frequently roost in cracks in tree trunks. Clinging woodpeckerlike to the bark and supported by the tail, they can withstand subfreezing temperatures even when partly covered by snow. When spring comes the pair begins to work intermittently on the nest, which may require a month to finish. Both sexes bring materials, but only the female does the construction. Observations in America suggest that only the female incubates, whereas in England it has been reported that the male participates in this to some extent. Both sexes also feed the young, which are ready to leave the nest in 13–14 days. Even though their short tail feathers do not provide them any support, the young fledglings are able to cling to vertical branches and move about like adults.

Suggested Reading: Bent 1948; Braaten 1975.

FAMILY CINCLIDAE
(DIPPERS)

Dipper

Dipper
Cinclus mexicanus

Breeding Status: Limited to the Black Hills of South Dakota, where it is an uncommon permanent resident.

Breeding Habitat: The dipper is associated with rapidly flowing mountain streams. In the Black Hills it is most common in Spearfish Canyon, but it also occurs elsewhere in the hills.

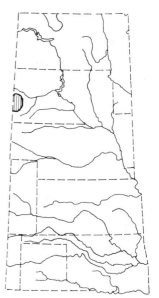

Nest Location: Nests are usually over water, either under bridges or under overhanging rock ledges. They have also been found among the roots of fallen trees, but in all cases they are made of woven mosses, usually with a roof, sides, and a front entrance. The nest may be from 8 to 12 inches in external diameter, thus being rather conspicuous, or may appear to consist of a small hole about 2 by 3 inches in a vertical wall of moss on the side of a cliff. The nest cup is of coarse grass, which effectively resists moisture.

Clutch Size and Incubation Period: From 3 to 6 white eggs, usually 4 or 5. The incubation period is 15–17 days, averaging 16. Single-brooded.

Time of Nesting: In the Black Hills the estimated breeding period is from late April through July, with eggs seen as late as July 5 and fledged young as early as June 13.

Breeding Biology: Dippers are very sedentary birds, and pairs tend to remain well separated. Studies in Montana indicate that by November the birds begin to establish winter territories, which are strongly defended through February and may include from 50 yards to as much as a half mile of stream. In spring the birds abandon these territories and begin to move upstream into breeding territories. During winter the birds also begin to sing, and singing increases in intensity to a peak in April. The songs of the two sexes are indistinguishable, and pair-formation is accompanied by loud singing as well as wing-quivering and chasing behavior. Both sexes participate in nest-building or in reconstructing an old nest, which seems to be more common than building entirely new nests. Incubation is entirely by the female, but the male frequently brings food to her during incubation. The young are hatched with a coating of down but grow relatively slowly, so that the nestling period is surprisingly long, about 19–25 days. The female broods regularly for about a week after hatching, and males rarely or never enter the nest. Instead, the older young poke their heads out the nest entrance to be fed. When the young birds leave the nest they are nearly as large as their parents and easily flutter out to a safe landing below. After the birds leave the nest, one or both of the parents typically removes the nest lining, presumably to prepare the nest for use in another year. The young birds soon learn to clamber about on the wet rock surfaces, and they remain in the vicinity of the nest for up to 15 days after fledging. It is likely that adult birds return to their same nesting areas each year, but in at least one case an adult bird had different mates in two successive years.

Suggested Reading: Bakus 1959; Hahn 1950.

FAMILY TROGLODYTIDAE
(WRENS)

Long-billed Marsh Wren

House Wren
Troglodytes aedon

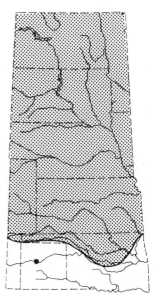

Breeding Status: Breeds commonly throughout the region from the Canadian border south to Oklahoma, where it reaches its southern limits in the panhandle (Cimarron County), south-central area (Cleveland County), and northeastern parts of the state. There is a single breeding record for the Texas panhandle (Randall County) but none for northeastern New Mexico.

Breeding Habitat: The house wren was originally associated with deciduous forests and open woods but now it is also city-adapted and nests in artificial structures.

Nest Location: Birds use natural cavities in trees, fenceposts, or stumps, as well as birdhouses or other artificial cavities with openings of the appropriate size (about 1 inch in diameter). The nest cavity is mostly filled with twigs, and the nest cup is formed of grasses, plant fibers, feathers, and other soft materials.

Clutch Size and Incubation Period: From 3 to 8 eggs (20 Kansas clutches averaged 5.8; 13 North Dakota nests had a mean of 6.5). The eggs are white with a fairly extensive covering of reddish to cinnamon dots. The incubation period is 12–15 days. Normally double-brooded.

Time of Breeding: Egg dates in North Dakota are from May 29 to July 22, and in Kansas they extend from April 11 to July 31. In Kansas, about 45 percent of the eggs are laid between May 11 and May 31.

Breeding Biology: As house wrens arrive on their breeding grounds in the spring, adults tend to precede immature birds, and males arrive about 9 days before females. An adult male that has nested previously returns to its old territory or establishes a new territory adjacent to it, and females also have a strong tendency to return to previous nesting areas. Males sing three kinds of songs, including a "territory song," a "mating song," and a "nesting song," and both sexes have a variety of call notes. Males typically have 2 or 3 possible nest sites within their territories and may have as many as 7, thus allowing the females considerable choice. When establishing nest sites, house wrens often destroy the eggs, nests, or young of their own or other species, and there is a good deal of territorial shifting owing to nest-site competition and to the frequent changing of mates between broods. The nestling period is approximately 15 days. In addition to mate-changing at this time, a second female may mate with a male and nest within his territory. In one study it was found that about 6 percent of the matings are polygynous and that about 40 percent of second matings are with the same mate. There is likewise about a 40-percent incidence of mating with the same individual in the following year, when both birds return to the same locality.

Suggested Readings: Kendeigh 1941; Bent 1948.

Bewick Wren
Thyromanes bewickii

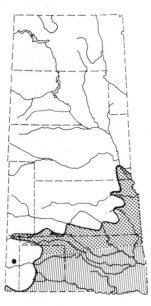

Breeding Status: Breeds from extreme southeastern Nebraska (one old nesting record for Otoe County, breeding also reported from Gage County and erroneously from western Nebraska) and presumably adjacent Iowa southward through northwestern Missouri (rare at Squaw Creek N.W.R.) and eastern Kansas, extending into western Kansas only south of the Arkansas River. It breeds locally in southeastern Colorado (Baca County), in the Oklahoma panhandle to Cimarron County, and in the eastern panhandle of Texas (west to Randall County). In northeastern New Mexico it breeds locally (Union and Quay counties).

Breeding Habitat: The Bewick wren is associated with open woodlands, brushy habitats, farmhouses, and towns.

Nest Location: Nests are in natural tree cavities, sheds and deserted buildings, old woodpecker holes, birdhouses, and similar cavities in manmade structures. Nests and nesting sites are like those of the house wren, and the two species usually conflict when in the same area. The house wren is dominant and can evict the Bewick wren.

Clutch Size and Incubation Period: From 4 to 8 eggs (averaging 5.5 in 12 Kansas clutches). The eggs are white, with dark spots and dots especially at the larger end. The incubation period is about 14 days. Double-brooded.

Time of Breeding: Kansas egg records are from March 21 to July 10, with first and second clutches usually in mid-April and mid-June. Active nests have been seen in Oklahoma from April 10 to June 25.

Breeding Biology: Bewick wrens are less migratory than house wrens and may move into areas used by house wrens when they leave in the fall, only to be forced out again in the spring. In some areas they avoid intense competition by breeding somewhat earlier and in thicker habitats than house wrens. Males establish breeding territories containing several potential nest sites, and some males appear to be bigamous or polygynous. Males advertise territories by singing, and pairing seems to occur rapidly after territories are established. At least in most instances, nests are not built until the female is present, and although both sexes may work on the nest the female does most of the construction or may even build alone. The nest is built in about 10 days, and the female does all the incubating. At this time the male apparently feeds her, and both sexes feed the young during the 14-day nestling period. For about 2 weeks after the young birds leave the nest they are cared for by the parents.

Suggested Reading: Miller 1941; Bibbee 1947.

Winter Wren
Troglodytes troglodytes

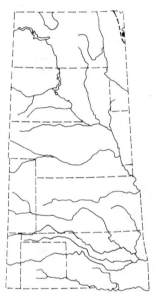

Breeding Status: Breeding is limited to northern Minnesota, Clearwater County being the only proved area of breeding; in Itasca State Park the birds are locally common and are known to have nested (*Loon* 49:85).

Breeding Habitat: The species is associated with heavy forests, usually coniferous. In Minnesota it is especially numerous in spruce and white cedar bogs.

Nest Location: Nests are most often in the upturned roots of a fallen tree but may also be amid the roots of a live tree or under a stump, in a rocky crevice, or rarely in an old woodpecker hole. However, they typically are not in an enclosed cavity like the nests of other wrens.

Clutch Size and Incubation Period: From 4 to 7 eggs, usually 5–6. The eggs are white, with reddish brown dots and spots often forming a wreath around the upper end. The incubation period is from 14 to 20 days, most commonly 16. Frequently double-brooded.

Time of Breeding: Egg dates are not available for Minnesota, but fledglings have been seen at Itasca Park as early as June 10, and parents with young have been seen in Minnesota as late as August 1.

Breeding Biology: This tiny wren is highly territorial, and in some areas males defend territories practically throughout the year. Territories are advertised by intensive singing, which is prolonged and melodious, frequently lasting more than 7 seconds and having remarkable acoustic complexity. The song consists of two similar major portions, each of which is terminated by a trill and totals well over 100 separate notes. Singing occurs from spring until as late as mid-August in northern Minnesota. Males are frequently polygynous and may have up to three mates simultaneously. They build several nests, averaging (in the European race) about 6 per bird, and may build as many as 12. Females rarely participate in building, but they do bring in lining materials. They apparently perform all the incubation, but males participate in feeding the young. In the European race, males help feed young in about 40 percent of the nests. The young remain in the nest for about 2–3 weeks, averaging about 17 days. A second clutch may be begun within 2 weeks of the fledging of the first brood.

Suggested Reading: Armstrong 1955; Bent 1948.

Carolina Wren
Thryothorus ludovicianus

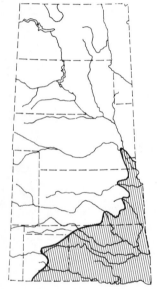

Breeding Status: Breeds in extreme southeastern Nebraska (periodically or locally north to Douglas County and west to Lancaster and Nuckolls counties), extreme northwestern Missouri (occasional at Squaw Creek N.W.R.), eastern Kansas (east of a line from Doniphan, Riley, and western Reno counties), Oklahoma west to the panhandle, and locally in the Texas panhandle (Randall County).

Breeding Habitat: Brushy forests, forest margins, cutover forests, cultivated areas with brush heaps or old buildings, and suburban parks and gardens.

Nest Location: Nests are in natural tree cavities, in woodpecker holes, under rocks, in overturned root cavities, in birdhouses, and in building crevices. They are rarely more than 10 feet above the ground, and the cavities are filled with a variety of soft, pliable materials. The lining is of fine grasses, hair, feathers, and the like.

Clutch Size and Incubation Period: From 3 to 8 eggs (9 Kansas nests averaged 4.2). The eggs are white, with heavy brown spotting around the larger end. The incubation period is 12–14 days. Normally double-brooded and sometimes raises three broods.

Time of Breeding: Kansas egg records are from April 11 to August 10, with a probable peak in mid-April and a possible initiation of nesting as early as late March. Oklahoma egg records are from March 18 to August 1.

Breeding Biology: The Carolina wren overlaps to some extent with the house wren in its breeding range but apparently does not conflict with it to the extent that the Bewick wren does. Being larger than either of these two species, it is probably socially dominant where contacts do occur. Males become territorial and sing persistently from late winter onward, although some song may occur almost throughout the year. Males apparently do not begin to build nests until they become mated. When mating occurs the pair begins to seek out suitable nest sites, and in one observed case they began building a day after a suitable nest basket had been hung. Both sexes build, and within 2 days the nest may be nearly completed. About 5 days after the start of nest-building the first egg is laid, and subsequent eggs are laid at daily intervals. The male feeds his mate to some extent during nest-building, incubation, and brooding, but the female is off the nest a surprisingly small amount of time during incubation. The nestling period is 13–14 days. The extent of double-brooding and polygyny or mate-changing between broods is still unreported in this species, but the male sometimes takes charge of the newly fledged brood so that the female can begin a second clutch in a new nest that the male has prepared.

Suggested Reading: Nice and Thomas 1948; Bent 1948.

Long-billed Marsh Wren (Marsh Wren)
Cistothorus palustris

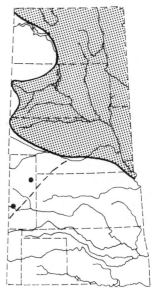

Breeding Status: Breeds throughout western Minnesota and North Dakota east of the Missouri, the eastern half of South Dakota, and most of Nebraska north of the Platte River, with local or infrequent breeding south of the Platte (*Nebraska Bird Review* 39:74). It also breeds locally in extreme northwestern Missouri and northeastern Kansas (Doniphan County, possibly others).

Breeding Habitat: Freshwater marshes or brackish tidal marshes with extensive tall emergent vegetation represent prime habitats, but the banks of tidal rivers, reservoir inlets, and similar habitats are also used.

Nest Location: Most nests are 3–5 feet above the marsh substrate, with early nests being lower and later ones higher. The nests are built in emergent vegetation (cattails, bulrushes, etc.) and are domed elliptical structures about 7 inches high and 5 inches across, with a lateral opening about 1¼ inches in diameter just above the equator. They are constructed of grass strips and lined with cattail down. In North Dakota, water depth at 19 nests averaged 12 inches, and 23 nest entrances averaged 16 inches above the water.

Clutch Size and Incubation Period: From 3 to 7 eggs (23 North Dakota clutches averaged 4.7). The eggs are cinnamon to brown with darker spots. The incubation period is 13 days. Double-brooded.

325

Time of Breeding: North Dakota egg dates are from May 26 to August 10, and nestlings have been reported from June 16 to August 6. In Kansas eggs are also laid between May and August.

Breeding Biology: Shortly after they arrive on their breeding areas males establish territories, which they advertise by persistent singing from all parts and by aerial displays above them. After a territory has been established, each male begins to build a number of "courting nests" (up to 5 or more), which are complete except for a lining. When a female selects a male as a mate she either accepts one of these nests for breeding and lines it or begins a new one, which is chiefly constructed by the male. After egg-laying has begun, the male moves to a new area in his territory and begins to advertise for additional mates. He may obtain as many as three mates, each of which incubates alone but is fed by the male. The male's role in feeding the young is often small or nil; instead, he continues to maintain the territory. The young leave the nest when about 14–16 days old but may be fed by adults for nearly 2 more weeks. Nest-building by the male increases during the nestling and fledgling period, in preparation for a second clutch.

Suggested Reading: Verner 1965; Welter 1935.

Short-billed Marsh Wren (Sedge Wren)
Cistothorus platensis

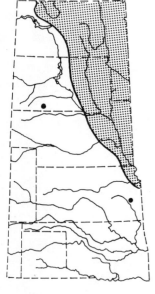

Breeding Status: Breeds through the western part of Minnesota and North Dakota east of the Missouri River southward through eastern South Dakota, western Iowa, and eastern Nebraska. It is uncommon in extreme northwestern Missouri, and in Kansas it is rare and irregular, with breeding records limited to a few eastern counties. There is a single breeding record for Oklahoma (Harper County, 1936).

Breeding Habitat: Wet meadows, especially those dominated by sedges, cottongrass, mannagrass, and reedgrass, are primary habitats in the northern plains, but the birds also utilize emergent vegetation associated with marshes as well as retired croplands and hayfields.

Nest Location: Nests are constructed over land or water in dense growing vegetation and are usually 1–3 feet above the substrate. They are globular structures about 4 inches in diameter, with a lateral entrance above the equator, lined with plant down, hair, or similar materials.

Clutch Size and Incubation Period: From 4 to 8 white eggs, often 7. The incubation period is 12–14 days. Reportedly double-brooded, but single-brooded in an Iowa study.

Time of Breeding: North Dakota egg dates are from June 7 to August 10. The few Kansas records indicate that eggs are laid in late July and August at that latitude.

Breeding Biology: Although this species is not nearly so well studied as the long-billed marsh wren, it is known that males regularly build numerous "dummy" nests, and thus a comparable pattern of pair-formation and evidently polygyny prevails. In favored habitats such as large meadows, the birds concentrate in high densities; about 35–40 singing males were counted in a Michigan meadow of only 10 acres. At the peak of the nesting period the male may spend as much as 22 hours a day singing, generally from 6–12 songs per minute. When a pair bond is formed the female selects or initiates a brood nest, which tends to be lower and harder to find than the courting nests. The female does all the incubating and most of the feeding of the young, with only occasional visits by the male. The young remain in the nest about 13 days, and presumably a second brood is often initiated shortly after the first one fledges.

Suggested Reading: Walkinshaw 1935; Crawford 1977.

Canyon Wren
Catherpes mexicanus

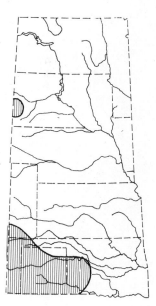

Breeding Status: Breeds locally in western South Dakota (Black Hills), southern Colorado (at least Baca County), the western panhandle (Cimarron County) and the southwestern corner of Oklahoma (Woodward to Caddo and Comanche counties), the Texas panhandle, and northeastern New Mexico.

Breeding Habitat: The species is associated with rocky canyons, river bluffs, cliffs, and occasionally cities.

Nest Location: Ledges in shallow caves or rocky crevices are favored, but birds also nest in buildings, utilizing rafters, chimneys, eaves, and similar locations. The nest is a haphazard accumulation of materials (in one case totaling 1,791 countable objects as well as half a pound of filling materials), lined with wool, feathers, or other soft material.

Clutch Size and Incubation Period: From 4 to 6 eggs, white with faint brownish spotting. The incubation period is about 12 days. Apparently double-brooded, at least in some areas.

Time of Breeding: In the Black Hills the nesting period is from late May through July, with fledglings observed as early as June 13. Active nests or eggs have been reported in Oklahoma from April 26 to July 5, and nestlings or fledgings have been seen from June 6 to July 14. Texas egg dates are from March 4 to July 7.

Breeding Biology: These birds are somewhat migratory in South Dakota and Colorado, arriving on their Black Hills breeding grounds in early April. By that time they are already actively singing and defending territories in southern Colorado, and females have begun to carry nesting material into crevices that may be the same sites as used in previous years. Males do not assist with incubation and have not been reported to feed incubating females. But when the nestlings are very small the male does feed his brooding mate, and soon both sexes begin to gather food. In one observed case the nestling period was at least 13 days. At least in Colorado, there seems to be a migration away from the nesting areas shortly after the young have fledged, although the birds are likely to return by early December.

Suggested Reading: Bailey and Niedrach 1965; Tramontano 1964.

Rock Wren
Salpinctes obsoletus

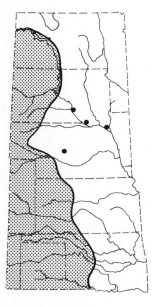

Breeding Status: Breeds in western North Dakota along the Missouri and Little Missouri drainages, southward through western South Dakota, western Nebraska (east to Cherry, Custer, and Lincoln counties), western Kansas (east to Decatur, Rooks, and Comanche counties), and western Oklahoma (east to Woods, Blaine, and Comanche counties and irregularly to Sequoyah County). It also breeds widely in eastern Colorado, northeastern New Mexico, and the Texas panhandle.

Breeding Habitat: Eroded slopes and badlands, rocky outcrops, cliff walls, talus slopes, and generally arid environments are preferred.

Nest Location: Nests are typically among rocks, in crevices of canyon walls, or sometimes in tree holes or cutbanks. There is often a small runway of stones leading to the nest, which is constructed of twigs and grasses, usually with a lining of hair or wool.

Clutch Size and Incubation Period: From 3 to 10 eggs, usually 4–5. The eggs are white with slight brown spotting. The incubation period is 14 days. Double-brooded at least in some areas.

Time of Breeding: North Dakota egg dates are lacking, but the probable egg-laying period is late May, since nestlings or dependent fledglings have been seen from June 19 to August 11. Kansas egg dates are from May 11 to July 20, with a mid-June peak. Active nests in Oklahoma have been seen from April 22 to June 13, and egg dates in Texas are from April 2 to July 3.

Breeding Biology: In Arizona, where rock wrens occur with canyon wrens, both species feed in a generalized fashion on similar foods, but rock wrens forage almost exclusively in open or relatively unvegetated situations, while canyon wrens forage mostly in secluded or covered habitats. The species also differ in favored nest sites, with this species using slopes of loose rocks and boulders rather than cliff or canyon walls. Eggs are laid at the rate of one a day, with incubation starting when the clutch is complete. Only the female incubates, but she is usually fed by the male, and both sexes feed the nestlings. When the young leave their nest (after about 14 days) the adults soon begin gathering nest material for their second brood or may begin a second clutch in the same nest.

Suggested Reading: Tramontano 1964; Bent 1948.

FAMILY MIMIDAE (MOCKINGBIRDS AND THRASHERS)

Brown Thrasher

Mockingbird
Mimus polyglottos

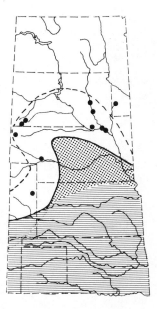

Breeding Status: The normal northern breeding limit of this species is probably northern Nebraska, although it is listed as a hypothetical breeding species for North Dakota and is rare but widespread in South Dakota. There are two Minnesota nesting records (*Loon* 49:229). In Nebraska the species breeds in the panhandle north to Sioux County, while in the Sandhills it occasionally breeds north to Thomas and Greeley counties and is most common in the southeast. In Kansas the species occurs throughout the state. In Colorado mockingbirds breed along the Platte Valley (Weld and Logan counties) and are locally common in the southern counties. They also breed through virtually all of Oklahoma and the Texas panhandle, as well as in northeastern New Mexico.

Breeding Habitat: Short, open woodlands, forest edges, farmlands, parks, cities and similar habitats are utilized, although treeless plains and deep forests are both avoided.

Nest Location: Nests are usually 2–10 feet above the ground, in trees, shrubs, or vines. They are rather loosely constructed of twigs, with an inner layer of leaves or grasses and a lining of rootlets and horsehairs. Small evergreen trees are favored nest sites, and evergreen vines are also used frequently.

Clutch Size and Incubation Period: From 3 to 5 eggs (27 Kansas clutches averaged 3.5). The eggs are greenish to bluish with heavy brown blotches or spots. Incubation lasts 12–13 days. Frequently double-brooded, and up to four broods have been reported.

Time of Breeding: Kansas breeding records are from April 21 to July 31, with most first clutches laid in early June. Breeding records in Oklahoma extend from April 9 (nest-building) to August 8 (nestlings), and dependent young have been observed as late as September 3.

Breeding Biology: Mockingbirds tend to be fairly long-lived, and mates frequently remain unchanged for several years. In fall the pair bonds are broken; the female leaves the nesting area and (in southern states) the male remains on his territory through the winter. Females establish their own nonbreeding territories at this time, but in spring they again seek out a male's territory. In addition to territorial advertisement by singing, nest-building is a part of the male's courtship. He carries nesting material to potential nest sites within the territory until he attracts a female and they establish a pair bond. She may then accept one of his nesting sites or select a new one. Both sexes participate in completing the nest, and a new nest is usually constructed for each brood. One case of simultaneous bigamy has been reported. The female incubates, and though the male continues to guard the territory he

333

does not feed her. However, both sexes feed the young birds, which have a fledging period of 10–13 days, and they may continue to be fed by one or both parents for nearly a month.

Suggested Reading: Laskey 1962; Adkisson 1966.

Gray Catbird
Dumetella carolinensis

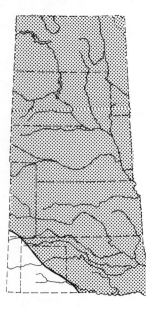

Breeding Status: Breeds in suitable habitats nearly throughout the region, becoming less common toward the west and south, and reaching its southern limits in southeastern Colorado (Baca County), the Oklahoma panhandle (Cimarron County), and the eastern panhandle of Texas (Hemphill and Wilbarger counties).

Breeding Habitat: Catbirds breed in thickets, woodland edges, shrubby marsh borders, orchards, parks, and similar brushy habitats. A combination of dense vegetation and vertical or horizontal "edge" is a major criterion for nesting habitats.

Nest Location: Small brushy trees, vine tangles, or dense thickets are preferred nesting sites, with the nest usually 2–10 feet above the ground. It has a bulky foundation of stick, weed stems, twigs, leaves, and grass, with a lining of rootlets and often horsehair.

Clutch Size and Incubation Period: From 3 to 5 greenish blue eggs (43 Kansas clutches averaged 3.3 eggs; 11 North Dakota nests averaged 4.3). The incubation period is 12–13 days. Frequently double-brooded.

Time of Breeding: North Dakota egg dates extend from June 3 to June 29, with dependent young seen as late as August 20. Kansas egg dates are from May 11 to July 31, with a peak of egg-laying in late May.

Breeding Biology: Although catbirds are distinctly territorial, active defense seems to be largely limited to the vicinity of the nest site, and much of the territorial proclamation is achieved by singing from within the dense vegetation the birds frequent. In one Michigan study, most nests were within 2 feet of the side or top of shrub cover, in sites providing good visibility for the sitting bird. Males frequently "point out" possible nest sites by sitting on branches with their wings spread and manipulating twigs or other objects as if nest-building. However, once a nest is begun the female does most of the actual building, although the male may bring her materials. The first egg is usually laid 2 days after the nest is completed, and thereafter eggs are laid daily until the clutch has been completed. Incubation is by the female alone, and the male apparently feeds her very little during this time. The young remain in the nest for an average of 11 days, and they are cared for by their parents for approximately 2 more weeks. In many cases the pair raises a second brood, but rarely if ever is a third brood successfully reared in central or northern states.

Suggested Reading: Nickell 1965; Bent 1948.

Brown Thrasher
Toxostoma rufum

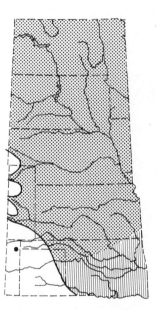

Breeding Status: Breeds in suitable habitats throughout most of the region, reaching its western limits in southwestern Nebraska, eastern Colorado (west at least to Adams and Crowley counties), the Oklahoma panhandle (rarely to Cimarron County), and probably the eastern parts of the Texas panhandle (nearest breeding record Wilbarger County). It has also bred at Clayton, New Mexico (Hubbard, 1978).

Breeding Habitat: The species frequents open brushy woods, or scattered patches or tracts of brush and small trees in open environments. Shelterbelts, woodlots, and suburban residential areas planted to shrubbery are also used.

Nest Location: Nests are in trees, shrubs, or vines in dense thickets, between 1 and 25 feet from the ground but usually less than 5 feet up. Ground nests are apparently common in New England but not in the Great Plains. The nest is loosely constructed of thorny twigs, leaves, and grasses with a deep cup lined with rootlets. The outside diameter of the nest averages 12 inches, and it is about 4 inches high.

Clutch Size and Incubation Period: From 2 to 5 greenish blue eggs (102 Kansas clutches averaged 3.6, and 6 North Dakota clutches averaged 4.3). The eggs are pale bluish white with small brown speckles. The incubation period is 12–13 days. Frequently double-brooded in southern states.

Time of Breeding: North Dakota egg dates are from May 20 to July 8, with dependent young seen as late as July 23. In Kansas, egg records span the period May 1 to July 20, with a peak in mid-May. Oklahoma egg records are from April 12 to July 10, with dependent young seen as late as August 24.

Breeding Biology: Males of this migratory species usually arrive on their breeding areas a few days ahead of females and apparently establish nesting territories almost immediately, although territorial singing may not begin for 10 days or more. Once a territory has been established, the males become very sedentary, and all the nests of the season are built within this territory.

Brown thrashers and catbirds have very similar territorial requirements, and at times thrashers will evict catbirds from their territory. Incubation is primarily by the female, and both birds also help brood the young, although males seem to be less efficient than females. The average nestling period is 11 days, but in some cases the female leaves the care of the young to the male soon after hatching and begins a second nest. In other cases the two parents may each take part of the brood after they fledge, later joining to begin a second nesting effort. Studies of banded birds have indicated that birds sometimes change mates between broods, even when the original mate is still available.

Suggested Reading: Bent 1948; Erwin 1935.

Curve-billed Thrasher
Toxostoma curvirostre

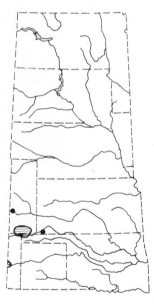

Breeding Status: Very rare or local breeder in southeastern Colorado (Baca County, *Colorado Field Ornithologist* 11:16), southwestern Kansas (Kearny County) and the Black Mesa of extreme northwestern Oklahoma (Cimarron County). In New Mexico it breeds in the Cimarron Valley of Union County and east to the Conchas Lake area of San Miguel County. This species is apparently now extending its breeding range eastward in Oklahoma (G. M. Sutton, pers. comm.), and bred in Morton County Kansas, in 1978.

Breeding Habitat: Arid semidesert to desert habitats, especially those with tall cacti, yuccas, and thorny brush, are favored.

Nest Location: With few exceptions, nests are in tall cholla cacti, 3–5 feet above the ground. The nests are bulky, about 10 inches in diameter, constructed of thorny twigs and lined with fine grasses, rootlets, and sometimes hair. Sometimes the old nests of cactus wrens are used as a foundation.

Clutch Size and Incubation Period: From 2 to 4 eggs, usually 3. The eggs are green to bluish with brown speckling or spotting. The incubation period is about 13 days. Regularly double-brooded.

Time of Breeding: The breeding season in Oklahoma is from April (heavily incubated eggs reported April 24) through the summer (newly fledged young found as late as October 2). Records from New Mexico also indicate a prolonged nesting season.

Breeding Biology: At least in many parts of their breeding range these thrashers are permanent residents, and pairs apparently remain mated throughout the winter. In some areas they coexist with Bendire thrashers, competing with them and generally being more successful, at least in the vicinity of Tucson, Arizona. Probably the birds maintain the same territory in successive years; at least as many as four old thrasher nests have been found in a single bush containing a new thrasher nest and five old cactus wren nests as well. Both sexes are said to participate in incubation and in the care of the young. The birds remain in the nest 14-18 days and do not become independent of their parents until they are about 40 days old. Studies on hand-raised birds indicate that young thrashers are apparently not innately able to recognize and respond appropriately to potential predators such as birds and reptiles before that age. Typically two broods are raised each season, with a second clutch begun about 2-3 weeks after the first brood has fledged. The birds either use the same nest or initiate a new one nearby.

Suggested Reading: Bent 1948; Ambrose 1963.

Sage Thrasher
Oreoscoptes montanus

Breeding Status: An uncommon to rare breeder in southwestern South Dakota (Buffalo Gap National Grassland) and possibly a very rare breeder in northwestern Nebraska (Sioux County, *Nebraska Bird Review* 40:71). It is considered a hypothetical breeding species in North Dakota, and in Colorado it is a common breeder in western areas, though specific breeding records seem to be lacking east of Fremont County. There is a single breeding record for southwestern Kansas (Morton County) and an early nesting record for Cimarron County, Oklahoma. There are no breeding records for northeastern New Mexico or the Texas panhandle.

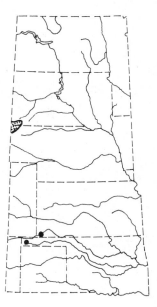

Breeding Habitat: The species is closely associated with sage-dominated grasslands or similar shrubby aridlands.

Nest Location: Nests are in low shrubs, especially sagebrush, or on the ground. They are typically from a few inches to about 3 feet above the ground, well hidden among the dense shrubbery. They are bulky structures of twigs with a lining of fine grasses, rootlets, and sometimes hair or fur. A few nests have had platforms of twigs above them, seemingly to provide protection from the sun.

Clutch Size and Incubation Period: From 3 to 7 eggs, usually 4-5. The eggs are blue to greenish blue with brownish spotting. The incubation period is 14-17 days, averaging 15. Single-brooded, but second nestings are known (*Auk* 95: 580-82).

337

Time of Breeding: Nests with eggs have been found in Colorado as early as May 13, and nestlings as early as June 24. The single Oklahoma egg record is for June 13.

Breeding Biology: Relatively little has been written on the breeding biology of this arid-adapted thrasher. Some early descriptions suggested a territorial song flight, with the bird zigzagging low over the ground, uttering a warbling song, and landing with upraised and fluttering wings. Apparently both sexes incubate, and incubation probably begins the day before the last egg is laid. The nestling period is 11–13 days. When bringing food to the young, the adults are highly secretive, landing on a sagebrush about 10 feet away, then approaching the nest while hidden from view. Pairs often remain mated during successive years, and the birds are sometimes rather long-lived, with one banded individual known to have reached 13 years.

Suggested Reading: Killpack 1970; Bent 1948.

FAMILY TURDIDAE
(THRUSHES, BLUEBIRDS,
AND SOLITAIRES)

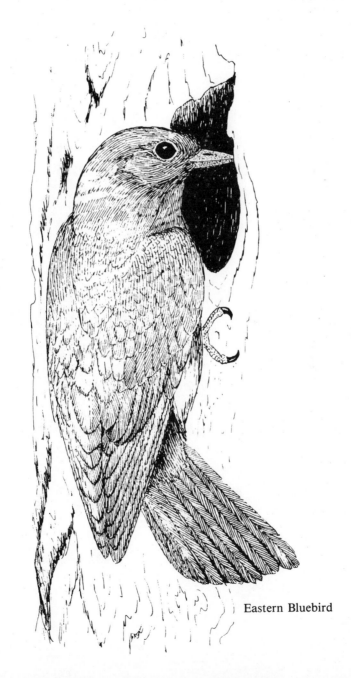

Eastern Bluebird

American Robin
Turdus migratorius

Breeding Status: Almost pandemic, but becoming rarer to the west and southwest, and increasingly limited there to urban areas. There seem to be no breeding records for northeastern New Mexico south of the Cimarron River Valley, the Oklahoma panhandle west of Beaver County, or for the Texas panhandle outside the Red River Valley.

Breeding Habitat: The original habitat was probably open woods, but now the species is most common in cities, suburbs, parks and gardens, and farmlands.

Nest Location: Nests are in tree forks, on horizontal branches, in shrubs, or on horizontal ledges of buildings, rarely more than 30 feet above the ground and usually 5–15 feet. They are constructed of mud into which grass and other vegetation has been worked while the mud is still wet, and are lined with fine grasses.

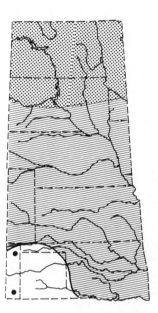

Clutch Size and Incubation Period: From 3 to 6 eggs (8 North Dakota clutches averaged 3.9; 57 Kansas clutches averaged 3.6). The eggs are uniformly blue green. The incubation period is 12–14 days. Multiple-brooded.

Time of Breeding: North Dakota egg dates range from April 23 to June 30, with nestlings seen from May 8 to July 15. In Kansas, egg dates are from April 1 to July 20, with about half the eggs laid between April 11 to 30. Full clutches have been found in Oklahoma as early as March 23, and nestlings have been seen as late as August 24.

Breeding Biology: A very early spring migrant, male robins tend to arrive on the breeding grounds slightly before females, and both sexes tend to return to the area where they were hatched. Males often establish essentially the same territory they held the previous year; the size of the territory seems to vary greatly with habitat and population density. The time clutches are begun is closely associated with latitude, and both sexes apparently help select the nest-site. The nest is sometimes completed in as little as 24 hours, with the male carrying much of the material and the female doing the shaping. However, most nests are built much more slowly, especially early ones, which often require 5–6 days. The eggs are laid at daily intervals, and incubation is done almost exclusively by the female. The fledging period is usually about 13 days, but varies from 9 to 16 days, and the young are cared for until they are about a month old. Even at the northern edge of its range the robin typically raises two broods, and pairs normally remain intact for the second brood. At times the same nest is used for the second clutch, but often a new one is constructed nearby.

Suggested Reading: Howell 1942; Bent 1949.

Wood Thrush
Hylocichla mustelina

Breeding Status: Breeds from west-central Minnesota (hypothetical breeder in extreme eastern North Dakota) southward through Iowa and eastern South Dakota, eastern Nebraska (west probably to Cherry, Lincoln, and Thomas counties, *Nebraska Bird Review* 34:18), eastern Kansas (west to Decatur and Edwards counties), and eastern Oklahoma (west to Kay, Oklahoma, and Murray counties).

Breeding Habitat: Associated with mature, shady forests, especially deciduous woods, and secondarily with wooded parks or gardens. Like veeries, wood thrushes prefer habitats with wet ground, running water nearby, and a dense understory, but apparently they also need tall trees for song perches.

Nest Location: Nests are usually 5–15 feet above the ground, rarely up to 50 feet, on a low horizontal fork or in a crotch of a tree. The nest is very similar to that of a robin but is somewhat smaller (less than 6 inches in diameter) and has a lining of leaves and rootlets rather than grass. Nests are typically well shaded from above and are well concealed.

Clutch Size and Incubation Period: From 2 to 4 blue or greenish blue eggs (9 Kansas clutches averaged 3.4). The incubation period is 13 days. Frequently double-brooded.

Time of Nesting: Egg dates in Kansas are from May 11 to August 10, with most eggs laid between May 21 and June 10. Egg dates in Oklahoma are from April 28 to July 21, and in Texas they are from April 29 to May 2.

Breeding Biology: Wood thrush males become territorial and obtain mates within a few days after they arrive on the breeding grounds; sometimes they are mated within a day after arrival, possibly to mates of the previous year. Although the male is able to influence the choice of a nesting site by calling or even incipient nest-building, the female makes the final choice and does the actual building. About 6 days are spent in nest-building; during that time sexual chases are frequent and copulation attempts are common. The first egg is laid 1–3 days after the nest is finished, and thereafter eggs are laid each day until the clutch is complete. Only the female incubates, but the male stands near the nest when his mate is absent. Both parents feed the nestlings, which fledge in 12–14 days. Thereafter both parents protect and feed the young for another 10–15 days, and typically a second clutch is begun soon afterward. New nests are constructed for the second nesting, usually somewhat lower than for the initial nesting effort. There is no evidence of mate-changing between broods.

Suggested Reading: Nolan 1974; Brackbill 1958.

Hermit Thrush
Catharus guttatus

Breeding Status: Breeds regularly in forests of north-central Minnesota (Clearwater and Roseau counties) and rarely in the Black Hills of South Dakota (*Wilson Bulletin* 78:321).

Breeding Habitat: Coniferous or mixed forests, especially shady, moist woods, are the favored habitat.

Nest Location: Nests are usually on the ground, well sunk in moss, but they may also be in the lower limbs of conifers. They are built of twigs, bark, grasses, and mosses and lined with materials such as grasses, mosses, and pine needles. In Colorado, nests are usually 3–10 feet above the ground, in spruces, and are close to a stream or spring.

Clutch Size and Incubation Period: From 3 to 4 very pale blue eggs, which rarely have a few spots. The incubation period is 12 days. Probably single-brooded but known to renest after nest failure.

Time of Breeding: Minnesota egg records are from May 30 to July 10, and nestlings have been reported as early as June 18. The Black Hills record was of a nest with eggs being incubated on June 19. Colorado egg records are from June 11 to July 5, with nestlings seen as late as August 3.

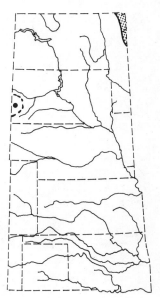

Breeding Biology: This is the first of the thrushes to arrive in northern areas in the spring and the last to depart in fall, a reflection of its adaptation to boreal nesting. It also is perhaps the most famous songster; sometimes its beautiful and complex song can even be heard in wintering areas, although it does not sing during migration. So far as is known, only the female incubates the eggs, but the male regularly feeds her while she is sitting. Additionally, he spends a good deal of time guarding the nesting territory, often standing on a sentinel post about 40 feet away from the nest itself. Both parents actively feed the young, which spend an average of 12 days in the nest. In the eastern states the range of egg dates, extending over about 3 months, is suggestive of double-brooding, but in Colorado and other areas of the region under consideration there seems to be no evidence of double-brooding.

Suggested Reading: Bent 1949; Bailey and Niedrach 1965.

Swainson Thrush
Catharus ustulatus

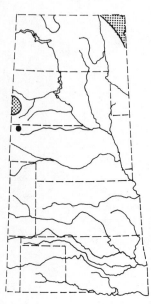

Breeding Status: Breeds in north-central Minnesota (Clearwater and Kittson counties) and is a hypothetical breeder in northeastern North Dakota. It is a common breeding species in the Black Hills of South Dakota and probably breeds regularly in adjacent northwestern Nebraska (one breeding record, Dawes County, *Nebraska Bird Review* 42:17).

Breeding Habitat: The species breeds in coniferous and occasionally in mixed forests in both South Dakota and Minnesota. In the Black Hills it is found in the higher, cooler spruce forests rather than in pines. Heavily shaded coniferous forests with brooks or springs and with a relatively open undergrowth allowing foraging on the forest floor are the favored areas.

Nest Location: Nests are usually in small trees, typically conifers, 2–20 feet above the ground, on horizontal branches near the main trunk. The nest is constructed of twigs, bark, leaves, mosses, and the like and lined with such things as lichens, fur, and fine strips of bark.

Clutch Size and Incubation Period: From 3 to 5 eggs, usually 3 or 4. The eggs are pale blue, with brown spotting around the larger end. The incubation period is 12–13 days. Probably single-brooded.

Time of Breeding: Minnesota egg records are from June 18 to July 2, and nestlings have been seen as early as June 24. The nesting season in the Black Hills is probably during June; nests or females ready to lay have been noted in early June, and dependent young have been observed in mid-July.

Breeding Biology: This species is the most arboreal of the North American *Catharus* thrushes, and the birds spend much of their time foraging in the foliage and catching flies. Territorial males sing a distinctive song that consists of almost continuous melodic phases that seem to spiral upward. Nests are built over a period of about 4 days, and eggs are laid daily. Apparently only the female incubates, and only she broods the young. They are fed by both parents and leave the nest 10–12 days after hatching. Unlike hermit thrushes, males of this species frequently begin to sing while still migrating, and on a territory they may average nearly 10 songs a minute, or more than 4,000 songs a day, from about 3:15 A.M. to 7:30 P.M.

Suggested Reading: Bent 1949; Dilger 1956.

Veery
Catharus fuscescens

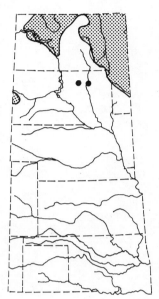

Breeding Status: Breeds over nearly all the wooded portions of northern and western Minnesota westward through comparable habitats of North Dakota to the Missouri and Little Missouri valleys. It is also a breeder in the Black Hills (especially Spearfish Canyon), as well as locally in northeastern South Dakota (Faulk and Spink counties).

Breeding Habitat: Moist deciduous or coniferous forests, river-bottom forests, moist wooded canyons or ravines, and tamarack swamps are favored. Sites with wet ground, nearby running water, and well-developed understories are especially preferred.

Nest Location: Nests are on the ground or in a low shrub, on a stump, in tangled vines, and in similar sites. Often the nest is sunk into the top of a mossy hummock, and it consists of twigs, bark, grasses, and moss, well concealed by surrounding vegetation. It is lined with grasses and rootlets; no mud is used.

Clutch Size and Incubation Period: From 3 to 5 eggs, usually 4 (4 North Dakota clutches averaged 4.2). The eggs are pale blue, usually unspotted. The incubation period is 11–12 days. Single-brooded.

Time of Breeding: North Dakota egg dates are from June 6 to June 15, but the probable breeding season extends from late May to late July. Minnesota egg dates are from May 24 to July 10, and nestlings have been reported as early as June 11.

Breeding Biology: Like the other North American *Hylocichla* and *Catharus* thrushes, males arrive in breeding areas before females and establish territories that they advertise by singing. When females arrive they intrude on these territories, and they are initially chased by the resident males, presumably because there are no plumage differences between the sexes in these species. When females are chased, they tend to remain in the male's territory, flying in circles. Ultimately the male accepts the female's presence and a pair bond is formed. Nest-building requires 6–10 days, and a clutch is begun soon thereafter. Although the male is strongly defensive of the nest, only the female incubates. The fledging period is 11–12 days, but the adults feed the young for some time thereafter. Veeries are frequently parasitized by cowbirds in spite of their well-concealed nests, and they typically accept the cowbird eggs.

Suggested Reading: Dilger 1956; Day 1953.

Eastern Bluebird
Sialia sialis

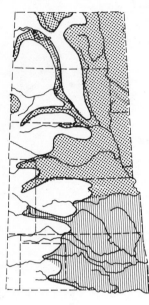

Breeding Status: Breeds throughout Minnesota, in North and South Dakota (west locally to their western borders), in eastern Nebraska (west locally to Dawes and Deuel counties), in eastern Colorado along the major river valleys, in eastern Kansas (rare and local west of Comanche County), in Oklahoma west to the panhandle, and in the eastern panhandle of Texas.

Breeding Habitat: The species frequents open deciduous woods especially where interspersed with or adjacent to grasslands. Upland and floodplain forest edges, city parks and gardens, shelterbelts, and farmsteads are all commonly used.

Nest Location: Nests are typically in old woodpecker holes or natural cavities of dead trees, dead limbs, or utility poles. In many areas birdhouses are used, especially where natural cavities are lacking. Nest boxes that are placed in open areas, 8 to 12 feet high, with entrances no larger than 1½ inches in diameter, are preferred by bluebirds. There should be a suitable tree perch with a view of the nest entrance, and if possible nests should face east or south, to avoid exposure to the westerly spring rains common to the Great Plains. The cavity is filled with weed stalks and grasses to form a loose cup, sometimes with a lining of finer grasses or a few feathers.

Clutch Size and Incubation Period: From 3 to 6 pale bluish eggs (276 clutches from Iowa, Minnesota, Nebraska, and South Dakota averaged 4.5). The incubation period is 12–18 days, which is surprisingly variable. Double-brooded in much of its range, with single-brooding usual in Canada and triple-brooding present in some areas.

Time of Breeding: Egg dates in North Dakota are from June 3 to July 22, and in Kansas they extend from April 1 to July 20, with first clutches peaking in late April and second clutches in early June. Oklahoma egg dates are March 27 to July 31.

Breeding Biology: Studies in Arkansas, where bluebirds are mostly permanent residents, indicate that wintering birds form pair bonds between November and the end of January, and courtship is closely associated with the visiting of nest boxes or other suitable nest cavities. Once pair bonds are established they seem to last throughout the year. A territory is established around a nesting site and is retained until the last brood fledges. Courtship feeding of the female by her mate is an important part of breeding activities; this starts before nest-building and continues into the nestling stage. Only females incubate, but males sometimes enter the nest box briefly when their mates are absent. The nestling period is usually 17–18 days, and at least in Arkansas

346

there are commonly three nesting attempts, sometimes as many as four, and up to three broods are reared.

Suggested Reading: Thomas 1946; Hartshorn 1962.

Mountain Bluebird
Sialia currucoides

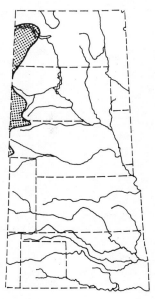

Breeding Status: Breeds in western North Dakota (east locally to the Souris River and Turtle Mountains), in the western third of South Dakota, (primarily the Black Hills), the western third of Nebraska (particularly the Pine Ridge area of Sioux to Sheridan counties and the Wildcat Hills of Banner County), and presumably eastern Colorado (no specific nest records for the region under consideration). There is one recent nesting record for Harmon County, Oklahoma, and some early records for Cimarron County. The only evidence for possible breeding in Kansas is a full-grown juvenile collected in 1911 in Hamilton County. There are no definite breeding records for the Texas panhandle or adjacent northeastern New Mexico.

Breeding Habitat: The species favors open woodlands, especially open stands of pine, pine forest edges, burned or cutover areas, and aspen clumps in open country.

Nest Location: Nests are in old woodpecker holes or other natural tree cavities, especially in aspens. The birds use cavities in both live and dead trees, as well as nest boxes and sometimes also cliff crevices. The openings are usually 4–10 feet above the ground, and the nest is built of whatever vegetation is locally available. Openings in nest boxes should be 1¾ inches across, since this species is slightly larger than the eastern bluebird.

Clutch Size and Incubation Period: From 4 to 8 pale blue or bluish white eggs. The incubation period is 13 days. Frequently double-brooded.

Time of Breeding: In North Dakota the probable breeding season is from early May to early August, with eggs reported as late as June 24. Nesting in South Dakota extends from early May through July, with nest-building reported as early as April 4 and nestlings seen from May 27 to August 5.

Breeding Biology: Bluebirds arrive relatively early in the central and northern plains and immediately begin searching for nesting sites. Paired birds often displace unmated males, which defend their territories only weakly, and when a nesting pair has established a territory they both defend it vigorously, the male defend-

347

ing the periphery and the female the actual nest site. One instance has been described of a male having two mates within his territory, nesting about 50 yards apart. Only the female builds the nest, which requires 4–6 days, and only the female incubates. Females brood their nestlings for about 6 days after hatching, and both parents actively feed the young. They fledge in 22–23 days, after which the female usually begins a second clutch and the male remains with the fledglings for about 10 days. Rarely, young of the first brood of mountain bluebirds have been observed feeding the second brood.

Suggested Reading: Power 1966; Haecker 1948.

Townsend Solitaire
Myadestes townsendi

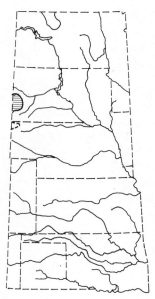

Breeding Status: Breeds commonly in the Black Hills of South Dakota and at least once bred in the pine forests of northwestern Nebraska (Sioux County). No breeding records for elsewhere in the region.

Breeding Habitat: In the Black Hills the species frequents pine and spruce forests and is generally found in rather dense montane forests, especially among conifers.

Nest Location: Nests are on the ground or close to it (rarely as high as 10 feet). The nest is usually well protected from above by overhanging vegetation or a rock and sunken into the earth or surrounded by roots that form a natural cavity. The nest is made of available materials and varies greatly in size to fit the cavity occupied; rarely, no actual nest is constructed, and the eggs rest on bare ground.

Clutch Size and Incubation Period: From 3 to 5 eggs, usually 4. The eggs are white with spots and blotches of varied sizes and colors. The incubation period is unreported.

Time of Breeding: In the Black Hills the birds probably nest from early May through June and probably July. An egg-laying female was collected in mid-May, and recently hatched young have been seen as early as June 2. In Colorado eggs have been found from the third week in May to the first of July, at elevations of 6,200 to about 12,000 feet.

Breeding Biology: This little-studied montane thrush in some respects acts more like a flycatcher. In the winter the birds move from the mountains to lower elevations. Thus, in California and Arizona it has been reported that the birds establish winter territories that they hold from late September until April and that are associated with the distribution of juniper berries, their main winter food in those areas. Although they compete with jays and robins for these berries, they do not attempt to defend their territories against these larger birds, though they do attack juncos, bluebirds, and nuthatches. During the breeding season, the remarkably beautiful song of the territorial male is the species' most attractive feature, and it sings not only while perched but also while hovering high in the air, after which it makes a spectacular plunging flight back toward earth. Remarkably little is known of nest-building or incubation; neither the relative roles of the sexes nor the incubation and fledging periods have been determined. Both sexes help care for the young, and the range of available egg dates suggests that two broods may be raised in a season, but this is not yet established.

Suggested Reading: Bent 1949; Lederer 1977.

FAMILY SYLVIIDAE (GNAT-CATCHERS AND KINGLETS)

Blue-gray Gnatcatcher

Blue-gray Gnatcatcher
Polioptila caerulea

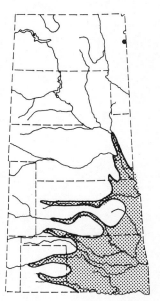

Breeding Status: Breeds in the Missouri Valley of eastern Nebraska and western Iowa southward through northwestern Missouri, eastern Kansas (west generally to Riley and Cowley counties, and locally to Wallace, Finney, and Grant counties), most of Oklahoma (one 1936 breeding record in Cimarron County is the only one for the panhandle), and the Texas panhandle (Randall and Armstrong counties). It has been reported in summer from Quay County, New Mexico, but there are no breeding records yet. There is a single breeding record (1955) for Itasca State Park, in Minnesota (*Loon* 49:93).

Breeding Habitat: Mixed and deciduous (especially oak) forests and pine woodlands are used in eastern areas, and deciduous brush, piñon pines, and oak-juniper woodlands are used at the western edge of its range.

Nest Location: Nests are 5–70 feet above the ground, but usually under 25 feet, and are often on dead horizontal branches or forks of deciduous trees. The small nest is beautifully constructed of plant fibers and down, oak catkins, and such, held together with spider webs and insect silk. It is lined with soft materials, and the outside is covered with lichens and plant down.

Clutch Size and Incubation Period: From 4 to 5 eggs, white to bluish with small reddish brown spots. The incubation period is 13–15 days. Often double-brooded.

Time of Breeding: Egg dates in Kansas are from April 20 to June 20, with a peak of egg-laying about May 10. In Oklahoma, nest-building has been seen as early as April 4 and nestlings have been seen as late as June 8. Texas egg dates are from April 4 to July 3.

Breeding Biology: Shortly after they arrive on their breeding grounds, male gnatcatchers initiate territories; the time depends on the abundance of foliage-dwelling arthopods in the locality. All the breeding activities occur within the territory, which is defended by the male and sometimes also by the female. Pair bonds may be established almost immediately after territoriality begins or may occur later. When a female appears on the territory of an unmated male he accompanies her to various nest sites, frequently perching in an upright posture and singing an elaborate but whispered song sequence. Both members of the pair build the nest, and they frequently obtain materials by dismantling old nests. When only new materials are used, they need about 2 weeks to complete a nest, but when materials are already available they may finish in 3–6 days. Both sexes incubate about equally, and both sexes brood the young, which remain in the nest for 12–13 days. They are at least occasionally fed by their parents for as

353

long as 19 days after leaving the nest. In one California study, 3 of 12 pairs raised one brood successfully, and 4 managed to raise two broods. The length of the breeding season there suggests that three broods may sometimes be raised in favorable years.

Suggested Reading: Root 1969; Fehon 1955.

Golden-crowned Kinglet
Regulus satrapa

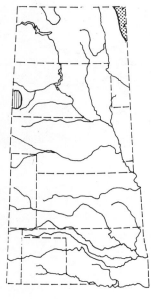

Breeding Status: Confined as a breeder to the Black Hills of South Dakota, where it is uncommon, and to the coniferous forests of north-central and northwestern Minnesota (Clearwater County; also occasionally present in summer at Agassiz N.W.R., Marshall County, and Tamarac N.W.R., Becker County).

Breeding Habitat: In the Black Hills the species is associated with spruce and occasionally pine forests, and it generally occurs near conifers throughout its range.

Nest Location: Nests are 6–60 feet above the ground in conifers, usually spruces. The semihanging nests are placed close to the trunks of large trees, attached either to twigs or to a horizontal branch. The nest is a nearly spherical mass of mosses and lichens, with a deep cup lined with rootlets, feathers, or strips of bark.

Clutch Size and Incubation Period: From 5 to 10 eggs, often 8–9, white to cream-colored with variable brown spotting. The incubation period is unreported but presumably is 12–16 days. Probably double-brooded.

Time of Breeding: In the Black Hills these birds probably breed in June and July; recently fledged young have been seen as early as July 8. In Minnesota, nest-building has been reported in May, and nestlings have been seen as early as June 18.

Breeding Biology: At least at the edges of their range, kinglets seem to be attracted to closed stands of conifers such as natural or planted groves of spruces that are spaced closely enough to produce a shaded undercover, providing a cool, moist environment for foraging. Most nests seem to be placed rather high in the tree, and they are constructed over a period of about a month, during which the female does all the gathering of materials and actual construction, while the male accompanies her. In one study where the nests of two pairs were destroyed after their clutches had been completed, each pair built three nests and laid a total of 26 eggs. Neither the incubation nor the nestling period has been reported for this species, but in a closely related European species, the goldcrest, the incubation period is 12–16 days and is performed by the female alone. Thereafter, both parents feed the young in the nest for another 15–17 days and are tended by the parents for another 2 weeks after fledging. At least in that species, two broods are regular.

Suggested Reading: Bent 1949.

Ruby-crowned Kinglet
Regulus calendula

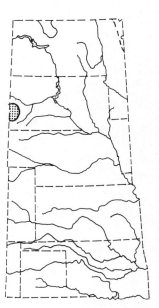

Breeding Status: Known to breed in our region only in the Black Hills, where it is a rare nester at higher elevations. However, it has been observed in summer at Itasca State Park and eastern Marshall County, Minnesota and probably also nests in that general area.

Breeding Habitat: The species is associated with spruce forests in the Black Hills and is generally limited to cool coniferous forests.

Nest Location: Nests are 2–100 (usually 6–30) feet above the ground in conifers, usually spruces, and are attached to hanging twigs below larger horizontal branches. The nest is built in thick foliage and is a deep, hanging cup of mosses, lichens, and other plant materials, lined with fur or feathers.

Clutch Size and Incubation Period: From 5 to 11 eggs, generally 7–9. The eggs are white with tiny brown spots. The incubation period has been estimated to be 12–15 days. Probably single-brooded.

Time of Breeding: Although there are no specific nesting records, the probable breeding season in the Black Hills is during June and July. Egg records for Colorado are as early as June 9, and young have been seen as early as June 21.

Breeding Biology: Apparently no modern studies have been done on either the golden-crowned or ruby-crowned kinglet. It is known that exhibition of their brilliant crown patches is an important part of aggressive and courtship display behavior. It is believed that only the female incubates, and that the young remain in the nest for about 12 days. Observations in Colorado suggest that the female broods the young and passes on to the young food that is brought in by the male.

Suggested Reading: Bailey and Niedrach 1965; Bent 1949.

FAMILY MOTACILLIDAE
(WAGTAILS AND PIPITS)

Sprague Pipit

Sprague Pipit
Anthus spragueii

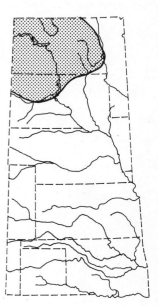

Breeding Status: Breeds primarily in North Dakota (except for the extreme northeast and southeast). Also breeds in adjacent Minnesota (all recent records from Clay County, earlier ones for Kittson, Pennington, and Marshall counties), and in northern South Dakota (south to Corson and possibly Dewey counties).

Breeding Habitat: The Sprague pipit is a prairie species associated with extensive areas of grassland dominated by grasses of medium height. The birds also breed in large alkaline meadows and locally in the meadow zones of larger alkali lakes.

Nest Location: Nests are in hollows on the ground, in clumps of grasses or grasslike plants. The nest is constructed entirely of grasses, circularly arranged, without any additional lining.

Clutch Size and Incubation Period: From 4 to 6 eggs (4 North Dakota clutches averaged 4.5). The eggs are grayish white with blotches of purplish brown. The incubation period is unreported, but in related species it is 13–14 days. Probably double-brooded.

Time of Breeding: North Dakota egg dates are from June 7 to June 30, but nestlings have been seen as late as August 2, suggesting that renesting or double-brooding may occur.

Breeding Biology: Like several other prairie-adapted species, this bird has inconspicuous plumage but a beautiful and spectacular song-flight display, in which it sings a series of two-note phrases as it rises in the air and is silent as it descends. Thus the song is heard periodically as the bird moves in a circular pattern over its territory. It is probable that the female does all the incubating, since she does all the brooding. It has been suggested, however, that the male takes charge of the first brood after they have left the nest at about 12 days of age, so that the female is able to begin a second clutch almost immediately. After the breeding season is over the young and adults gather in large flocks, often with horned larks and longspurs, and gradually begin to move southward.

Suggested Reading: Roberts 1932; Bent 1950.

FAMILY BOMBYCILLIDAE
(WAXWINGS)

Cedar Waxwing

Cedar Waxwing
Bombycilla cedrorum

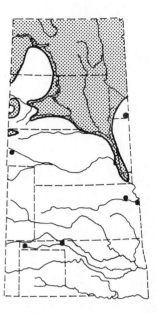

Breeding Status: Breeding occurs rather commonly throughout Minnesota and North Dakota except in southwestern North Dakota. In South Dakota the species breeds locally in the Black Hills and more generally in the eastern half of the state. Breeding in Nebraska is apparently confined to the Missouri Valley, the Pine Ridge area, and perhaps rarely occurs elsewhere, such as the Bessey Division of Nebraska National Forest and near Scottsbluff (*Nebraska Bird Review* 39:16, 45:3). It is reported to be an uncommon breeder at Squaw Creek N.W.R., Missouri. There are a few breeding records for northeastern Kansas and some old nesting records for northwestern Oklahoma.

Breeding Habitat: The species is generally associated in North Dakota with semiopen deciduous woodlands, including floodplain forests, wooded hillsides, and sometimes farmsteads, parks, or residential areas. It sometimes also breeds in stands of cedar in western North Dakota and in other conifers elsewhere.

Nest Location: Nests are placed 6-20 feet above the ground in small trees and usually are elevated about 10 feet. The nest is placed on a horizontal limb and is loosely constructed of grasses, weeds, twigs, and fibrous materials, lined with softer and finer materials.

Clutch Size and Incubation Period: From 2 to 6 eggs, usually 4-5. The eggs are pale grayish with scattered brownish spots. The incubation is 11-13 days. Sometimes double-brooded.

Time of Breeding: North Dakota egg dates are from June 24 to August 2, and dependent fledglings have been seen as early as June 30. In Kansas, eggs have been found in mid-June and nestlings reported as late as July 22.

Breeding Biology: Not much is known of the courtship behavior of this rather common and highly gregarious species, which remains in flocks for much of the year. Adult birds often may be seen passing berrries back and forth, but whether this is courtship feeding is questionable. Mutual breast-preening and bill-clicking are probable courtship activities. During the period of nest-building the female does perform begging behavior and is fed by her mate. Territoriality is virtually absent in cedar waxwings. The nests are frequently situated colonially, and breeding seems to correspond with the period when berries and fruit ripen. Both sexes build the nest, which requires 2-6 days. The female does the incubating, but she is frequently fed by her mate, and she also broods the young for several days after hatching. The young birds leave the nest when about 16 days old and may remain in the nest vicinity for about a month.

Suggested Reading: Lea 1942; Bent 1950.

FAMILY LANIIDAE
(SHRIKES)

Loggerhead Shrike

Loggerhead Shrike
Lanius ludovicianus

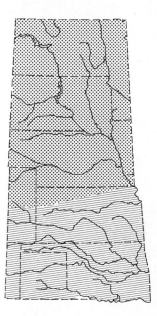

Breeding Status: Breeds pandemically throughout the region in suitable habitats.

Breeding Habitat: The species is associated with open-country habitats that include scattered or clustered shrubs or small trees, such as shelterbelts, cemeteries, farmsteads, hedgerows, and the like. It is uncommon to rare in heavily wooded areas such as the Black Hills and northern Minnesota.

Nest Location: Nests are in woody vines, shrubs, and small trees, usually 4–20 feet above the ground in dense foliage. Most nests are less than 10 feet high, in isolated trees or in small thickets. They are bulky, constructed mostly of sticks and twigs, with a soft lining of rootlets, feathers, cottony materials, and the like. Thorny trees seem to be especially favored sites, probably because they provide protection from predation.

Clutch Size and Incubation Period: From 4 to 7 eggs (9 North Dakota clutches averaged 5.4 and 32 Kansas clutches averaged 5.3). The eggs are white to grayish with gray to brown spots and blotches. The incubation period is 16 days. Reportedly double-brooded, yet recent Colorado studies suggest that renesting but not double-brooding is typical in that area.

Time of Breeding: North Dakota egg dates are from April 18 to June 19, and Kansas dates extend from April 1 to June 30, with a peak in mid-April. In Oklahoma, egg dates are from March 20 to about June 28, with nestlings seen as late as July 13.

Breeding Biology: On a study area in north-central Colorado, shrikes arrived in early April and had established territories by the first of May. Both sexes help build the nest. Nests in Colorado were separated by at least 400 meters, and nest sites of previous years were often used. The female incubates, but her mate feeds her and both sexes bring food to the young. The fledgling period is normally 17 days but may require up to 20 days. The young birds gradually learn how to wedge or impale food items in forks or on thorns or other sharp objects; the development of impaling seems to include both learned and innate components. Impaling is generally recognized as a means of short-term food storage and is thus most likely to occur in shrikes that are not hungry.

Suggested Reading: Wemmer 1969; Porter et al., 1975.

1. Common loon, adult, upright aggressive display.

2. Eared grebe, adult incubating (photo courtesy of Nebraska Game and Parks Commission).

3. Black-crowned night heron, adult.

4. Snowy egret, adult foraging.

5. Trumpeter swan, adults and cygnets.

6. Goshawk, adult.

7. Prairie falcon, adult (photo courtesy of Nebraska Game and Parks Commission).

8. Swainson hawk, adult.

9. Greater prairie chicken, male booming before female.

10. Sharp-tailed grouse, adult males in territorial confrontation.

11. Greater sandhill crane, adult incubating.

12. American coot, adult feeding young.

13. Mountain plover, adult (photo courtesy of Ed Schulenberg).

14. Upland sandpiper, adult.

15. Long-billed curlew (photo courtesy of Nebraska Game and Parks Commission).

16. American avocet, adults foraging.

17. Roadrunner, adult.

18. Burrowing owl, adult (photo courtesy of Ed Schulenberg).

19. Short-eared owl, adult.

20. Red-headed woodpecker, adult.

21. Rough-winged swallow, adult.

22. Long-billed marsh wren, male singing.

23. Mountain bluebird, adult male.

24. Yellow-headed blackbird, adult male.

25. Western meadowlark (photo courtesy of Nebraska Game and Parks Commission).

26. Western tanager, adult male.

27. Le Conte sparrow, adult (photo courtesy of Phil and Judy Sublett).

28. Vesper sparrow, adult (photo courtesy of Phil and Judy Sublett).

29. Clay-colored sparrow, adult (photo courtesy of Phil and Judy Sublett).

30. Field sparrow, adult (photo courtesy of Phil and Judy Sublett).

FAMILY STURNIDAE
(STARLINGS)

Starling

Starling
Sturnus vulgaris

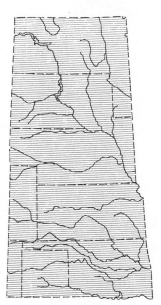

Breeding Status: Introduced in the eastern United States; now breeds pandemically throughout the region as a result of range expansion.

Breeding Habitat: The starling is generally associated with human habitations such as cities, suburbs and farms or with wooded areas having suitable nest sites.

Nest Location: Nests are in natural or artificial cavities, including birdhouses, old woodpecker holes, natural tree cavities, or other sites. Birdhouses with entrances less than 1½ inches in diameter effectively exclude starlings, but they can easily take over martin houses.

Clutch Size and Incubation Period: From 4 to 7 pale bluish to greenish eggs (19 Kansas clutches averaged 5.2). The incubation period is 11–13 days. Usually double-brooded.

Time of Breeding: In North Dakota, breeding extends from mid-April to early July, with a peak from late April to mid-June. Kansas egg records are from March 1 to June 30, with first and second clutches usually laid in mid-April and in early June. Oklahoma breeding dates are from March 16 (nest-building) to June 2 (nearly fledged brood).

Breeding Biology: Although starlings are migratory, adults tend to return in the spring to the areas where they previously bred; some females are known to have nested in the same site for 3–4 consecutive years. Territorial defense in starlings is limited to the nest hole itself and a few inches immediately around it. Pair-formation is achieved when the male locates a suitable nest site and defends it from intrusion by other birds while simultaneously singing and attempting to attract females to it. Males are normally monogamous, although a few instances of polygyny have been noted. The male sometimes helps to incubate both clutches but usually aids in the brooding and feeding of only one of the broods. Normally, both members of a pair participate in nest-building and incubation, and usually both parents assist in brooding and feeding the young. However, if either of the parents should die or desert the brood, the remaining member of the pair has been known to successfully raise the family alone. The young usually leave the nest when 21 days old and rarely remain until their 25th day. Almost as soon as the first brood is fledged a second clutch is begun; in one case the first egg was laid only a day after the first brood was fledged.

Suggested Reading: Kessel 1957; Bent 1950.

FAMILY VIREONIDAE
(VIREOS)

Red-eyed Vireo

Black-capped Vireo
Vireo atricapilla

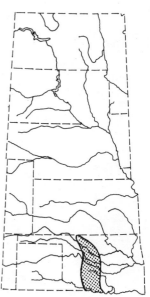

Breeding Status: Currently limited as a breeder to central Oklahoma (mainly Caddo, Blaine, and Dewey counties, less often seen in Cleveland, Major, and Comanche counties). Previously the range extended appreciably farther north into central Kansas and it was also sighted rarely in southeastern Nebraska, but it apparently is now extirpated from Kansas.

Breeding Habitat: Scrubby and clumped tree growth such as is found in prairie ravines, hilltops, or slopes, or overgrazed grasslands seems to represent this species' primary habitat. In Oklahoma it uses clumps of oaks, junipers, and similar trees in rather sloping topography, but not streamside or bottomland thickets. Thus it avoids contact with both the Bell and white-eyed vireos.

Nest Location: Nests are in rather small trees, often scrubby oaks (51 of 70 Oklahoma nests were in black-jack oaks and averaged only about 3 feet above ground). Nearly all are above sloping ground, and they often are near the base of a cliff or steep bank. The nest is typically vireolike, a deep and suspended cuplike structure.

Clutch Size and Incubation Period: From 3 to 5 white eggs, usually 4. The incubation period is 14–17 days. Sometimes double-brooded.

Time of Nesting: The nesting season in Oklahoma extends from late April (uncompleted nest seen April 26) to late July (eggs July 23, nestlings as late as August 2).

Breeding Biology: Males arrive in breeding areas about a week before females and immediately establish territories. As soon as the females arrive they form pair bonds, apparently choosing the male and territory as a unit. Remating with males of the previous year is known, and one pair probably remated three years. The pair looks for nest sites together, and unmated males sometimes even begin nests, but most building is done by the female. Nests usually take 4–5 days to complete, and incubation begins with the laying of the second or third egg. Both sexes incubate, although the male lacks a brood patch. Although the young may leave the nest as early as 10 days after hatching, they usually leave at 11–12 days. Fledglings are cared for 4–7 days, after which the female may begin a second nesting. She may either remate with a second male, leaving her first mate to care for the original brood, or undertake the incubation and brooding of the second brood alone.

Suggested Reading: Graber 1961; Bent 1950.

White-eyed Vireo
Vireo griseus

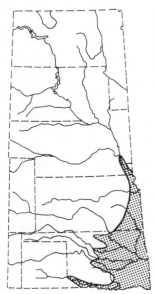

Breeding Status: Breeding occurs in the Missouri Valley as far north as Sarpy County, Nebraska, and extends southward through eastern Kansas (specific nesting records limited to Doniphan County and Kansas City region) and the eastern half of Oklahoma (principally in the eastern third, but reported west to Alfalfa, Payne, Oklahoma, Caddo, and Love counties).

Breeding Habitat: This species is associated with the dense understory of bottomland woodlands and favors thickets near water. It does not extend out into prairie areas along small streams like the Bell vireo, nor does it occur in scattered hillside tree clumps like the black-capped vireo.

Nest Location: Nests are in shrubs or low trees, up to 8 feet above the ground, but averaging only about 3 feet. Like other vireo nests, they are suspended from horizontal forks and are held together with cobwebs. They are less cuplike than those of the red-eyed or black-capped vireo, and are more conelike in shape.

Clutch Size and Incubation Period: From 3 to 5 eggs, usually 4. The eggs are white with a few scattered blackish spots. The incubation period is 14–15 days. Single-brooded.

Time of Nesting: Oklahoma breeding dates from mid-April (several nests April 13–17) to July 1 (1 vireo egg and 2 cowbird eggs in nest). Texas egg records are from April to July 4.

Breeding Biology: Although few detailed observations on this species are available, it is apparently much like the other vireos in its breeding biology. It is known that the male participates fully in the nesting, sharing both nest-building and incubation with his mate. He also defends the nest fearlessly, judging from an early account by A. A. Saunders, who stated that he sometimes had to lift one male off the nest by the bill in order to examine the eggs.

Suggested Reading: Bent 1950.

Bell Vireo
Vireo bellii

Breeding Status: Breeds from the Missouri Valley of south-central North Dakota (rarely) southward through central South Dakota (west rarely to the Black Hills and east to the Big Sioux River), Nebraska (west to Dawes County and the Colorado border), western Iowa and northwestern Missouri, eastern Colorado

376

(probably at least Logan and Yuma counties), all of Kansas, Oklahoma except for the western panhandle and southeastern areas, and the eastern panhandle of Texas, possibly extending rarely or locally to the New Mexico border.

Breeding Habitat: This vireo is generally associated with thickets near streams or rivers and with second-growth scrub, forest edges, and brush patches.

Nest Location: Nests are usually less than 5 feet high, in various shrubs or trees, typically suspended in terminal or lateral forks of small tree branches or in the periphery of shrub thickets. When nests are built in low shrubs, the site is always overhung with the foliage of a much taller shrub or a tree. They are constructed of a variety of plant materials, supplemented on the outside with silk from spider webs and cocoons and lined with fine grasses.

Clutch Size and Incubation Period: From 3 to 6 eggs (33 Kansas clutches averaged 3.4), white with scattered blackish spots. The incubation period is 14 days. Sometimes double-brooded and a persistent renester; frequently parasitized by cowbirds.

Time of Breeding: Kansas egg records extend from May 1 to July 20, with a major peak in the last part of May and a second smaller peak in mid-June, reflecting renesting efforts. Oklahoma egg dates are from May 4 to July 23.

Breeding Biology: On the basis of a Kansas study, it seems that male Bell vireos become territorial shortly after they arrive in spring, and they establish territories averaging about an acre. They rather quickly attract females, which they initially attack, then recognize as females by their submissive behavior. Nest-building is an important component of courtship in this species; the female probably selects the site, but the male constructs the initial suspension stages, interspersing the work with intensive courting behavior. About 4–5 days are spent in constructing the nest, and the first egg is laid a day or two later. The two sexes share incubation, although only the female has a brood patch and she does all the night time incubation. Brooding is likewise done by both sexes, and both sexes bring in food, although the female seems to take the major responsibility for this. The young birds remain in the nest 9–12 days, but feeding of fledglings has been reported as long as 35 days after fledging. At least in Kansas, double-brooding is frequent, and in four cases of pairs that succeeded in their first nesting all began nesting within a day or two after fledging their first brood.

Suggested Reading: Barlow 1962; Nolan 1960.

Gray Vireo
Vireo vicinior

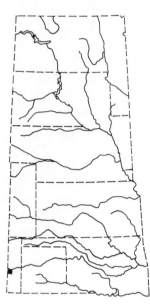

Breeding Status: Very rare or accidental breeder; with a breeding pair observed in Cimarron County, Oklahoma, in 1937 and a breeding colony found near Montoya, Quay County, New Mexico, in 1903. The species has also been collected during May in Prowers County, southeastern Colorado.

Breeding Habitat: The species is associated with wooded foothills and canyons, or dry chaparral, in the southwestern states. Dry chaparral, forming a continuous zone of twigs 1–5 feet above the ground, is the most favored habitat.

Nest Location: Nests are built in dense or thorny bushes or trees, usually 4–6 feet above the ground. Like other vireo nests they are built in forks, but often the fork is nearly upright rather than horizontal, and the nest lacks a lichen covering and instead may be decorated with sagebrush leaves.

Clutch Size and Incubation Period: From 3 to 5 eggs, white with sparse dark spotting near the larger end. The incubation period is unreported. Probably double-brooded.

Time of Breeding: In Oklahoma, nest-building was observed in late May, and two active nests were found in mid-June in northeastern New Mexico.

Breeding Biology: Unlike most other vireos, which tend to be adapted to arboreal foraging on foliage, this species is a dry-slope forager and thus extends well out into nonforested regions. Little has been written on the prenesting behavior of the gray vireo, and the relative roles of the sexes in nest-building, incubation, and brooding have not been established. It is likely that this species often raises two broods, since fully fledged young were found in Arizona on the same day that a pair was seen starting to build a nest, and the female was judged to have previously been brooding. In another case some newly fledged young were found under the care of a male parent, also suggesting that the female may have been occupied with a second nesting.

Suggested Reading: Johnson 1972; Bent 1950.

Yellow-throated Vireo
Vireo flavifrons

Breeding Status: Breeds from the eastern half of North Dakota (west to the Souris and Missouri rivers) and western Minnesota southward through eastern South Dakota, western Iowa, eastern Nebraska (west locally to Brown, Garfield, and Hall counties), northwestern Missouri, eastern Kansas (west to Shawnee and Woodson counties), and eastern Oklahoma (west probably to Tulsa and Cleveland counties).

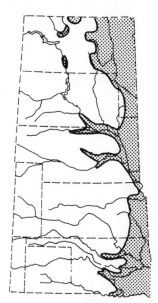

Breeding Habitat: This bird is associated with mature, moist deciduous forest, especially river-bottom forest or north-facing slopes; it is less frequently found in wooded residential areas. This species occupies a broader habitat range than red-eyed and warbling vireos and overlaps with both, but it favors more open woodlands than the red-eyed vireo and forages in the interior of trees, whereas warbling vireos forage on their periphery.

Nest Location: The nest is well above the ground (3–60 feet, usually above 30 feet), near the trunk of a deciduous tree, in the horizontal fork of a slender branch. It is suspended and typically vireolike, with spider silk and attached lichens and mosses, and has a deep, slightly incurved nest cup lined with fine grasses.

Clutch Size and Incubation Period: From 3 to 5 eggs, usually 4. The eggs are white, with brown spots around the larger end. The incubation period is 14 days. Probably double-brooded.

Time of Breeding: In North Dakota, active nests have been seen in June and early July, and dependent young as late as August 4. Oklahoma breeding dates extend from April 25 (nest ready for eggs) to July 22 (fledged young).

Breeding Biology: Studies on this species in Michigan suggest that males tend to establish territories in either of two general sites— either high in tall trees, totally above the ground, or low and including much ground area, with the two types so well separated that at times the territories of adjacent pairs overlap. Both sexes help gather nesting material, though the female does most of the actual building. About a week is required to complete the nest. Like other vireos, the male participates in incubation, frequently singing even while on the nest. He also brings food to the female while she is incubating. Both sexes feed the young, which fledge in about 15 days. Apparently the young birds remain together as a brood until they leave the area in the fall, but opinions differ on the frequency with which this species raises second broods.

Suggested Reading: Sutton 1949; James 1973.

Solitary Vireo
Vireo solitarius

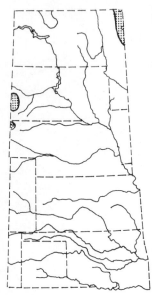

Breeding Status: A common breeder in the Black Hills of South Dakota, locally in Sioux County of extreme northwestern Nebraska, and in the wooded portions of north-central Minnesota (breeding records for Becker and Clearwater counties).

Breeding Habitat: In the Black Hills this species is found in pine forests and occasionally in mixed pine and aspen forests. In Minnesota it is also associated with conifers, especially spruce and tamarack swamps. More generally, it seems to prefer open mixed forests with considerable undergrowth.

Nest Location: Nests are usually 5–10 feet above the ground, rarely as high as 20 feet, in conifers or less frequently in hardwoods. Like those of other vireos, the nest is suspended in a horizontal fork and is generally basketlike. It is highly variable in materials but is often externally "decorated" with bits of lichen or "paper" from wasp nests.

Clutch Size and Incubation Period: From 3 to 5 eggs, usually 4. The eggs are white with scattered blackish spots. The incubation period is uncertain, but probably is 13–14 days. Probably single-brooded in our area, but possibly double-brooded farther south.

Time of Breeding: In the Black Hills, this species breeds from late May through July, with nestlings seen as early as June 16 and as late as July 2. In Minnesota, eggs have been found as early as May 25 and as late as July 7, and nestlings have been seen as late as July 17.

Breeding Biology: The territories of this species seem rather diffuse; the resident male frequently wanders about and does not confine his singing to a particular tree, and territories often overlap with those of red-eyed vireos. Both sexes are said to assist with nest-building; the male primarily brings materials to the female, and she shapes the nest. Both sexes also share in incubation, but the female probably does most. As in other vireos, the male sometimes sings while on the nest. The male continues singing until the young are 5–6 days old, then stops until the young have been out of the nest for about a week, when he may begin again, presumably signaling the start of another nesting cycle.

Suggested Reading: Barclay 1977; James 1973.

Red-eyed Vireo
Vireo olivaceus

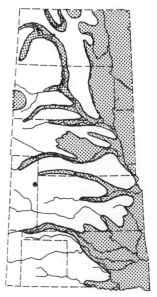

Breeding Status: Breeds in suitable habitats throughout Minnesota, North Dakota, South Dakota, western Iowa, Nebraska (west at least to Dawes, Thomas, and Deuel counties), eastern Kansas (west rarely to Decatur, Trego, and Morton counties), and eastern Oklahoma (west to Woods, Ellis, and Tillman counties; probably occurs in Cimarron County but no breeding record). There is a single state record of a territorial male collected in Union County, New Mexico (*Condor* 68:213).

Breeding Habitat: The species is associated with deciduous forests, especially those with semiopen canopies. It is found in both upland and river-bottom forests and sometimes also in older tree-claims.

Nest Location: Nests are usually 10–40 feet above the ground in large trees, in horizontal forks of small branches. Like other vireo nests they are suspended structures, with deep cups lined inside with fine grasses and covered outside with spider webs, wasp-nest "paper," and lichens.

Clutch Size and Incubation Period: From 3 to 5 eggs (5 Kansas clutches averaged 4.0). The eggs are white, with scattered blackish spots near the larger end. The incubation period is 12–14 days. Apparently single-brooded in our region; double-brooding is said to be occasional in eastern states.

Time of Breeding: In North Dakota, egg dates are from June 10 to June 18, and dependent fledglings have been seen from August 4 to August 27. Kansas egg dates are from May 21 to July 31, with a peak in the first week in June. Oklahoma egg dates range from May 18 to June 25, and nestlings have been seen as late as August 6.

Breeding Biology: Studies in Michigan suggest that these birds prefer to establish nesting territories in open woods where there are tall trees for singing perches, scattered clusters of hardwood saplings such as maples for nest concealment, a small clearing near the nest site, loose, fibrous bark for nest material, and abundant leaf-eating larvae for food. Apparently males do not assist with nest-building but only accompany females during that period and often attempt to mate with them. About 4–5 days are spent in nest-building, and the first egg is laid 1–3 days later. Although males have often been reported to assist in incubation, this was not observed in recent Michigan or Ontario studies, and the female also did all the brooding. Fledging occurs at about 12 days, and at least in Michigan there does not seem to be any double-brooding, although some singing may again occur at this time.

Suggested Reading: Southern 1958; Lawrence 1953*a*.

381

Philadelphia Vireo
Vireo philadelphicus

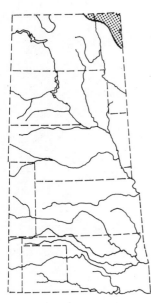

Breeding Status: Breeds locally in wooded areas of north-central Minnesota (Becker County is the only breeding record) and in northeastern North Dakota (Turtle Mountains and Pembina County).

Breeding Habitat: In North Dakota this species is restricted to mature stands of quaking aspens with relatively closed canopies. More generally it is associated with second-growth willow, poplar, or alder woods, deserted farmsteads, and rarely with village shade trees.

Nest Location: Nests are 10-40 feet high, in forks of horizontal twigs. They are deeply cupped and suspended, with an inner lining of fine grasses, and the outside is covered with spider webs and attached birch bark, *Usnea*, or other camouflaging materials.

Clutch Size and Incubation Period: From 3 to 5 eggs, usually 4. The eggs are white with sparse dark spotting on the larger end. The incubation period is 14 days.

Time of Breeding: No egg dates are available for our region, but a dependent fledgling has been observed in North Dakota in early August.

Breeding Biology: As soon as males arrive on their breeding areas they establish territories and begin to patrol them, sometimes forming pair bonds within a few days of arrival. As in most other vireos, the female soon chooses a nest site while the male attends her, and in this species the female does all the nest-building. However, male Philadelphia vireos do participate in incubation, and both sexes share in feeding the nestlings and fledglings. This species overlaps extensively with the red-eyed vireo, and there is extensive co-occupation of habitats in these two forms, as well as similarities in foraging ecology. It is thus not surprising that they exhibit interspecific territoriality, facilitated by the close similarities in their territorial songs. The young are brooded by both parents, and leave the nest at 13–14 days.

Suggested Reading: Barlow and Rice 1977; Bent 1950.

Warbling Vireo
Vireo gilvus

Breeding Status: Breeds in suitable habitats virtually throughout the two Dakotas and western Minnesota southward through most of Nebraska (west to Dawes, Thomas, and Deuel counties), eastern Colorado (at least to Logan County, *Nebraska Bird Review* 29:9), eastern Kansas (west locally at least to Decatur County and probably to the Colorado line), Oklahoma (west locally to Cimarron County), and the Texas panhandle (suspected but unproved nesting).

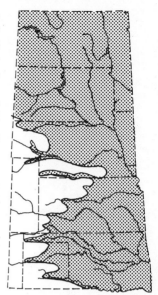

Breeding Habitat: This vireo is associated with open stands of deciduous trees, including streamside vegetation, groves, or scrubby hillside trees, and residential areas. Although warbling and red-eyed vireos occur in the same areas, this species favors more open woodlands and tends to be a treetop forager.

Nest Location: Nests are in deciduous trees at heights of 4–90 feet. They are often below 20 feet in our region, but generally tend to be rather high. They are on horizontal branches well away from the trunk, and frequently in aspens. Like other vireo nests they are suspended, deeply cupped, and covered outside with spider webs, lichens, and the like.

Clutch Size and Incubation Period: From 3 to 5 eggs, usually 4. The eggs are white with sparse dark spotting. The incubation period is 12 days. Single-brooded.

Time of Breeding: North Dakota egg dates are from June 9 to June 19, but the breeding period probably extends from late May to late July. Kansas egg dates are from May 1 to June 20, with a peak in early June. The recorded breeding season in Oklahoma is from late April (nest-building) to June 22 (eggs), but the nestling period probably extends into July.

Breeding Biology: In spite of its widespread occurrence, this species has not been extensively studied, perhaps because many of its breeding activities take place high in trees. An early observation by Audubon indicated that both sexes helped build the nest, which is unusual in vireos, and that 8 days were required to complete it. Both sexes also incubate, and males often sing while on the nest. The fledging period is said to be 16 days. Adult birds continue to sing well into the summer, and young males learn to sing fairly well before they leave for the South in fall. In this respect, this species resembles the red-eyed and yellow-throated vireos.

Suggested Reading: Sutton 1949; Dunham 1964.

FAMILY PARULIDAE
(WOOD WARBLERS)

Black-and-white Warbler

Black-and-white Warbler
Mniotilta varia

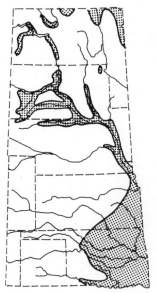

Breeding Status: Breeds in partially wooded areas of northwestern Minnesota (records for Roseau, Clearwater, and Becker counties), North Dakota (especially the Little Missouri Valley), South Dakota (probably including the Black Hills), Nebraska (west to the Pine Ridge area, commonest along the Niobrara and Missouri rivers), eastern Kansas (west to Doniphan, Coffey, and Sedgwick counties), and eastern Oklahoma (west at least to Alfalfa, Caddo, and Love counties). There are no breeding records for the Texas panhandle.

Breeding Habitat: The species is generally found in semiopen upland stands of deciduous or coniferous forest, especially those composed of immature or scrubby trees such as oaks and junipers and hillside or ravine groves with thin understories. It is also found in riverside forests adjacent to grasslands.

Nest Location: Nests are normally on the ground, at the base of a tree or stump, under a fallen tree branch or log, near a rock, or otherwise well protected and concealed from above. The nest is built of dried leaves and lined with various softer materials. Infrequently it is placed on top of a low stump, in a rotted cavity.

Clutch Size and Incubation Period: From 4 to 5 eggs, white with brown spots and blotches over most of the surface. The incubation period is 11–12 days. Single-brooded.

Time of Breeding: In North Dakota, dependent young have been seen in late July and August. In Kansas, eggs are probably laid in May and June, and in Oklahoma eggs are probably laid as early as late April, and dependent young have been seen as late as July 20.

Breeding Biology: This warbler is adapted for gleaning arthropods from the trunks of trees in creeper fashion. It is not dependent on foliage and thus tends to migrate well before most other warblers. Males are the earliest to arrive on their breeding grounds and soon begin to announce territories with their high-pitched seesawing song. The female incubates alone, and presumably she also does all the brooding. The fledging period is said to be 8–12 days.

Suggested Reading: Bent 1953; Smith 1934.

Prothonotary Warbler
Protonotaria citrea

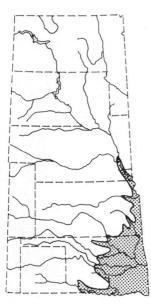

Breeding Status: A rare breeder in the Missouri Valley of southeastern Nebraska (extending north to Sarpy County) as well as adjacent southwestern Iowa and northwestern Missouri, and southward through eastern Kansas (breeding records for Doniphan, Douglas, Linn, and Cowley counties) and the eastern half of Oklahoma (probably west to Alfalfa, Canadian, Greer, and Tillman counties).

Breeding Habitat: The species is restricted to moist bottomland forests and wooded swamps or periodically flooded woodlands, in the vicinity of running water or pools.

Nest Location: Nests are in natural cavities or in old woodpecker holes, 5-20 feet high, often over water. Both tree stumps and trees are used, and there have been some records of nesting in birdhouses, gourds, and even tin cups. The cavity is lined with a variety of grasses, leaves, rootlets, and twigs, forming a deep hollow.

Clutch Size and Incubation Period: From 3 to 6 eggs (15 Kansas clutches averaged 4.5). The eggs are white with extensive brown spotting. The incubation period is 12-14 days. Frequently double-brooded, at least in Oklahoma.

Time of Breeding: Kansas egg records are from May 11 to July 10, with a peak in early June, and 75 percent of the records fall between June 1 and June 20. In Oklahoma, breeding records are from May 5 (nest-building) to early August (dependent young).

Breeding Biology: As soon as the males return to their breeding areas in spring, they become territorial and particularly try to stake out areas along riverbanks or other well-shaded water edges. The presence of suitable nest sites, preferably downy woodpecker holes only 5-6 feet above the water or ground, is critical, and males begin to carry moss into the cavities even before the females arrive. Apparently it is not typical for these birds to mate with the partner from the previous year. The female completes the nest and performs all the incubation. Fledging occurs after 10-11 days, and a second clutch may be begun in the same nest.

Suggested Reading: Walkinshaw 1953; Bent 1953.

Swainson Warbler
Limnothlypis swainsonii

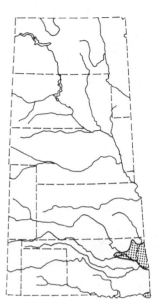

Breeding Status: Breeding in our region is limited to eastern Oklahoma, including the extreme southeast (McCurtain County), and the northeastern area of Arkansas River drainage (Delaware, Washington, Tulsa, and Payne counties).

Breeding Habitat: In Oklahoma the species is associated with moist, bottomland woods and woods with a brushy cane thicket or briar cover. It is generally most abundant in canebrake swamps, wooded ravines, or brushy thickets that provide a combination of deep shade, moderately dense undergrowth, and dry land within a general floodplain forest situation.

Nest Location: Nests are 2-12 feet high, in shrubs, vines, or small trees, often far from water and sometimes at the edge of or even outside the male's advertised territory. The nest is usually in rather thin cover but is always close to a denser thicket and is well concealed. It is a bulky structure of twigs, needles, and leaves, lined with fine grasses.

Clutch Size and Incubation Period: From 3 to 5 white eggs, usually 3. The incubation period is 14-15 days. Probably single-brooded.

Time of Breeding: In Oklahoma, eggs have been found as early as May 29 and as late as June 20, and fledged but dependent young have been seen in late June.

Breeding Biology: During the breeding season, these warblers can be found as unmated males, isolated pairs, or clustered breeding groups or "colonies." Typically, males establish territories immediately upon arrival and defend them with vigorous singing and chases. Singing is done from a variety of perches, but usually from below 30 feet. Pairs forage on the ground within these territories and begin nest-building 2-4 weeks after the males arrive. The nest is built by the female and takes 2-3 days to complete. After another 2 days, the first egg is laid. Apparently the male does not visit the nest during incubation but feeds the female some distance away. However, both sexes visit the nest when feeding the young, which fledge in 10-12 days. Singing by the males becomes sporadic after late June, but until that time they are persistent singers, often averaging 200 songs an hour.

Suggested Reading: Meanley 1969, 1971.

Worm-eating Warbler
Helmitheros vermivorus

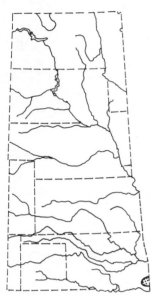

Breeding Status: A very rare breeder in our region, with few actual records. Periodically seen in southeastern Nebraska, but no breeding records exist for the state. In Kansas it is a possible resident in the east, but a territorial male reported in Doniphan County is apparently the only suggestion of breeding there. In Oklahoma the species has been seen in several eastern counties, but the only definite records are for 1961 and 1962, in McCurtain County, when a breeding pair and a pair with one young were seen.

Breeding Habitat: Wooded hillsides covered with medium-sized deciduous trees and an undergrowth of small shrubs and saplings are the preferred habitat of this species, especially when there are streams or swampy areas nearby.

Nest Location: The nest is in a hollow on the ground, usually on steep or sloping hillsides. It is constructed of partly skeletonized leaves and lined with mosses, and is often placed under a canopy of dead leaves at the base of a tree or under bushes where it is well concealed.

Clutch Size and Incubation Period: From 3 to 6 eggs, usually 4–5. The eggs are white with dark speckles. The incubation period is 13 days. Probably single-brooded.

Breeding Biology: Rather little has been written on the breeding biology of this woods-adapted species, partly owing to the heavy cover it frequents and perhaps also because the male's song is easily mistaken for that of the chipping sparrow. The female apparently incubates and broods the young alone, but she is fed on the nest by her mate. In one case, the young left the nest after 10 days.

Suggested Reading: Bent 1953.

Golden-winged Warbler
Vermivora chrysoptera

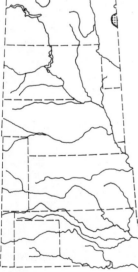

Breeding Status: Limited as a breeder in our region to north-central Minnesota (breeding records for Becker and Clearwater counties), north at least to Itasca State Park.

Breeding Habitat: The species is associated with openings in deciduous forest or forest edges, where there is a dense understory of ferns and other moisture-loving plants. Also found in hillside thickets, overgrown pastures, and brushy fields. At Itasca Park the birds occupy both upland forest such as aspens and lowland areas that provide edge habitat or openings in cover.

Nest Location: Nests are on the ground or close to it, usually at the forest edge but within its shade, and may be in grass clumps, at the bases of trees, or on a substrate of dead leaves. The cup of the nest is lined with grasses, bark strips, and sometimes hair.

Clutch Size and Incubation Period: From 3 to 6 eggs, usually 4–5. The eggs are white, with brown spotting on the larger end. The incubation period is 10–11 days. Probably single-brooded.

Time of Breeding: Minnesota egg records are from May 30 to June 24, and fledged but dependent young have been seen from early July to early August.

Breeding Biology: The blue-winged and golden-winged warblers are closely related species that have come into contact in recent times and that hybridize locally, as well as competing strongly with one another. In these species and their hybrids males with similar plumage and songs or with similar plumage but dissimilar songs have nonoverlapping territories. Those with dissimilar plumage and songs or with similar songs but dissimilar plumage have overlapping territories. Both species do have very similar courtship displays and form pair bonds very rapidly, which helps to explain the occasional hybridization in areas of overlap. In the golden-winged warbler the female incubates alone, with the male starting to feed her some days before the eggs hatch and continuing for a few days thereafter in one observed instance. The young leave the nest when about 8 days old, and a few days later they are able to fly fairly well.

Suggested Reading: Eyer 1963; Ficken and Ficken 1968*a*.

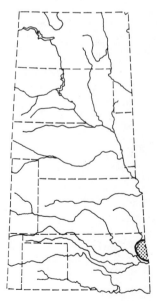

Blue-winged Warbler
Vermivora pinus

Breeding Status: Restricted as a breeder in our region to eastern Oklahoma (breeding record only from Delaware County; probably also breeds in Adair County).

Breeding Habitat: The primary habitat in the Great Plains consists of woodland edges, overgrown pastures, and thickets. The species is also more generally associated with rank growth near the borders of swamps or streams but is usually not found in heavy forests. This species uses slightly moister habitats than the golden-winged warbler, although in some areas the species overlap and hybridize.

Nest Location: Nests are on the ground, usually attached to the stems of weeds or grasses. The nests closely resemble those of the golden-winged warbler, but in this species the eggs are slightly less heavily marked.

Clutch Size and Incubation Period: From 3 to 6 eggs, usually 4–5. The eggs are white, with limited brown spotting toward the larger end. The incubation period is 10–11 days. Probably single-brooded.

Time of Breeding: No egg dates are available for our region, but nestlings in Oklahoma have been seen during late May and dependent fledglings in mid-June.

Breeding Biology: Adaptations for foraging in this species and in other species of the genus include a very straight and sharply pointed bill tip that allows for efficient probing of buds and leaf clusters and for feeding in flowers. The more typical warbler method of gleaning food from flat surfaces is used less frequently. The breeding biology of *Vermivora* warblers fits the general pattern for the family, with the female doing most or all of the nest-building and also all of the incubating. Both sexes feed the young birds, and the female does the brooding at night. The nestling period is normally about 10 days, rarely as short as 8.

Suggested Reading: Ficken and Ficken 1968*b*; Murray and Gill 1976.

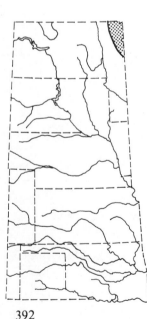

Tennessee Warbler
Vermivora peregrina

Breeding Status: Breeds in north-central and northwestern Minnesota, south to Itasca State Park, Clearwater County. Regarded as a hypothetical breeder in eastern North Dakota (singing males reported in Cass, Bottineau, and Pembina counties).

Breeding Habitat: In Minnesota this species is associated with open spruce and tamarack or white cedar (arbor vitae) bogs, where sphagnum moss is abundant. More generally it is also associated with deciduous forests, forest clearings, and brushy hillsides.

Nest Location: Nests are on the ground, typically in sphagnum-covered hummocks. They are deeply cupped and overhung with grasses and sedges and are lined with fine grasses. Infrequently, nests are placed on rather dry hillsides, under the cover of shrubs or saplings.

Clutch Size and Incubation Period: From 4 to 7 eggs, usually 6. The eggs are white with extensive brown blotches or spotting. The incubation period is unreported but probably is 11–12 days. Presumably single-brooded.

Time of Breeding: Specific egg dates are not available for this region, but territorial males have been found throughout June and into early July. Singing has been heard as late as August 16 in northern Minnesota.

Breeding Biology: Remarkably little has been learned of the breeding biology of this species. An early study in New Brunswick indicates that the birds are most abundant in low areas of streams and boggy ground, in spruce and balsam woodlands that have been partially cut, leaving areas of second growth not much higher than shrubs. In such boggy and open habitats the birds were found to be numerous, with up to five territorial males in a single clearing. Although it is known that only the female incubates, the incubation period and the details of nestling development have not been established.

Suggested Reading: Bowdish and Philipp 1916; Bent 1953.

Nashville Warbler
Vermivora ruficapilla

Breeding Status: Breeding in our region is restricted to north-central and northwestern Minnesota (breeding records for Becker and Clearwater counties).

Breeding Habitat: In Minnesota this species is especially typical of tamarack-ash and tamarack-spruce-white cedar swamps, and less frequently occurs in second growth and heavy timber in upland situations. Generally, it also occurs in swales, slashings, and undergrowth of mixed forests.

Nest Location: Nests in Minnesota are usually in sphagnum hummocks, well concealed by overhanging vegetation. The nest is a shallow cup of leaves and rootlets, lined with fine grasses, moss stems, needles, and hair.

Clutch Size and Incubation: From 4 to 5 eggs, white with scattered or concentrated brown spotting. The incubation period is 11–12 days. Probably single-brooded.

Time of Breeding: Minnesota egg dates are from May 30 to July 14, and nestlings have been reported as early as June 20.

Breeding Biology: Apparently male Nashville warblers become territorial immediately upon arrival, and pairing occurs soon afterward. The nests are usually so hard to find that few observations on them are available, but in one case nest-building took 7–9

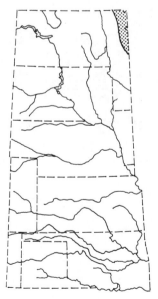

days, with the female presumably doing most of the work while the male perched nearby and sang. The female does all the incubating and occasionally leaves the nest for foraging, accompanied by the male. The female also broods the young, but both sexes bring food to them. They fledge in 8–11 days and thereafter seem to leave the nesting area.

Suggested Reading: Lawrence 1948; Bent 1953.

Northern Parula
Parula americana

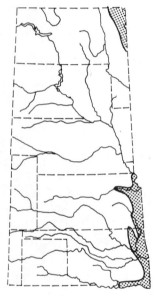

Breeding Status: Breeds in north-central and northwestern Minnesota (breeding records for Becker and Clearwater counties) and from northwestern Kansas southward through eastern Kansas (west locally to Riley and Montgomery counties) and eastern Oklahoma (records for Washington, Delaware, and McCurtain counties; probably breeds west at least to Marshall county, since there is a breeding record for Cooke County, Texas).

Breeding Habitat: The species is primarily associated with swampy woods, especially those where there are mosslike lichens or *Tillandsia* ("Spanish moss"). In Minnesota, tamarack and spruce swamps with abundant *Usnea* lichens are prime habitat.

Nest Location: Nests are suspended from tree branches at heights averaging about 10 feet (range 6–100). In Minnesota they are woven of *Usnea*, as they are in Oklahoma; where lichens or Spanish moss (*Tillandsia*) are not available, other materials may be used. In all cases the nest is a woven pouch, with a lateral or dorsal entrance.

Clutch Size and Incubation Period: From 3 to 7 eggs, usually 4–5. The eggs are white with brown spotting. The incubation period is at least 12 days. Single-brooded.

Time of Breeding: Minnesota egg records are for early June, with nestlings seen in late June and dependent young as late as August 7. In Kansas, eggs are apparently laid at least from mid-May to mid-June, and in Oklahoma nest-building has been reported in late April, with eggs in mid-May and nestlings as late as early July.

Breeding Biology: The distinctive nests of this species are far less easy to find than one might suspect; not only are they constructed of the abundant hanging lichens, but at times the female builds a concealing curtain of *Usnea* above and around the nest site before starting the nest itself, which may thus be seen only from directly below. About 4 days are required to construct the nest, and laying may begin the next day. Only the female incubates, and she leaves the nest rather often for foraging. Both sexes feed the young, but females do so more actively than males. In one observed case a young bird left the nest at 10 days of age, but this was the result of disturbance and apparently was premature.

Suggested Reading: Graber and Graber 1951; Bent 1953.

Yellow Warbler
Dendroica petechia

Breeding Status: Breeds pandemically in suitable habitats throughout the region.

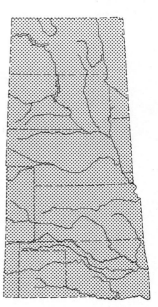

Breeding Habitat: The species prefers moist habitats, such as brushy bogs, edges of swamps, marshes, or creeks, but is also found in dry sites such as roadside thickets, hedgerows, orchards, and forest edges. It avoids both heavy forests and grassland environments lacking shrubs or trees.

Nest Location: Nests are 2–12 feet above the ground in upright forks or crotches of shrubs and trees. The nest is a compact cup of grasses, milkweed fibers, and cottony materials, lined with plant down and fine grasses.

Clutch Size and Incubation Period: From 3 to 5 eggs (29 Kansas clutches averaged 4.2, and 12 North Dakota nests averaged 3.6). The eggs are white, with gray to brown spotting near the larger end. The incubation period is 11–12 days. Single-brooded.

Time of Breeding: North Dakota egg dates are from June 2 to June 30, with dependent young seen as late as July 29. In Kansas, egg dates are from May 11 to June 20, with a peak in late May. Oklahoma records extend from May 11 (eggs) to late July (dependent young), and Texas egg records are from May 17 to July 13.

Breeding Biology: This widespread and abundant warbler seems to have rather generalized territorial needs, including suitable nest sites, tall singing posts, concealing cover, and foraging areas in shrubs and trees, within an area of about two-fifths of an acre. Territorial behavior begins soon after the males arrive, and pairs may be formed in 1–4 days. Approximately 4 days are needed for nest construction, which is done mostly or entirely by the female. The female also does all the incubating, often beginning before the clutch has been completed. Both sexes feed the young, and occasionally males have been seen brooding them, but this seems to be atypical. The young fledge in 9–12 days and remain in the general vicinity for another 7–10 days.

Suggested Reading: Schrantz 1943; Frydendall 1967.

Magnolia Warbler
Dendroica magnolia

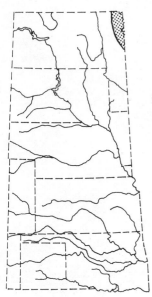

Breeding Status: Breeding is limited to northern Minnesota (south rarely to about Itasca Park, Clearwater County) and confined to coniferous forest habitats.

Breeding Habitat: Spruce and fir forests with low trees and open coniferous bogs dominated by white cedar or other coniferous species are preferred. Dense thickets of spruce and fir, second-growth following logging, edges of taller coniferous forests, and similar bush and sapling habitats are used.

Nest Location: Nests are usually in small conifers, 1–8 feet above the ground and generally are well concealed in the foliage near the tip of a horizontal branch. Occasionally hardwood tree or shrub species such as cherries, rhododendrons, or birches are also used, but hemlocks and spruces are favored. The nest is a loosely constructed cup of twigs, grasses, and weeds, lined with black rootlets.

Clutch Size and Incubation Period: From 3 to 5 eggs, usually 4. The eggs are white, with extensive markings of brown, reddish, or purplish near the larger end. The incubation period is 11–13 days. Probably single-brooded.

Time of Breeding: Few specific egg dates are available, but young have been found in Minnesota as early as June 22, and there have been several sightings of young during July. A nest at Itasca State Park hatched on June 7 and fledged on June 16 or 17 (*Loon* 49:199).

Breeding Biology: This species maintains territories by songs, visual displays, and chases, with favorite singing posts about 10–45 feet up in a tree. Males display toward other birds with wing- and tail-spreading and persistently follow females that enter the territory. Apparently both birds help build the nest, which may require 4–6 days. Only the female incubates and broods the young, although both sexes feed them. They fledge in about 10 days, and thereafter the young birds wander about, being fed by their parents until about 3 weeks after fledging.

Suggested Reading: Kendeigh 1945; Bent 1953.

Cape May Warbler
Dendroica tigrina

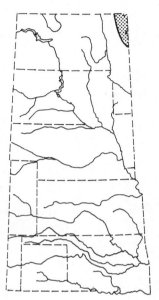

Breeding Status: Breeding in our region is limited to north-central and probably northwestern Minnesota (Clearwater County is the only specific record, and it is rare even at Itasca State Park).

Breeding Habitat: The species is associated with fairly open stands of tall conifers or the edges of coniferous forests, especially if birches or hemlocks are present.

Nest Location: Nests are in thick foliage near the top of tall spruces, usually 30–40 feet above the ground, and generally are invisible from the ground. The nest is built of mosses, twigs, and grasses and lined with soft vegetation, feathers, and hair.

Clutch Size and Incubation Period: From 4 to 9 eggs, usually 6–7. The eggs are white with reddish brown spots or blotches. The incubation period is not known but presumably is about 12 days. Probably single-brooded.

Time of Breeding: No definite egg dates or dates of young birds are available from Minnesota, but nesting probably occurs during June.

Breeding Biology: Because of the species' tendency to nest so high in conifers, few nests have been found, and little is known of the nesting biology. In one observed case, only the female gathered nest materials, which she collected from two places within 60 feet of the nesting tree. The nest was built mostly of sphagnum around the exterior, interwoven with vine stems, generally resembling a green ball of moss. Apparently only the female incubates, but almost nothing else is known of the breeding biology, since the female is very secretive about approaching and leaving her nest.

Suggested Reading: Bent 1953; Walley 1973.

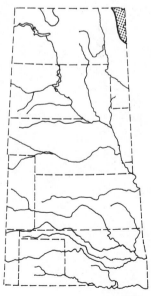

Black-throated Blue Warbler
Dendroica caerulescens

Breeding Status: Breeding in our region is restricted to north-central and perhaps northwestern Minnesota (Itasca State Park, Clearwater County, is the only definite record).

Breeding Habitat: In Minnesota this species is found in mixed coniferous forests as well as hardwood forests; heavy deciduous forest with an ample undergrowth of saplings and evergreen or deciduous shrubs seems to be preferred. Small conifers, creeping yew, or species of the heather family are usually present in such habitats and often serve as nesting cover.

Nest Location: Nests are in low coniferous seedlings or evergreen shrubs such as rhododendron, laurel, or yew, and usually are elevated less than 2 feet. They are often well concealed and are built of a mixture of twigs, bark, and leaves with a lining of hair and black rootlets.

Clutch Size and Incubation Period: From 3 to 5 eggs, usually 4. They are white, with spotting or blotching of browns and grays toward the larger end. The incubation period is 11–13 days, usually 12. Probably single-brooded.

Time of Breeding: Minnesota records of eggs or nestlings are all no earlier than July, but almost certainly eggs are laid by June, and singing by territorial males occurs in May.

Breeding Biology: Like the other species of its genus, the black-throated blue warbler is specialized for gleaning. It forages in the shrub, subcanopy, and lower canopy layers of forests, up to about 20 yards above the ground, mostly in the foliage and outer tree branches. Males tend to forage at higher levels than females and sing from territorial posts rather high in the trees. Most nest-building is done by the female, but sometimes males bring in materials, and in one case a nest was completed in 4 days. Incubation and brooding are done by the female, but both sexes feed the young with equal intensity. The young are fledged in about 10 days but are fed by the adults for some time thereafter.

Suggested Reading: Black 1976; Harding 1931.

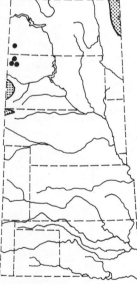

Yellow-rumped Warbler (Myrtle and Audubon Warblers)
Dendroica coronata

Breeding Status: Breeds commonly in north-central and north-western Minnesota (records for Clearwater, Marshall, and Roseau counties) and in the Black Hills of South Dakota. Also breeds commonly in the Pine Ridge area of Nebraska (Sioux and probably Dawes counties, *Nebraska Bird Review* 40:41), and rarely in North Dakota (Slope County).

Breeding Habitat: In the Black Hills, breeding occurs both in pine and in spruce forests, and in Minnesota it occurs rather generally in coniferous forests. Scattered evergreens, thickets near a stream or lakeshore, or open plantings are more typically used than dense, mature stands.

Nest Location: Nests are almost invariably in conifers, elevated 5–40 feet, and are usually well out on horizontal branches where they are screened from above by clumps of needles. Sometimes they are built in the topmost clump of needles of small conifers or, rarely, in a hardwood species. The nest is built mostly of twigs and bark fibers, lined with plant down and almost always some feathers.

Clutch Size and Incubation Period: From 3 to 5 eggs, usually 4. The eggs are white, with brown blotching near the larger end, often forming a continuous wreath. The incubation period is 12–13 days. Probably single-brooded, but double-brooding has been suspected.

Time of Breeding: Nesting in the Black Hills occurs from the last week of May to mid-July. In Minnesota, fledged young have been reported as early as June 13, and fledged but dependent young have been seen as late as late July.

Breeding Biology: This warbler is among the first of its family to move northward in spring, often becoming fairly abundant as the first tree leaves appear. Little has been written on the early stages of its nesting, but it is believed that only the female incubates and broods the young. One early account suggests that rarely the male may incubate too and may even sing while on the nest. The young fledge at about 12–14 days.

Suggested Reading: Ficken and Ficken 1966; Hubbard 1969.

Black-throated Green Warbler
Dendroica virens

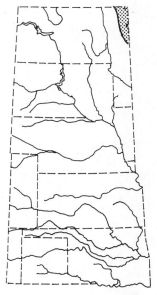

Breeding Status: Restricted as a breeding species in our region to north-central and perhaps northwestern Minnesota (Clearwater County is the only specific breeding record).

Breeding Habitat: In Minnesota, this species nests in rather open coniferous and mature stands of mixed forests. It is generally associated with large trees but occasionally is found in second-growth timber as well as scattered trees in pastures and on hillsides. Stands of pines, especially white pines, seem to be preferred habitat.

Nest Location: Nests are elevated 2–80 feet, in horizontal forks or branches of trees, well away from the trunk in rather thick

399

foliage. Conifers are the usual sites, but various hardwoods are also used at times. The nest is constructed of a variety of materials that often include birch bark, and feathers or hair are frequently part of the lining.

Clutch Size and Incubation Period: From 4 to 5 eggs, white, with brown to purple spotting around the larger end. The incubation period is 12 days. Two broods may occasionally be raised.

Time of Breeding: In Minnesota, eggs have been found in late June, and nestlings have been seen as early as July 4. Dependent but fledged young have been seen to the end of July.

Breeding Biology: In a study done in Michigan, it was found that a pair of black-throated green warblers spent 4 days building a nest; both sexes worked initially, but the bulk of the work was done by the female. In an earlier study only the female built, and 8 days were spent in this phase of breeding. Incubation is apparently done only by the female; early reports of males incubating have not been substantiated. Likewise, all brooding is by the female, but both sexes feed the young. Fledging occurs 10 days after hatching, and young birds have been observed being fed as late as early September in Massachusetts.

Suggested Reading: Pitelka 1940; Nice and Nice 1932.

Cerulean Warbler
Dendroica cerulea

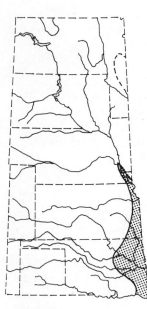

Breeding Status: Probably breeds in west-central and north-central Minnesota (no specific records) and is a hypothetical breeder in southeastern North Dakota (territorial male noted in Richland County). Also breeds in the Missouri Valley, but the northern limits are uncertain; several nests have been found in Sarpy County, Nebraska, and it presumably breeds south to the Kansas border. However, it is not known to breed at Squaw Creek N.W.R., Missouri, and there are no definite breeding records for Kansas, where it probably is a rare nester. It is known to breed in eastern Oklahoma (Delaware and McCurtain counties), and it has been seen as far west as Washington, Tulsa, and Latimer counties.

Breeding Habitat: The species is found in moist deciduous bottomland forests as well as shady and mature upland woods. Rather open forests, with large elms, maples, and ashes and fairly free of undergrowth, seem to be preferred.

Nest Location: Nests are usually fairly high (20–60 feet) in tall trees, usually well away from the trunk on horizontal branches, which are usually substantial and free of vegetation below. The nest is compactly rounded, made of bark fibers, spider silk, and grasses and decorated outside with brownish lichens and mosses. It resembles a large knot.

Clutch Size and Incubation Period: From 3 to 5 eggs, usually 4. They are white, with brown spotting near the larger end. The incubation period is unreported but probably is about 12 days. Probably single-brooded.

Time of Breeding: No egg dates are available for Minnesota, but young birds have been seen in late July. In Oklahoma, fledglings have been seen in late June, and there is one egg record for April 29.

Breeding Biology: Rather little is known of the early stages of the breeding cycle, during the period of territorial establishment and nest-building. Only the female is known to incubate; during incubation the male rarely visits the nest and apparently is forcibly kept away by the female, as is typical of several other species of warblers. Nothing specific has been reported on the fledging period or other posthatching stages of the breeding biology.

Suggested Reading: Trapp 1967; Bent 1953.

Blackburnian Warbler
Dendroica fusca

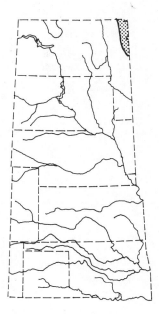

Breeding Status: Breeding in our region is restricted to north-central and perhaps northwestern Minnesota (records for Becker and Clearwater counties).

Breeding Habitat: The species is associated with mature coniferous forests, especially heavy timber and swampy areas where *Usnea* lichens are abundant. Deciduous or mixed second-growth timber is used secondarily, but mature stands of spruces, hemlocks, and pines are the primary breeding habitat.

Nest Location: Nests are 5–90 feet above the ground, usually at least 25 feet high, on horizontal branches of large conifers, commonly hemlocks. The nest is built of twigs, into which *Usnea* is normally woven, concealing the nest, and the cup is usually lined with hair, grasses, and black rootlets.

Clutch Size and Incubation Period: From 4 to 5 eggs, usually 4. They are white, with extensive brownish spotting or blotching around the larger end, usually forming a wreath. The incubation period is unknown but presumably is 12 days. Probably single-brooded, but attempted renesting has been seen.

Time of Breeding: The few available egg records from Minnesota are for July, but it is likely that most eggs are laid in June.

Breeding Biology: Studies of several pairs in Ontario indicate that only females do the nest-building and that 3–6 days may be spent in this phase of breeding. Likewise, only the female incubates and broods the young; in some instances the female is fed on the nest by the male and in other cases she leaves the nest briefly for foraging. The period to fledging has not been established. By late July the adults go into molt, and singing stops abruptly. At that time territories are abandoned and the adults begin to disperse, often followed by fledglings still begging to be fed.

Suggested Reading: Lawrence 1953*b*; Bent 1953.

Yellow-throated Warbler
Dendroica dominica

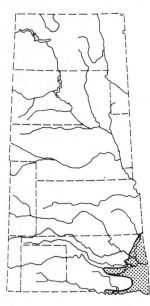

Breeding Status: Current evidence indicates that breeding in our region is restricted to Oklahoma, where the species is resident in the eastern part of the state, extending west at least to Tulsa and Cleveland counties and along the Red River perhaps to Love County, since breeding has been reported across the river in Cooke County, Texas. There are no breeding records for Kansas, but in 1968 the collection of possible breeding birds was reported for Cherokee County (*Kansas Ornithological Society Bulletin* 20:7). Reported without apparent justification to be a rare breeder in southeastern Nebraska.

Breeding Habitat: In Oklahoma this species occupies swampy hardwood forests and upland forests of mature pines and oaks. Over most of its range it is associated with forests with abundant Spanish moss.

Nest Location: Nests are most frequently in oak trees, at heights of 10–100 feet, averaging about 35 feet. Where Spanish moss is available the nest is constructed of it and is hung from a horizontal branch a considerable distance from the trunk. Where this material is lacking the nest is usually saddled over a horizontal branch, and is a more typically warbler nest of bark strips, grasses, and weeds lined with plant down and feathers.

Clutch Size and Incubation Period: From 4 to 5 eggs, usually 4. The eggs are white, with extensive spotting or blotching of browns and grays often forming a wreath around the larger end. The incubation period is probably 12–13 days. Believed to be double-brooded in some areas, but not in Oklahoma.

Time of Breeding: In Oklahoma, nest-building has been reported in late April and early May, and recently fledged broods have been seen from early June to mid-July.

Breeding Biology: Throughout most of the southern states this species has an ecological affinity for *Tillandsia*-draped forests of oaks, pines, and cypresses. On the eastern coast, its northern limit coincides with the northern limit of loblolly pine, and its major foraging adaptation is a narrow skull and an elongated bill that allow it to probe deeply into the cones of these pines. The birds often forage creeperlike on tree trunks as well, in both pines and oaks. The female does most of the nest-building and all the incubating. The fledging time has apparently not been established. The birds are persistent renesters after clutch-loss; in one case a pair built four nests and laid complete clutches of 4 eggs in each nest between late April and early June.

Suggested Reading: Ficken, Ficken, and Morse 1968; Bent 1953.

Chestnut-sided Warbler
Dendroica pensylvanica

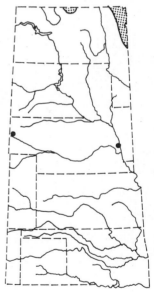

Breeding Status: Regular breeding in our region is probably restricted to north-central and northwestern Minnesota (records for Clearwater, Polk, and Roseau counties) and to northeastern North Dakota (Turtle Mountains of Bottineau and Rolette counties and Pembina Hills of Cavalier County). Elsewhere it has apparently been known to nest only near Omaha, Nebraska, in 1894 and during 1975 in Scottsbluff County (*Nebraska Bird Review* 44:10). There is an 1895 nest record from Dickinson County, Kansas (*Kansas Ornithological Society Bulletin* 28:32).

Breeding Habitat: The species is associated with low shrubbery, briar thickets, forest clearings and edges, overgrown pastures, and similar rather open and dry habitats having some woody vegetation in the form of shrubs and small trees. The nesting requirements seem to be trees or shrubs for singing perches and shade, moist situations with dense vegetation for about 3 feet above the ground to provide the nest sites and foraging, and sufficient nesting materials.

Nest Location: Nests are in bushes or saplings, 1–4 feet above the ground. They are loosely constructed structures made of grasses, bark strips, and weed stems and are lined with fine grasses, rootlets, and plant fibers.

Clutch Size and Incubation Period: From 3 to 5 eggs, usually 4. The eggs are white with dark spotting or blotching of various tones. The incubation period is 11-12 days. Probably single-brooded.

Time of Breeding: Minnesota egg records extend from May 27 to July 2. Nestlings have been reported as late as July 22, and dependent young out of the nest have been seen as early as June 27.

Breeding Biology: Nests are apparently built only by the female, although her mate often follows her about during this 3-4 day period. From 1-6 days may follow before the first egg is laid, and subsequent eggs are laid daily. Only the female incubates, but the male sometimes will approach the nest with food or feed her while she is away from the nest. Only the female broods, but both sexes tend the young birds, which remain in the nest 8-10 days.

Suggested Reading: Tate 1970; Cripps 1966.

Bay-breasted Warbler
Dendroica castanea

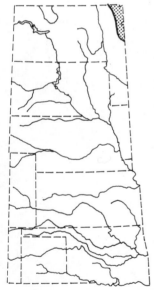

Breeding Status: Breeding in our region is probably restricted to north-central and northwestern Minnesota, but there is still only indirect evidence of breeding in the area (Itasca Park, Clearwater County).

Breeding Habitat: The species is associated mainly with coniferous forests, especially with rather low forests in swampy areas containing some birches and maples, and with mixed coniferous and deciduous forests having clearings or edge areas.

Nest Location: Nests are in conifers, at heights of 4-40 feet, but usually less than 25 feet. They are generally well away from the trunk on horizontal branches, well concealed by foliage. Nests are rather loosely constructed of twigs and grasses and are lined with rootlets, hair, and grasses. They are often built so that they are nearly invisible from directly below and may also have a limb or foliage directly above.

Clutch Size and Incubation Period: From 4 to 7 eggs, usually 5. They are white, with extensive spotting or blotches of brown and gray around the larger end. The average incubation period is slightly over 12 days. Probably single-brooded.

Time of Breeding: No Minnesota egg dates are available, but adults apparently tending eggs or young have been seen in July, and dependent young have been observed in late July.

Breeding Biology: The nest in this species is begun with a few stems of hay placed in the fork of several twigs on a flat limb;

spruce twigs are then anchored to this base, and the nest is gradually built up of more of these components. Only the female incubates, but she is often fed on the nest by her mate and often seems quite fearless of humans. The female also broods the young, which fledge in about 11 days, and may remain in the general area for a week or more.

Suggested Reading: Mendall 1937; Bent 1953.

Pine Warbler
Dendroica pinus

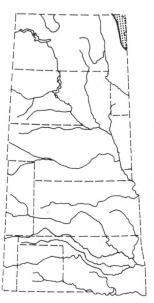

Breeding Status: Breeding occurs in north-central and perhaps northwestern Minnesota (records for Clearwater and Beltrami counties), and also in extreme eastern Oklahoma (from Cherokee and Mayes counties south through Pittsburg, Pushmataha, and McCurtain counties).

Breeding Habitat: The species is associated with open pine woods and "pine barrens," including especially jack pine in Minnesota, and upland southern pines such as shortleaf pine in eastern Oklahoma. It generally avoids very tall, moist, and dense coniferous forests.

Nest Location: Nests are in pines (rarely other conifers) at heights of 8–80 feet, averaging about 40 feet, out near the tips of horizontal branches and well hidden in the foliage. They are constructed of twigs, bark, pine needles, and spider webs and lined with such materials as hair, pine needles, and feathers.

Clutch Size and Incubation Period: From 3 to 5 eggs, usually 4. The eggs are white, with extensive brownish spotting or blotches around the larger end. The incubation period is unrecorded but probably is about 12 days. At least two broods are produced in some southern areas, probably including Oklahoma.

Time of Breeding: Minnesota breeding records extend from early June (nests or nest-building) through early July (fledged but dependent young). In Oklahoma, sitting birds have been seen as early as April 27, and nest-building as late as June 28. Fledged but dependent young have been noted as late as August 7.

Breeding Biology: This species occurs only in pine forests throughout its range and tends to forage creeperlike over pine trunks. It occasionally also hawks for insects in the air or feeds on the ground, picking up surface invertebrates. The role of the sexes in nest-building is evidently not known, but several observers have indicated that males help with incubation. The young are fed by both parents as well, but the fledging period has not been established.

Suggested Reading: Bent 1953; Ficken, Ficken, and Morse 1968.

Prairie Warbler
Dendroica discolor

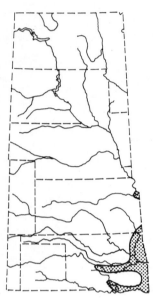

Breeding Status: Breeding in the Great Plains is almost entirely limited to eastern Oklahoma, where the species nests from Ottawa County south to McCurtain County, and locally west to Mayes, Wagoner, Cleveland, Caddo, and Murray Counties. In Kansas breeding is known only from Wyandotte, Johnson, and Cherokee counties. There is no good evidence of any recent sightings in Nebraska, where reports of breeding (Dakota and Richardson counties) in the late 1800s now seem likely to have been the result of misidentification.

Breeding Habitats: In Oklahoma, brush-grown fields, openings in pine woods, dense or scattered oak saplings, and scrubby oaks and junipers on canyon slopes have all been described as breeding localities. In Kansas, edge habitats between oak-hickory forests and old fields are used. Generally, dry habitats with brush or low trees seem to serve the primary needs of this species. Recently, Christmas tree farms have provided a new habitat for prairie warblers in eastern states.

Nest Location: Nests are elevated 1–10 feet, in brush or small trees, in crotches or on limbs. They are constructed of bark shreds, leaves, plant fibers, and down, lined with hair, grasses, and feathers. The incubation period is 12–13 days. Single-brooded.

Time of Breeding: Oklahoma egg records are lacking, but fledged young or recently abandoned nests have been found from mid-June to mid-July, and in Kansas eggs are laid at least during June.

Breeding Biology: The breeding biology of this species evidently is much like that of the other *Dendroica* species. Furthermore, it has been intensively studied in one unusual aspect, "anticipatory food-bringing" behavior by the male. In this and several other warbler species the male may bring food to the nest before the young hatch. If the female is absent from the nest the male may eat the food himself, and if she is present she may either eat it or ignore it. This behavior seems to allow the male to determine the stage of hatching and thus to stimulate a prompt onset of parental feeding.

Suggested Reading: Nolan 1978; Bent 1953.

Palm Warbler
Dendroica palmarum

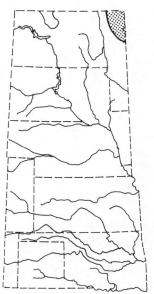

Breeding Status: Breeding in our region is restricted to local areas in north-central and northwestern Minnesota (at least eastern Marshall County and probably Roseau County).

Breeding Habitat: The species is largely confined to open to dense boggy areas dominated by tamarack, black spruce, and white cedar (arbor vitae). Dry and open forests of spruce or jack pine are also reportedly used in some areas, but the Minnesota breeding records are all associated with swampy or boggy sites. In Nova Scotia, territories were found at the periphery of open heath bogs rather than in the open areas themselves.

Nest Location: Nests are usually on the ground, nearly buried in sphagnum mosses, and are built of coarse grasses and rootlets, usually with feather lining. In some areas nests have also been found in low branches of conifer saplings, such as on hummocks of ground junipers.

Clutch Size and Incubation Period: From 4 to 5 eggs, white, with brownish spots or blotches forming a wreath around the larger end. The incubation period is 11 days. Probably normally single-brooded; a few unsuccessful second clutches have been found.

Time of Breeding: The only available egg record for Minnesota is for mid-June (hatching nest), but nestlings or dependent fledglings have been seen from mid-June through late July.

Breeding Biology: In a study area in Nova Scotia, it was found that male territories were usually less than 5 acres in area, and as many as 30 song posts may be used. Males prefer high trees but also sing from some low posts such as shrubs. The female incubates, but she is fed by her mate, and monogamy is typical. Among 10 males studied in Nova Scotia, 3 were unmated, 1 had two mates, and the other 6 were monogamous. Young leave the nest at about 10 days and are able to fly well about 3 days later. The male's participation in feeding the young appears to be variable, but he usually helps feed the fledged young, which are tended for at least 11 days after leaving the nest.

Suggested Reading: Bent 1953; Welsh 1971.

Ovenbird
Seiurus aurocapillus

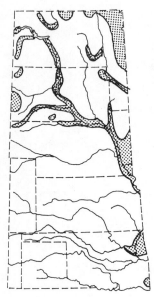

Breeding Status: Breeds in suitable habitats from the Missouri and Little Missouri valleys of North Dakota eastward through western Minnesota and southward through the wooded portions of South Dakota including the Black Hills, northern Nebraska (Pine Ridge, the Niobrara Valley west at least to Cherry County, and the Missouri Valley), northwestern Missouri, and probably adjacent northeastern Kansas (no actual breeding records). Also breeds in northeastern Oklahoma (Delaware, Mayes, Adair, McCurtain, and possibly Washington counties).

Breeding Habitat: Breeding is mostly confined to well-drained bottomland deciduous forests and to well-shaded and mature upland forests, often on north-facing slopes or in ravines.

Nest Location: Nests are on the ground, in a depression among dead leaves or under a tuft of grass so as to be invisible from above. The structure has a lateral entrance and somewhat resembles an oven. The nest is built of grasses, twigs, leaves, and similar materials, lined with rootlets, hair, and plant fibers.

Clutch Size and Incubation Period: From 3 to 6 eggs, usually 4–5. The eggs are white, with reddish brown to lilac spots or blotches around the larger end. The incubation period is from 11 to 14 days, averaging about 12. Single-brooded.

Time of Breeding: There is a single North Dakota egg record for July 1. Minnesota egg records extend from May 17 to June 28, nestlings have been noted between June 20 and July 4, and fledglings have been seen as early as June 17. In Oklahoma, nests have been found in late April and early May.

Breeding Biology: Both sexes of ovenbirds display a strong tendency to return to their breeding areas of the previous season, and thus remating by old pairs sometimes occurs. While the male patrols and advertises his territory, the female selects a nest site and constructs the nest. Incubation begins on the day after the last egg is laid and is done entirely by the female. The male begins to bring food to the nest at the time of hatching or slightly before, and both sexes actively feed the young. In one instance a male was found to maintain two mates simultaneously, and in another two males were observed to carry food to the young of one nest. The young leave the nest in 8–10 days but may continue to be fed with decreasing intensity by the adults until they are nearly 5 weeks old.

Suggested Reading: Hahn 1937; Bent 1953.

Northern Waterthrush
Seiurus noveboracensis

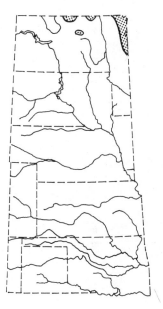

Breeding Status: Breeding is restricted to northeastern North Dakota (Turtle Mountains, Pembina Hills, Devils Lake area, rarely elsewhere) and to north-central and northwestern Minnesota (relatively rare, no specific breeding records).

Breeding Habitat: The species is associated with ponds, lakes, and streams having woody borders, with brushy bogs and swamps, and with second-growth swamp forests in North Dakota. Standing-water habitats are favored, rather than swiftly running water as in the case of the Louisiana waterthrush.

Nest Location: Nests are in the sides of overhanging banks beside water, among the roots of living trees, or under the upturned roots of fallen trees. The nest is mostly composed of mosses and dead leaves, lined with mosses, grasses, and sometimes hair.

Clutch Size and Incubation Period: From 3 to 6 eggs, usually 4–5. The eggs are white, with variable spotting and blotching of brownish near the larger end. The incubation period is 12 days. Probably single-brooded.

Time of Breeding: In North Dakota, incubated eggs have been found in mid-July, and young have been collected in late July. In Minnesota, dependent young or parents carrying food to young have been seen in late June.

Breeding Biology: In contrast to the Louisiana waterthrush, and like the ovenbird, males of this species frequently perform flight songs over their territories, which tend to be rectangular rather than linear. These differences seem to be related to habitat similarities of the ovenbird and northern waterthrush and to associated difficulties in pairs' maintaining visual contact. Only the female builds the nest, and only she incubates. In one case the young left the nest when 9 days of age; for the first few days after leaving, the young are extremely secretive and hard to locate. Soon they begin to perch on low branches and beg for food, and within another week they are able to fly well.

Suggested Reading: Eaton 1957; Bent 1953.

Louisiana Waterthrush
Seiurus motacilla

Breeding Status: Breeds from the Kansas River Valley of Kansas (west to Riley County; one report of nesting in Decatur County in

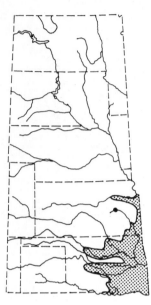

1910), southward along the eastern edge of the state to Oklahoma, where it apparently extends westward locally to Major, Caddo, and Love counties. There was an apparent breeding in Fremont County, Iowa, in 1977 *(Iowa Bird Life* 47:130).

Breeding Habitat: The species is usually found near swift streams, in wooded and hilly country, but also breeds around lagoons and swamps where trees grow to the shoreline.

Nest Location: Nests are usually close to water, either in holes in steep banks or under the roots of nearby trees, well hidden by overhanging vegetation. The nest is made of dead leaves that are well packed and intermixed with twigs, and the nest cup is lined with grasses, rootlets, and so forth. There may be a pathway of leaves leading to the nest.

Clutch Size and Incubation Period: From 4 to 6 eggs, white with extensive spotting or blotching of gray and brown. The incubation period is 12–14 days. Single-brooded.

Time of Breeding: In Kansas eggs are laid in May and June, and in Oklahoma fledglings have been seen as early as May 12, while nestlings have been noted as late as June 25.

Breeding Biology: Males establish territories immediately after arriving in the spring and begin to advertise them from song posts in trees. Territories are long and narrow, centered on fast-flowing streams, and pairs typically occupy about 200–400 yards of stream habitat. The nest is constructed by both sexes, with the female doing most of the work. The main structure takes a day or two, and 2 or 3 more days are needed for lining it and for laying the first egg. The female incubates alone, and the male rarely is seen near the nest until the young are about to hatch, when he may begin to bring food. The young leave the nest when 10 days old, and within another month they begin to wander about unattended by the adults.

Suggested Reading: Eaton 1958; Bent 1953.

Kentucky Warbler
Oporornis formosus

Breeding Status: Breeds in the Missouri Valley from southeastern Nebraska (north to Sarpy County) southward through northwestern Missouri (uncommon at Squaw Creek N.W.R.), eastern Kansas (Doniphan to Labette counties, and west locally to Riley County), and eastern Oklahoma (west locally to Kay, Payne, Oklahoma, Caddo, Grady, Murray, and probably Love counties).

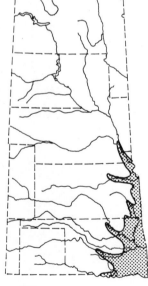

Breeding Habitat: The species is associated with shrubby woodland borders and the understory of damp or shady woods, especially moist ravines and bottomlands.

Nest Location: Nests are usually on sloping ground, at the base of a tree or bush, under the branches of a fallen limb, or just above the ground in a shrub. The nest is slightly above the ground on a foundation of dead leaves and is built of grasses, rootlets, and plant fibers, lined with fine grasses and rootlets.

Clutch Size and Incubation Period: From 4 to 5 eggs, rarely 6. The eggs are white, with dark spotting or blotching at the larger end. The incubation period is 12–13 days. Single-brooded.

Time of Breeding: In Kansas, records indicate that eggs are laid in May and June. In Oklahoma unfinished nests have been found as early as May 6, and eggs have been found as early as May 12 and as late as June 3. Nestlings have been seen from May 26 to July 10, and dependent young as late as August 5.

Breeding Biology: The breeding biology of this ground-foraging warbler is much like that of other warblers, judging from the rather small amount of published information. Unlike most warblers, males are accomplished and loud singers; their wrenlike vocalizations are uttered about once every 12 seconds at the peak of the breeding period and carry for as much as 150 yards. Apparently only the female incubates the eggs and broods the young, and she also does most of the feeding of the young. The young are fed in the nest for 8–9 days and thereafter are tended by the parents for as long as 17 days.

Suggested Reading: Bent 1953; De Garis 1936.

Mourning Warbler
Oporornis philadelphia

Breeding Status: Breeding in our region is restricted to north-central and northwestern Minnesota (records for Otter Tail, Clearwater, Marshall, and Roseau counties) and to northeastern North Dakota (Turtle Mountains, also locally in Pembina and Cavalier counties).

Breeding Habitat: Brushy areas such as clearing, slashings, roadsides, and forest edges are favored, especially those near woods and where thickets of raspberries or other briars occur. The species is often found in areas undergoing succession after forest fires.

Nest Location: Nests are on the ground, or infrequently may be as high as 30 inches above the ground, in tangles of thorny

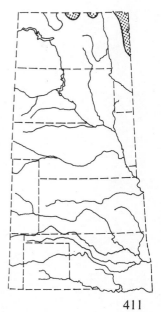

411

shrubs, ferns, grass tussocks, or other rather rank growth. The nest is a fairly bulky structure of weeds, grasses, and vines with a cup of finer grasses, rootlets, and hair.

Clutch Size and Incubation Period: From 3 to 5 eggs, usually 4. The eggs are white, with brown spots and blotches, especially at the larger end. The incubation period is 12 days. Probably single-brooded.

Time of Breeding: North Dakota records indicate a breeding season from early June to early August, and in Minnesota eggs have been reported between May 26 and July 17, with nestlings seen as early as June 20 and as late as July 17.

Breeding Biology: Studies of this species at Itasca State Park indicate that territories average about 2 acres and occur where there is a partially open canopy and a mixture of shrubby and herbaceous ground vegetation. The nest is placed in a relatively open part of the territory, and incubation may begin before the clutch has been completed. Only the female incubates, but the male may feed her both at the nest and away from it. The young leave the nest at 8–9 days and for about 3 weeks may remain in the general vicinity of the nest. The young are scarcely able to fly until they are about 2 weeks old, and the adults begin to molt about the time the young become independent.

Suggested Reading: Cox 1960; Bent 1953.

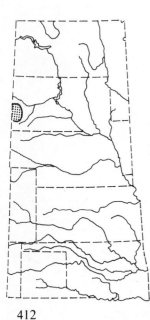

MacGillivray Warbler
Oporornis tolmiei

Breeding Status: Apparently confined as a breeding species to the Black Hills of South Dakota, where it is uncommon. It is rare in the Nebraska Pine Ridge area and is not known to nest there.

Breeding Habitat: In the Black Hills this species is primarily found near streamside thickets and other low deciduous growth in canyons or gulches. Less often it occurs in dense stands of deciduous woods or in mixed woodlands on upland slopes, and also in river-bottom stands of cottonwoods and associated riparian vegetation.

Nest Location: Nests are in grass clumps virtually on the ground or, more frequently, elevated from a few inches to about 2 feet in the upright stalks of shrubs. They are rather well concealed but are often poorly constructed, built of bark shreds, weeds, and grasses with a lining of rootlets, grasses, and sometimes hair.

Clutch Size and Incubation Period: From 3 to 6 eggs, usually 4. Eggs are white, with brown spots, scrawls, and blotches, primarily at the larger end. The incubation period is about 13 days. Probably single-brooded.

Time of Breeding: In the Black Hills the estimated breeding period is from late May to mid-July. Eggs have been reported in the first half of June, and dependent fledglings have been noted in early July.

Breeding Biology: This western counterpart of the mourning warbler has been much less studied than that species but presumably has a breeding biology much like it. Apparently the female undertakes virtually all the parental duties, including incubation and brooding the young. The male also is said to participate only slightly in feeding the young, which leave the nest 8–9 days after hatching.

Suggested Reading: Bent 1953; Griscom et al. 1957.

Common Yellowthroat
Geothlypis trichas

Breeding Status: Pandemic, breeding throughout the region in suitable habitats.

Breeding Habitat: This species is primarily found near moist ground, at aquatic sites, and among associated lush vegetation, including tall grasses and often shrubs and small trees. Pond or river margins, or swampy areas, are especially utilized, but occasionally it is found in upland thickets of shrubs and small trees, in poorly tended orchards, retired croplands, or weedy residential areas.

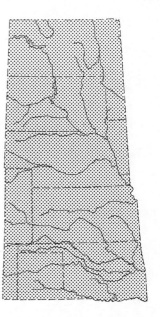

Nest Location: Nests are near or, rarely, on the ground, in heavy growth of weeds, grasses, or low shrubs generally in vegetation less than 2 feet high. They are bulky, made of coarse grasses, leaves, and sometimes mosses and lined with grasses, bark fibers, and hair. Nests are well concealed unless the vegetation is parted, and are usually no more than a few inches above the ground, rarely as high as 3 feet.

Clutch Size and Incubation Period: From 3 to 6 eggs, normally 4. The average of 16 first nestings in Minnesota and Michigan was 3.9 eggs, and 8 subsequent nestings averaged 3.8. The eggs are white, with brown speckling near the larger end. The incubation period is 12 days. Frequently double-brooded in favorable habitats and known to be a persistent renester.

Time of Breeding: North Dakota egg dates are from mid-June to mid-July, and nestlings have been seen as early as July 10. In

413

Kansas, egg records are from May 11 to June 10, and in Oklahoma eggs have been found from May 1 to June 8.

Breeding Biology: As soon as they arrive in spring, males establish territories that usually are less than 2 acres in area, but in some instances may exceed 3 acres (one bigamous male was found to occupy an area of 3.4 acres). Nests are usually built in the drier and more open parts of the territory, but water is always nearby. The female builds the nest over a period of 2 to 5 days and also performs all the incubation. Feeding of the young is done by both sexes, and the young may leave the nest when only 7–8 days old. However, they cannot fly until they are 11–12 days old and they do not begin feeding on their own until about 3 weeks old. They may be tended by one or both parents until they are 4–5 weeks old. Most or all females attempt to raise second broods, but few are successful. At least some females built at least 3 nests, and mate-changing between broods apparently is infrequent.

Suggested Reading: Stewart 1953; Hofslund 1959.

Yellow-breasted Chat
Icteria virens

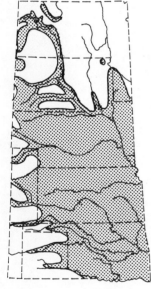

Breeding Status: Breeds in suitable habitats from western North Dakota (Missouri, Little Missouri, and Souris valleys) southward through South Dakota (west locally to the Black Hills and east along the Big Sioux drainage into western Iowa and southwestern Minnesota), Nebraska (virtually statewide), eastern Colorado (west locally to the foothills), all of Kansas, Oklahoma (except possibly Cimarron County), the Texas panhandle (rare in the southern panhandle), and locally in northeastern New Mexico (casual in Union County).

Breeding Habitat: Ravine or streamside thickets are favored, especially those with small trees and tall shrubs, also forest edges, dense stands of tree saplings, and clumps of shrubs in overgrazed pastureland.

Nest Location: Nests are from a few inches to 6 feet above the ground, in forks or crotches of shrubs, low trees, or vines. They are relatively bulky (about 5 inches wide), built of leaves, stems, and grasses and lined with finer grasses and stems.

Clutch Size and Incubation Period: From 3 to 5 eggs (21 Kansas clutches averaged 3.9). The eggs are white, with brown to lilac spotting over most of the egg, but especially at the larger end. The incubation period is 11 days. Probably single-brooded.

414

Time of Breeding: In North Dakota the probable breeding season is from early May to early August, and in the Black Hills of South Dakota the nesting period is apparently from June through early July. Kansas egg records extend from May 11 to July 20, with the first ten days of June the peak period of egg-laying. Oklahoma egg dates are from May 11 to July 14, and fledged young have been seen as early as June 10.

Breeding Biology: This species is one of the most aberrant of all the warblers and quite possibly should be removed from the family, although its relationships are still obscure. Males are unusual in their remarkable diversity of vocalizations, which often have 6–10 different song phrases, which are almost randomly uttered and vary greatly in loudness, pitch, and duration. So far as is known, only the female incubates the eggs and broods the young, which leave the nest in 8–11 days.

Suggested Reading: Bent 1953; Griscom et al. 1957.

Hooded Warbler
Wilsonia citrina

Breeding Status: The probable regular breeding range in our region is confined to southeastern Oklahoma (McCurtain and Le Flore counties). Additionally, there is a single breeding record for eastern Kansas (Anderson County) and specimens have been collected during the breeding season from Leavenworth and Shawnee counties.

Breeding Habitat: Associated with thick bottomland woods in Oklahoma, and wet, open woods in eastern Kansas. More generally, it is found in the undergrowth of deciduous forests and in wooded swamps, especially in the south.

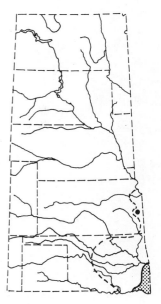

Nest Location: Nests are from a few inches to 6 feet above the ground, in bushes, vines, saplings, and diverse other sites. The nest is often in the lowest branch of a sapling and invariably is extremely inconspicuous, initially resembling a random accumulation of leaves lodged in the vegetation. It is composed of dead leaves, grasses, bark strips, and various other materials, lined with fine grasses, hairs, and rootlets.

Clutch Size and Incubation Period: From 3 to 5 eggs, often only 3. The eggs are white, with brown to lilac markings toward the larger end. The incubation period is 12 days. Usually single-brooded, but sometimes double-brooded in Florida.

Time of Breeding: The single Kansas nesting record is for a clutch completed about May 23, and the young fledged on June 22. In

Oklahoma, fledged young have been seen in late June and early July.

Breeding Biology: Males begin to establish and advertise territories as soon as they arrive on the breeding grounds in spring, and they soon acquire mates. Both sexes defend the territories, which are often spaced at 50–100-yard intervals in favored swamp habitats. In spite of the almost haphazard appearance of the nest, it is carefully constructed and may take up to a week to complete. It has been questionably reported that the male participates in incubation, which would be highly unusual for wood warblers, but it is not surprising that both sexes participate in feeding the young, which remain in the nest for about 8 days. In an Ohio study, it was reported that only about one-seventh of the pairs were successful in raising their first brood, but second and sometimes even third nesting efforts are made by unsuccessful pairs.

Suggested Reading: Bent 1953; Griscom et al. 1957.

Canada Warbler
Wilsonia canadensis

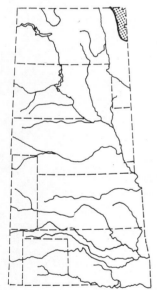

Breeding Status: Breeding in our region is apparently confined to north-central and perhaps northwestern Minnesota (Clearwater County is the only breeding record). The species is also a hypothetical breeder in northern North Dakota, since adults have been reported in the Turtle Mountains during July.

Breeding Habitat: The species frequents heavy undergrowth in mature mixed or deciduous woodlands, the borders of cedar swamps, and second-growth deciduous forests. In Minnesota, moist spruce and birch forests that have a thick brushy understory are favored. Habitats similar to those of the mourning warbler are used but tend to be considerably moister.

Nest Location: Nests are on or very near the ground, often in a mossy hummock, under upturned tree roots, in a bank cavity, or in a rotted, moss-covered stump. The nest is built of leaves, weed stems, grasses, and bark fibers on a foundation of large leaves and is lined with hair, rootlets, and plant down.

Clutch Size and Incubation Periods: From 3 to 5 eggs, usually 4. The eggs are white, with brown spotting or blotching toward the larger end. The incubation period is not established but is probably about 12 days. Single-brooded.

Time of Breeding: There is only a single egg record (June 4) for Minnesota, but dependent young have been seen through most of July. A nest at Itasca State Park fledged on July 8 (*Loon* 49:199).

Breeding Biology: One of the few nesting studies of this warbler was done in Michigan, but so far no observations on the early phases of nest-building and early incubation are available. Apparently only the female incubates, and the male rarely comes to the nest but communicates with his mate by vocalizing. In the Michigan study the male always brought food to the nest during his visits. The female was never on the nest at the time, however, and the eggs were still unhatched. Such anticipatory food-bringing has also been seen in various other warblers. There is no information on the development of the young.

Suggested Reading: Krause 1965; Bent 1953.

American Redstart
Setophaga ruticilla

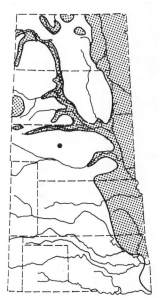

Breeding Status: Breeds in suitable habitats from western Minnesota and most of North Dakota (Turtle Mountains, Pembina Hills, and major river valleys) southward through western Iowa, South Dakota (major river valleys and the Black Hills), Nebraska (the Pine Ridge area, the Niobrara and Missouri valleys, and sporadically in the Platte Valley west at least to Adams County), eastern Kansas (east of Cloud and Sumner counties), and eastern Oklahoma (west regularly to Washington, Tulsa, and Pushmataha counties).

Breeding Habitat: The species frequents moist bottomland woods, usually deciduous but sometimes mixed or coniferous, especially young or second-growth stands, and the margins or openings of mature forests.

Nest Location: Nests are 4–30 feet above the ground, in vertical forks or crotches of shrubs or trees, usually hardwood saplings. The nest is compactly built of bark strips, grasses, rootlets, and so forth, and is lined with finer grasses, hair, and sometimes feathers. The outside is often wrapped with spider webs, to which lichens or other materials are attached.

Clutch Size and Incubation Period: From 2 to 5 eggs, usually 4. The eggs are white, with gray or brown spotting and blotching, especially at the larger end. The incubation period is 12–13 days. Single-brooded.

Time of Breeding: The breeding season in North Dakota is probably from late May to late July, and eggs have been found as late as June 27. In Kansas eggs are probably laid in May and June, and in Oklahoma nest-building has been noted as early as May 12.

417

Breeding Biology: The American redstart is one of the few warblers whose social behavior has been extensively analyzed, including both courtship and more general aggressive behavior. Males arrive on the breeding grounds before females and immediately become territorial. Females typically arrive at night and may obtain a mate by the following morning, presumably being attracted to the males by their singing. Up to 60 hours may elapse between the formation of the pair bond and the start of nest-building, during which time the female investigates the territory and begins to restrict her activities to the eventual nest site. The female selects the site alone and immediately begins to build a nest there, which may require about 3 days. Incubation and brooding are performed by the female alone, although males are typically highly attentive to the young during the nestling period, which usually lasts 8–9 days.

Suggested Reading: Ficken 1962; Bent 1953.

FAMILY PLOCEIDAE (OLD WORLD SPARROWS AND WEAVER FINCHES)

House Sparrow

House Sparrow (English Sparrow)
Passer domesticus

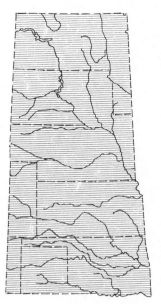

Breeding Status: Introduced and now pandemic, breeding throughout the region.

Breeding Habitat: The species is associated everywhere with humans, breeding in cities, suburbs, and around farm buildings.

Nest Location: Nests are placed in almost any kind of cavity or crevice, including those provided by buildings, dense vines growing against walls, old nests of barn swallows and cliff swallows, billboard braces, and tree cavities. The size of the nest varies with the size of the cavity, which is haphazardly filled with grasses and weeds and lined with feathers, hair, or other soft materials that happen to be available.

Clutch Size and Incubation Period: From 3 to 7 eggs, usually 4 or 5 in our region. The eggs are white with gray and brown spots and dots. The incubation period is usually 11–13 days, averaging 12.2. Normally multiple-brooded in our region.

Time of Nesting: In North Dakota the principal breeding period is from early April to mid-September. Kansas egg dates are from March 20 to August 7, with first and second clutches usually laid in early April and early May. In Texas, breeding occurs throughout the year but mainly extends from early February to late July.

Breeding Biology: Although it has been reported on the basis of European studies that mated house sparrows remain permanently faithful to their nest site and to each other, recent studies in Mississippi do not support this view and indicate that mate-changing by males is rather prevalent, even between broods during a single breeding season. In this study there was also little indication of return to a previous year's nest site by adults, particularly among females. Both sexes participate in nest-building, and although males do not develop brood patches they sometimes relieve females for as long as 20 minutes. Both sexes feed the young, and there may also be a high incidence (63 percent of 254 nests in Mississippi) of feeding by "helper" birds. The posthatching nest period varies greatly, from 12–24 days, and usually averages about 14–18 days, depending on the region. Nests are usually reused about twice in a season, and up to 4 nesting cycles have been noted in a small percentage of birds.

Suggested Reading: Sappington 1977; Summers-Smith 1963.

FAMILY ICTERIDAE (MEADOWLARKS, BLACKBIRDS, AND ORIOLES)

Red-winged Blackbird

Bobolink
Dolichonyx oryzivorus

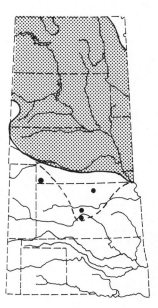

Breeding Status: Breeds in suitable habitats throughout western Minnesota and North Dakota southward through South Dakota (rarely in the Black Hills), most of Nebraska (locally common west to Sioux and Garden counties), western Iowa, northwestern Missouri, and locally or sporadically in northern Kansas (records for Stafford, Cloud, and Barton counties). There are no nesting records for Oklahoma and none for the portion of Colorado covered by this book.

Breeding Habitat: Tall-grass prairies, ungrazed to lightly grazed mixed-grass prairies, wet meadows, hayfields, retired croplands, and sometimes small-grain croplands are used for breeding.

Nest Location: Nests are simple hollows scraped into the ground or natural depressions suitable in depth and size, such as that made by a horse's hoof. The nest is invariably concealed in dense vegetation such as tall grasses, clover, or other thickly growing plants. The hollow is filled with grasses and weeds and lined with fine grasses.

Clutch Size and Incubation Period: From 4 to 7 eggs (4 North Dakota clutches averaged 4.8). The eggs are gray to cinnamon, rather irregularly blotched and spotted with browns, and sometimes are almost entirely brown. The incubation period is 13 days. Single-brooded.

Time of Breeding: In North Dakota the breeding season extends from late May to mid-August, with egg dates ranging from June 4 to June 27 and fledglings seen from July 8 to August 16. Minnesota egg dates are from May 27 to June 22, and limited information of birds at the southern edge of their range in Kansas indicates that there the eggs are laid in June as well.

Breeding Biology: Males arrive on their breeding areas about a week before females and quickly spread out, although specific territorial establishment and defense seems to be weak or lacking. Although the nests are well scattered, males tolerate other males surprisingly near the nest site. The female incubates alone; the male seldom visits her and apparently never feeds her. But males do help feed the young. Broods usually remain in the nest for about 10–14 days but have been reported to leave when only 7–9 days old, or well before they are able to fly. Males often acquire second mates after their first mate has begun nesting. These secondary mates tend to lay smaller clutches than the primary mates, perhaps because they often are young birds or are renesting. This smaller clutch size of secondary mates is adaptive, since males less frequently assist in feeding their second broods, and unassisted females are more likely to be able to tend smaller broods.

Suggested Reading: Kingsbury 1933; Martin 1971.

Eastern Meadowlark
Sturnella magna

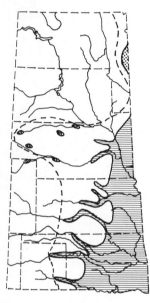

Breeding Status: Breeds commonly through Minnesota west to the Red River Valley and south to the northwestern corner of Iowa. South of that area the species is common only in Iowa, southeastern Nebraska, northwestern Missouri, and the eastern portions of Kansas and Oklahoma, but local breeding apparently extends west to the western edge of Nebraska (Garden and Sioux counties), eastern Colorado (no definite records), western Kansas (Jewell, Edwards, and Comanche counties), the Oklahoma panhandle (Beaver County), and the western panhandle of Texas (Deaf Smith County). There has been a recent western expansion into the plains along rivercourses (*Transactions of the Kansas Academy of Sciences* 75:1-19).

Breeding Habitat: The species is associated with tall-grass prairies, meadows, and open croplands of small grain, as well as weedy orchards and other open, grass-dominated habitats. At the western edge of its range, where the western meadowlark also occurs, it is predominantly limited to low and rather moist situations, such as wet meadows and the edges of sandhill marshes (*Ecology* 37:98-108).

Nest Location: Nests are on the ground in scrapes or natural depressions, well hidden in grass clumps and with a canopy woven into the surrounding vegetation. There is a lateral entrance to the nest, which is constructed of coarse grasses and lined with finer grasses, and there is sometimes also a visible trail leading to the nest through the adjacent vegetation. Nests are often placed on sloping ground and often face toward the east or north, away from the prevailing winds.

Clutch Size and Incubation Period: From 3 to 6 eggs (26 Kansas clutches averaged 5.2). The eggs are white with brown to lavender spotting, particularly around the larger end. The incubation period is 13–15 days, averaging 14. Normally double-brooded.

Time of Breeding: Minnesota nesting dates extend from late April (nests) to June 24 (hatching eggs), and fledged young have been seen as early as May 28. In Kansas, egg records are from April 10 to July 20, with most from May 1 to May 20. Oklahoma egg dates are from April 25 to June 26, and nestlings have been seen as late as July 19.

Breeding Biology: Nest-building begins early in eastern meadowlarks and evidently is performed by only the female over a period of 3-8 days, the shorter periods being typical of later nests. Polygyny is fairly frequent in this species; about 50 percent of the males in one New York study had 2 mates, and one had 3. Only the female incubates, and she leaves the nest infrequently once it has begun. The young birds normally leave their nest 11-12 days after hatching but may leave sooner if disturbed. Within 2-3 days of nest departure, the female may start a second clutch, with the male remaining to tend the young of the first brood. The female does, however, periodically feed the first brood until she begins incubating again, when the young are about 3 weeks old.

Suggested Reading: Roseberry and Klimstra 1970; Bent 1958.

Western Meadowlark
Sturnella neglecta

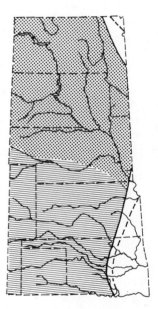

Breeding Status: Breeds over virtually all of North Dakota and western Minnesota except the most forested portions of northwestern Minnesota, all of South Dakota, western Iowa, all of Nebraska except possibly the extreme southeastern corner (occasionally present but not known to nest in adjacent northwestern Missouri), all of Kansas except for the southeastern corner, and the western half of Oklahoma (east probably to Payne, Oklahoma, Cleveland, and Marshall counties).

Breeding Habitat: In our region the species is associated with tall-grass and mixed-grass prairies, wet meadows, hayfields, weedy borders of croplands or retired croplands, and to a limited extent with short-grass prairies and sage prairies. In arid areas it is limited to moist lowland situations, but where it overlaps with the eastern species at the eastern edge of its range it occupies dry upland sites whereas the eastern meadowlark is found in moister areas (*Ecology* 37:98-108). A very limited amount of mixed pairing has been reported in such areas of habitat sharing (*Southwestern Naturalist* 10:307).

Nest Location: Nests are in the same situations as described for the eastern meadowlark, and neither the nest nor the eggs can be distinguished from those of that species.

Clutch Size and Incubation Period: From 3 to 6 eggs (16 Kansas clutches averaged 4.3 and 18 North Dakota clutches averaged 5.3). The eggs are white, with variable amounts of brown to lavender spotting near the larger end. The incubation period is 13-15 days. Probably generally double-brooded in our region.

427

Time of Breeding: In North Dakota the breeding season extends from late March to early August, with egg dates ranging from May 5 to July 6. In Kansas, the egg dates are from April 10 to July 30, with peaks in early May and early June for the first and second broods. Texas egg dates are from March 27 to July 24.

Breeding Biology: The breeding biology of this species is virtually identical to that of the eastern meadowlark, and these two species provide an interesting problem in terms of their ecology and evolutionary relationships. Where the two species occur together they are sometimes intermediate in their primary songs, but this does not prove frequent hybridization; their call notes are more diagnostic and indicative of ancestry. One area of apparent hybridization is the Platte Valley of Nebraska, where intermediate birds are several times more frequent than elsewhere in the Great Plains. In areas of overlap there has been no evolutionary convergence in song types, but apparently some of the aggressive displays of the two species do exhibit convergent elements.

Suggested Reading: Lanyon 1957; Rohwer 1971.

Yellow-headed Blackbird
Xanthocephalus xanthocephalus

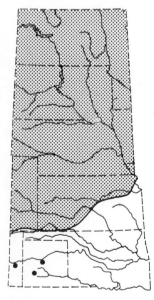

Breeding Status: Breeds in suitable habitats throughout North Dakota and western Minnesota and southward through South Dakota (except the Black Hills), western Iowa, Nebraska, northwestern Missouri, eastern Colorado, and northwestern Kansas, reaching its normal limits north and west of a line drawn from Meade to Douglas counties, Kansas. There is only a single old breeding record for Oklahoma, from Cimarron County, and likewise a single recent breeding record for Texas, for Castro County in 1978. There are no breeding records for northeastern New Mexico.

Breeding Habitat: Favored habitats are deep marshes, the marsh zones of lakes, and shallow river impoundments where there are stands of cattails, bulrushes, or phragmites. Where it breeds on the same marshes with red-winged blackbirds, this species occupies deeper areas, while the red-winged blackbirds establish territories around the marsh edges.

Nest Location: Nests are usually clustered in stands of emergent vegetation, most frequently (in North Dakota) in hardstem bulrush or cattails, with other emergent plants used relatively little. In a sample of 79 nest sites, the water depth ranged from 3 to 32 inches and averaged 18 inches, and the height of the nest rim above water also averaged 18 inches. By comparison, 28 nests of red-winged blackbirds were in water averaging only 9 inches deep, and 8 more nests were in terrestrial sites.

Clutch Size and Incubation Period: From 3 to 7 eggs (109 North Dakota clutches averaged 3.7). The eggs are off-white with spots and dots of browns and grays. The incubation period is 12–13 days. Single-brooded, but males are often bigamous.

Time of Breeding: North Dakota egg dates are from May 10 to July 13, with nestlings seen as early as May 27 and dependent young as late as August 10. Kansas egg dates are from May 20 to June 30, with a probable peak in early June.

Breeding Biology: The displays of the yellow-headed blackbird are very similar to those of the red-winged blackbird, but the species differs ecologically in that the males normally participate in brood care, are more dependent on emerging aquatic insects such as damselflies, and thus are more dependent on marshes than are redwings. In both species, the males' conspicuous and prolonged displays seem to be related to the importance of territorial size and quality in attracting the maximum number of females. As in the red-winged blackbird, only the female incubates, but males often help feed the young, particularly those of their first female. The young leave the nest at 9–12 days.

Suggested Reading: Willson 1964; Orians and Christman 1968.

Red-winged Blackbird
Agelaius phoeniceus

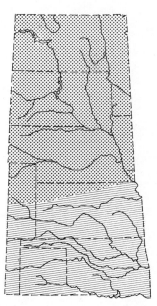

Breeding Status: Pandemic, breeding throughout the region in suitable habitats.

Breeding Habitat: Habitats range from deep marshes or emergent zones of lakes and impoundments through progressively drier habitats such as wet meadows, ditches, and brush patches in prairie, hayfields, and weedy croplands or roadsides.

Nest Location: Nests are in herbaceous or woody vegetation that is usually in or near water, but they may also be in weeds, bushes, or trees some distance from water, and, rarely, up to 14 feet above the substrate. Of 48 North Dakota nests, hardstem bulrush provided the most common vegetational support, and the rims of 28 nests built over water averaged 13 inches above the water surface. Eight terrestrial nests averaged 19 inches above ground. The nest is built of leaves of grasses and sedges woven together and bound to adjacent vegetation and is lined with fine grasses.

Clutch Size and Incubation Period: From 2 to 7 eggs (38 North Dakota clutches averaged 3.6, and 243 Oklahoma clutches averaged 3.4). The eggs are pale bluish green, with scrawls and spots of dark tones, mostly toward the larger end. The incubation period is 10–12 days. Frequently double-brooded, and males are often polygynous, with up to six females per male.

Time of Breeding: North Dakota egg dates are from May 15 to July 13, and nestlings are seen from June 9 to July 28. Kansas egg dates are from May 1 to July 30, with nearly three-fourths of the eggs laid between May 11 and June 10. Oklahoma breeding dates are from April 22 (eggs) to July 31 (nestlings).

Breeding Biology: This is one of the commonest and most thoroughly studied of all North American songbirds. Adult males arrive on their breeding marshes well before females and begin to advertise their territories by flight song and "song-spread" displays, both of which prominently exhibit the red upper wing coverts. Experiments with surgically muting males or painting these red markings black before they acquire mates result in the loss of territories by such altered males, although later alteration has no obvious effect. Pair bonds last only during the breeding season, and most territorial males manage to acquire at least two females. In one Wisconsin study, it was found that experienced males tend to return to their old territories in successive years and that first-year males are usually unable to hold territories long enough to breed. In that study, no more than three females were mated to a single male, but a few instances of double-brooding were found. The young birds leave the nest at 10–11 days but are dependent for some time thereafter.

Suggested Reading: Nero 1956; Peek 1971.

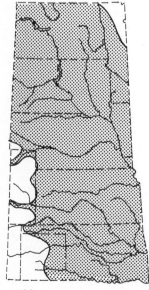

Orchard Oriole
Icterus spurius

Breeding Status: Breeds in suitable habitats virtually throughout North Dakota and western Minnesota except in the most forested areas, southward through South Dakota, western Iowa, Nebraska, northwestern Missouri, eastern Colorado (west at least to Morgan and Yuma counties), Kansas (throughout), Oklahoma (to Cimarron County), and the eastern panhandle of Texas (breeding records for Childress and Wilbarger counties). There is no breeding evidence for northeastern New Mexico.

Breeding Habitat: The species is associated with lightly wooded river bottoms, scattered trees in open country, shelterbelts, farmsteads and residential areas, and orchards. Open rather than closed woodlands are preferred, and even grassland habitats may be used if suitable nest sites are available nearby.

Nest Location: Nests are 5–70 feet (averaging about 20 feet) above the ground, in the forks or crotches of a wide variety of broadleaved trees and shrubs. The nest is a hanging, basketlike structure almost entirely made of woven grasses and is usually wider than deep, thus differing from that of the northern oriole.

Clutch Size and Incubation Period: From 3 to 7 eggs (41 Kansas clutches averaged 4.1). The eggs are pale bluish white, with spots, blotches, and scrawls of dark colors around the entire egg or concentrated near the larger end. The incubation period is 12–14 days. Single-brooded, but known to renest.

Time of Breeding: North Dakota egg records are from May 21 to July 3; Minnesota records are from May 20 (eggs) to early July (nestlings and fledglings). Kansas egg records are from May 11 to August 11, with nearly half the eggs laid during the first ten days of June.

Breeding Biology: There is a marked separation of the sexes and ages in migration, with old males arriving in breeding areas first, followed by females and, finally, first-year birds. Territories are gradually established by singing and display, and there seems to be little direct competition between orchard and northern orioles. Trees supporting kingbird nests are often used by orioles for nest sites. The nest is built over a period of about 3–6 days, and both sexes participate in its construction. The male has also been reported by an early observer to assist with incubation, but this seems unlikely, and probably his major contribution is feeding the incubating female. The male also helps feed the nestlings, which leave the nest in 11–14 days.

Suggested Reading: Bent 1958; Dennis 1948.

Northern Oriole (Baltimore and Bullock Orioles)
Icterus galbula (including *I. bullockii*)

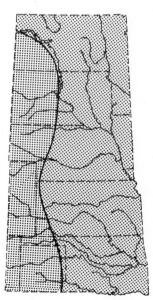

Breeding Status: Collectively, the two forms are pandemic. The eastern race *galbula* ("Baltimore oriole") generally occurs east of the 102d meridian, and the western form *bullockii* ("Bullock oriole") occurs west of this line, but there is a zone of frequent hybridization extending approximately 100 miles on either side of the line of contact (*Condor* 66:130–50; *Systematic Zoology* 19: 315–51).

Breeding Habitat: Favored habitats of the eastern race *galbula* include wooded river bottoms, upland forests, shelterbelts, and partially wooded residential areas and farmsteads. It is absent from pure coniferous forests but colonizes these areas after clear-

431

ing and the development of deciduous second-growth. In our region, *bullockii* is largely limited to river-bottom stands of willows and cottonwoods (in the north) and mature mesquite trees in flat uplands (in the southwest).

Nest Location: Nests are usually about 25 feet above the ground (range 9–70 feet) in rather large trees, especially elms and cotton-woods growing in open spaces below, and are deep woven baskets of plant fibers including grasses but not exclusively made of them as in the orchard oriole. The nest is deeper than it is wide and sometimes has a lateral rather than an upper opening. There are some regional variations in placement and structure (*Condor* 78:443–48).

Clutch Size and Incubation Period: From 2 to 6 eggs (57 Kansas clutches averaged 4.7). The eggs are pale grayish white, with spots, blotches, and scrawls of dark brown or black, especially around the larger end. The incubation period is 12–14 days. Single-brooded.

Time of Breeding: North Dakota egg dates are from June 11 to June 28, with nestlings seen from June 27 to July 6. Kansas egg records are from May 11 to July 10, with two-thirds of the eggs being laid between May 21 and June 10. Oklahoma breeding records are from May 3 (nest-building) to August 2 (nestlings).

Breeding Biology: The remarkable pendant nests of this species are built mostly by the female, sometimes in as little as 4½ days, though usually a week or so is needed. No true knots are tied in the process, but a loose tangle of fibrous materials is gradually pulled together and tighted, forming a woven structure. A new nest is made each year, but certain trees or territories from previous years seem to be favored, as the remains of old nests are often found near new ones. Incubation is by the female, who is fed on the nest by her mate. The nestling period is approximately 2 weeks, and the young are dependent on the adults for another 2 weeks. Until recently, the western race ("Bullock oriole") was regarded as a species distinct from the eastern race, but extensive hybridization in the Great Plains favors the view that they are a single species. However, recent evidence (*Condor* 79:335–42) suggests that hybrids are becoming less frequent in the overlap zone.

Suggested Reading: Bent 1958; Sibley and Short, 1964.

Brewer Blackbird
Euphagus cyanocephalus

Breeding Status: Breeds in suitable habitats virtually throughout North Dakota and western Minnesota, except perhaps the extreme southwestern corner, and much of South Dakota except for the southeastern and south-central prairie areas. In Nebraska the species is a common breeder only in the northwestern corner, but there have been sporadic breedings elsewhere (Hall, Lancaster, and Johnson counties). A reported rusty blackbird breeding in Hall County, Nebraska (*Nebraska Bird Review* 41:7), is presumed to be in error and probably involves this species. There are no definite breeding records for Kansas or northwestern Missouri, and there is only a single old record for Oklahoma, in Cimarron County. Nevertheless, a rather extensive area of breeding occurs in eastern Colorado (Kingery and Graul, 1978).

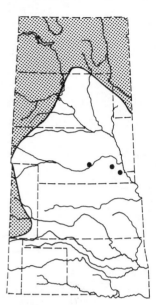

Breeding Habitat: The species is generally associated with low-stature grasslands, such as mowed or burned areas along roadsides and railroads, as well as with residential areas of towns or farmsteads. Shrubby marsh edges are also favored. Grassy areas that have small trees or shrubs for nesting and that are fairly near water seem to be used extensively.

Nest Location: Nesting may be in colonies or singly, and nests may be either on the ground or in shrubs or trees. Elevated nests are usually within 3 feet of the ground. The nest is built of twigs and grasses, usually supported by mud or dung, and is lined with grasses, rootlets, and horsehair.

Clutch Size and Incubation Period: From 3 to 7 eggs (6 North Dakota nests averaged 4.8). The eggs are grayish to pale greenish, with dark dots, spots, and blotches over the entire egg or concentrated near the larger end. The incubation period is 12–14 days. Sometimes double-brooded.

Time of Breeding: Egg records in North Dakota are from May 25 to July 1, and fledged but dependent young have been seen from June 19 to July 25. In South Dakota fledged but dependent young have been reported as early as June 16 and as late as August 6.

Breeding Biology: Brewer blackbirds are often colonial nesters, but unlike red-winged and yellow-headed blackbirds they are more frequently monogamous than polygynous. In contrast to these species, pair formation begins when the birds are still in winter flocks, and frequently mates of the previous season will form pair bonds again. The female builds the nest, although the male may often accompany her as she gathers material. From 10 to 14 days are spent in building the nest and laying the clutch. Males do not assist in incubation, but, rarely, do visit the nest to feed the incubating female. Both sexes feed the young, which leave the nest in about 13 days, and fledglings may be cared for by their parents for at least 3 weeks. At least in some areas, double-brooding is fairly frequent, and as many as three nesting attempts may be made in a single season by an unsuccessful pair.

Suggested Reading: Williams 1952; Horn 1970.

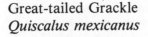

Great-tailed Grackle
Quiscalus mexicanus

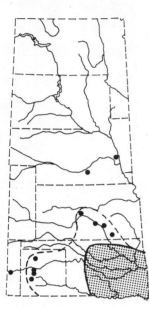

Breeding Status: Until the late 1960s this species was limited to eastern Oklahoma, which it colonized from the south in the 1950s, quickly spreading westward and northward to Caddo, Woods, Garfield, Payne, Tulsa, and Mayes counties. By 1970 the species had spread into central Kansas, breeding at least in Sedgwick, Reno, and Barton counties. Finally, in 1977 the species bred in Douglas and Adams counties, Nebraska *(Nebraska Bird Review* 45:18), suggesting that the rapid northern expansion of its range has not terminated. It has also spread into the Texas panhandle, with nesting observed in Potter, Randall, Moore, and Castro counties (K. Seyffert, pers. comm.), and has bred at Tucumcari, New Mexico.

Breeding Habitat: The species is associated with a wide variety of habitats but generally is found near standing water and open ground; thus it is especially common in farmed prairie or other manmade habitats provided by the planting of shade trees and irrigated crops (*Condor* 63:37–86). The very similar and closely related boat-tailed grackle (*Quiscalus major*) is largely confined to coastal marshes in Texas.

Nest Location: Nests are elevated from 2 feet above the water in marshes to about 50 feet in tall trees; generally they are as high as the surrounding vegetation will permit. Tree nests in Texas range from 5 to 15 feet above the ground. Where trees are not available, utility poles or other artificial structures may be used. The nest is a bulky mass of grasses, rushes, and so forth, varying greatly in size and lined with mud followed by fine grasses and weed stems.

Clutch Size and Incubation Period: From 2 to 5 eggs, usually 4. The eggs are light blue to bluish gray, with scrawls or spots of darker tones over much of the surface. The incubation period is 13 days. Single-brooded in our area, but renesting is likely.

Time of Breeding: In Nebraska, eggs were found in mid-May, and adults taking food to presumed nestlings were seen as early as May 12. In Oklahoma eggs have been found from May 7 to July 1, nestings seen from late May to late June, and fledged young as early as May 25. Egg dates in Texas are from April 1 to July 19.

Breeding Biology: The recent recognition that the "boat-tailed grackle" consists of two different species has confused the earlier literature on these birds, but the males of this species are fairly easily recognized by their yellow iris coloration. In a Texas study, it was found that nesting occurred from mid-May through early August, but this prolonged period was the result of renesting rather than multiple brooding. The females build the nest in 5–10 days, without any help from the males, which apparently are polygynous or promiscuous. Nests are in colonies, within which the birds refrain from destroying eggs of their own or other species. However, beyond the colony limits the eggs and young of virtually all species up to the size of coots are sought out and eaten. The nestlings fledge at 20–23 days and thereafter follow their mothers about for some time, begging at every opportunity.

Suggested Reading: Tutor 1962; Pruitt 1975.

Common Grackle
Quiscalus quiscula

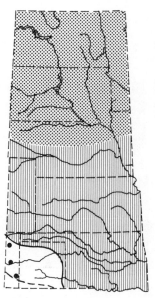

Breeding Status: Nearly pandemic throughout the region, but becoming progressively rarer in the more arid southwestern parts. Its southern limits probably include all of Oklahoma, but it is uncommon and local in the panhandle. In the Texas panhandle it is becoming widespread (possible breeding in Armstrong County is the only nesting record), and is apparently colonizing the area. In northeastern New Mexico it is local and mostly confined to planted trees in residential areas, having bred in the Cimarron Valley, Clayton, Des Moines and Tucumcari.

Breeding Habitat: The common grackle generally frequents woodland edges or areas partially planted to trees, such as residential areas, parks, farmsteads, shelterbelts, and the like. Both coniferous and deciduous tree plantings are used, and tall shrub thickets near croplands or marshes are also favored.

Nest Location: Nests are 1–60 feet above the ground or water, averaging about 20 feet, in coniferous or deciduous trees, shrubs, birdhouses, cavities, ledges, or even cattails. Higher elevations are preferred when available. The nest is a bulky structure of grasses and weeds, sometimes reinforced with mud, lined with finer grasses and sometimes feathers. Nests are often in loose colonies.

Clutch Size and Incubation Period: From 3 to 6 eggs (33 Kansas clutches averaged 4.5). The eggs are usually pale greenish with dark brown or purplish spots and scrawls. The incubation period is 13–14 days. Single-brooded.

Time of Breeding: North Dakota egg dates are from May 7 to June 22, with dependent young seen as late as July 23. Kansas egg records are from April 11 to June 30, with two-thirds of the eggs laid between May 1 and May 20. In Oklahoma, egg dates are from May 6 to June 15, with nestlings seen as late as July 21.

Breeding Biology: Males of this colonial-nesting species usually arrive on their breeding areas well before females and remain in flocks until the females arrive. There is only a gradual breakup of

435

migratory flocks as pairs are gradually formed. A major component of pair formation is a flight involving a single female and up to five males, which follow her while keeping their tails strongly keeled. After pairing occurs, the females begin to select nest sites, and their mates defend only a small area of the nesting tree. The female gathers most of the material and does all the actual construction, which sometimes takes about a week and sometimes may be spread out over several weeks. The female incubates alone and also does all the brooding. Both sexes feed their young, which leave the nest at 10–17 days.

Suggested Reading: Maxwell and Putnam 1972; Maxwell 1970.

Brown-headed Cowbird
Molothrus ater

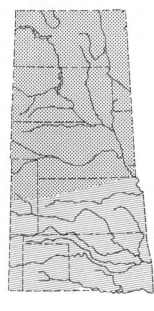

Breeding Status: Pandemic, breeding throughout the region.

Breeding Habitat: The species is generally associated with woodland edges, brushy thickets, and other habitats where low or scattered trees are interspersed with grassland vegetation. Originally presumed to associate with bison, it more recently has adapted to feeding near cattle and has spread to both coasts as deforestation and cattle-raising have become more widespread.

Nest Location: No nests are constructed; the species is an obligate parasite. Although more than 100 species have been found to serve as hosts, the most important host groups are the tyrant flycatchers, the vireos, the warblers, and the finches. The thrush and icterid families are also frequent targets of parasitism.

Clutch Size and Incubation Period: The "clutch" probably consists of about 6 eggs, laid one per day in different nests. After several days a second clutch may be laid, and a female may lay up to about 20 eggs in a single season. The eggs are white, with small brown spots, especially at the larger end. The incubation period is usually 11–12 days and is performed by the host species. Studies in North Dakota indicated that 1–6 cowbird eggs occurred in 69 parasitized nests, averaging 2.0. Presumably most of these multiple layings were by more than one female.

Nesting Period: North Dakota egg dates range from May 10 to July 15, with nestlings seen from June 15 to July 25, and include more than 30 host species (Stewart 1975). In Kansas, egg dates are from April 21 to July 20 and include 40 hosts (Johnston 1964). Eggs in Oklahoma have been seen from April 13 to July 23, and at least 39 species have served as hosts (Sutton 1967). Species that have served as hosts in all three states include the red-winged

436

blackbird, brown thrasher, chipping sparrow, dickcissel, and lark sparrow. Common hosts in Nebraska, Kansas, and Missouri include the Bell vireo, meadowlarks, cardinal, indigo bunting, and field sparrow (*Bird-Banding* 48:358).

Breeding Biology: This is the only species of North American bird that is an obligatory nest parasite. Before colonization the species was largely limited to the Great Plains, but recently it has come into contact with many new potential host species that have not had time to evolve defensive mechanisms. These include many forest-adapted songbirds, including the rare Kirtland warbler. There is no egg-mimicry in this species, but apparently each cowbird normally lays only one egg in each host's nest, usually during the host's egg-laying period. The female sometimes but not always removes a host's egg from the nest as well, but only when at least two eggs are already present. The eggs do not always hatch at the same time as or before the host's eggs, but the nestling period is roughly 10 days, about the same as for many of its hosts. However, the young grow much more rapidly than the host nestlings and thus the amount of food available for the young of the host is correspondingly reduced.

Suggested Reading: Mayfield 1965; Bent 1958.

FAMILY THRAUPIDAE
(TANAGERS)

Western Tanager

Western Tanager
Piranga ludoviciana

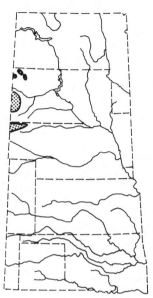

Breeding Status: Mostly limited to the Black Hills of South Dakota, where it is a common resident. Also breeds uncommonly in the Pine Ridge area of northwestern Nebraska, possibly extending east in the Niobrara Valley far enough to come into contact with the scarlet tanager (*Nebraska Bird Review* 29:19).

Breeding Habitat: In the Black Hills and Pine Ridge areas this species is primarily associated with pine forests, and it secondarily uses deciduous woods along rivers or in gulches and canyons. In more mountainous country it extends higher, into the Douglas fir zone, inhabiting relatively dense and mature stands.

Nest Location: Nests are 6-50 feet above the ground, usually on horizontal branches of conifers, 3-20 feet out from the trunk. Infrequently, deciduous trees are used as nest sites. The nest is substantial but rather crude, constructed of twigs and rootlets lined with rootlets, hair, or other soft materials.

Clutch Size and Incubation Period: From 3 to 5 eggs. The eggs are pale bluish, with specks, spots, or blotches of brown, usually concentrated near the larger end. The incubation period is 13 days. Probably single-brooded.

Time of Breeding: In the Black Hills, breeding occurs during June and July, with eggs reported as early as June 10 and dependent fledglings seen as late as August 20.

Breeding Biology: Surprisingly little has been written on the breeding biology of this beautiful species, but it presumably closely resembles that of the scarlet and summer tanagers. In spite of the bright coloration of the males, breeding pairs are not conspicuous, since the olive-colored female remains high in the trees, and they tend to be very elusive during nesting. The female incubates alone, and the male evidently rarely if ever approaches the nest during this period. After hatching, he does help feed the young, which probably fledge in about 2 weeks, judging from what is known of the other tanagers.

Suggested Reading: Bent 1958.

Scarlet Tanager
Piranga olivacea

Breeding Status: Breeds locally in North Dakota (Pembina Hills, the Cheyenne, Red, and lower Missouri valleys, rarely the Turtle

441

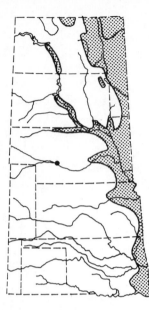

Mountains), western Minnesota, eastern and central South Dakota, western Iowa, eastern Nebraska (west to Cherry County along the Niobrara River), northeastern Kansas (west at least to Clay County), northwestern Missouri, and eastern Oklahoma (west to Mayes and Pushmataha counties).

Breeding Habitat: In our region the species is restricted primarily to mature hardwood forests in river valleys, hill slopes, and valleys; it is less frequently found in coniferous forests and in city parks and orchards.

Nest Location: Nests are usually in tall trees, often oaks, 8–75 feet above the ground, but usually between 35 and 50 feet high, well out on horizontal limbs. The nest is rather small and loosely constructed of twigs and rootlets, lined with grasses and weed stems.

Clutch Size and Incubation Period: From 3 to 5 eggs, usually 4. The eggs are pale greenish or bluish, with brown specks, spots, or blotches, especially at the larger end. The incubation period is 13–14 days. Single-brooded.

Time of Breeding: In North Dakota, active nests have been found between mid-June and mid-July. Kansas egg records are from May 11 to June 20, and in Oklahoma active nests have been found in early June and early July.

Breeding Biology: The relatively late spring arrival of this species, combined with its typical foraging characteristics of remaining high in the canopy of mature trees, keeps most of its behavior obscured from normal view, much to the disappointment of bird watchers. Further, the nests are usually inaccessible for easy observation. It is known that the female incubates alone and in some cases the male participates very little even in feeding of the young. The young leave the nest in about 15 days and remain with their mother in the general vicinity of the nest for 10 days or more.

Suggested Reading: Bent 1953; Prescott 1965.

Summer Tanager
Piranga rubra

Breeding Status: Breeds from the Missouri Valley of southeastern Nebraska (north to Sarpy County) southward through northwestern Missouri (uncommon at Squaw Creek N.W.R.), eastern Kansas (at least to Doniphan, Shawnee, and Montgomery counties), and the eastern half of Oklahoma (locally west to Cleveland and Comanche counties).

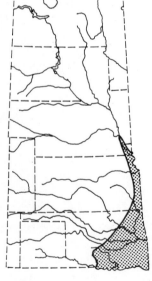

Breeding Habitat: The species habitat is not greatly different from that of the scarlet tanager, namely upland forests including hardwoods, mixed woods, and also open coniferous forests. It perhaps favors somewhat lower and more open forests than those used by scarlet tanagers, but at the northern edge of its range it is likewise limited to rich bottomland forests.

Nest Location: Nests are 10–35 feet above the ground, usually in deciduous trees such as oaks, placed well out from the trunk on horizontal branches. The nest strongly resembles that of the scarlet tanager, and the eggs are likewise very similar.

Clutch Size and Incubation Period: From 3 to 5 eggs, usually 4. The eggs are pale green to bluish, with brown specks, spots, and blotches that tend to be slightly heavier than those of the scarlet tanager. The incubation period is 11–12 days. Single-brooded.

Time of Breeding: Kansas egg records are from May 21 to July 20, with a peak of egg-laying around June 5. Oklahoma egg dates are from May 8 to July 17, and newly fledged young have been seen as late as early August.

Breeding Biology: A study of this species in northeastern Kansas indicated that territories were centered in areas of thick second-growth deciduous forest. The pair members seemed to remain in contact by uttering clicking notes, and in one case both usually stayed within an area of about 300 by 600 feet. The nest was built by the female, with the male standing guard, and required about 2 weeks. Incubation was by the female alone, but both sexes fed the young. At the age of only 1 week one brood left the nest, and the young fell to the ground. After another 3 days they were able to fly into nearby trees and were nearly feathered. They gradually moved away from the nest site but remained in the parents' territory for nearly 3 weeks more. A study in North Carolina indicated that courtship feeding of the female is probably an important part of this biology, since it seems to strengthen the pair bond and also probably prepares the male for feeding the young. In that study the young birds left the nest on the 9th or 10th day after hatching, and one of the young was singing a "primitive version" of the male's song when only 2 weeks old.

Suggested Reading: Potter 1973; Fitch and Fitch 1955.

FAMILY FRINGILLIDAE (GROSBEAKS, FINCHES, SPARROWS, AND BUNTINGS)

Cardinal

Cardinal
Cardinalis cardinalis (*Richmondena cardinalis*)

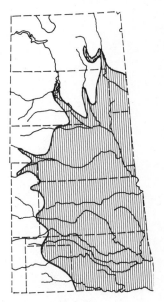

Breeding Status: Breeds from the Red River Valley of southeastern North Dakota (north to Cass County) southward through adjacent southwestern Minnesota, eastern South Dakota (north along the Missouri River at least to Dewey County), most of Nebraska (west to Cherry, Thomas, and Deuel counties), eastern Colorado (at least to Morgan County), virtually all of Kansas (local and infrequent west of Seward, Hamilton, and Cheyenne counties), Oklahoma (west locally at least to Texas County), and the eastern panhandle of Texas (west at least to Potter County), The northern range limits of this species have expanded considerably in this century (*Wilson Bulletin* 68:111-17).

Breeding Habitat: The species is associated with forest edges or brushy forest openings, parks and residential areas planted to shrubs and low trees, and second-growth woods, and with river bottom gallery forests in grasslands.

Nest Location: Nests are usually 3–8 feet (rarely to 40 feet) in dense shrubbery, small trees, vines, or briar thickets. The nest is constructed of twigs, vines, and other materials in a loose cup, lined with fine grasses and hair. The nests are usually well concealed in forks or in mats of vine stems but at times are placed very close to human traffic patterns.

Clutch Size and Incubation Period: From 3 to 6 eggs (65 Kansas clutches averaged 3.5). The eggs are grayish to bluish white, with brown dots, spots, and blotches varying greatly in extent. The incubation period is 12–13 days. Regularly multiple-brooded in our region.

Time of Breeding: Nests with eggs have been found in North Dakota and Minnesota from late April to early June. In Kansas the range of egg dates is from April 1 to September 20, with a peak in initial clutches about May 1, followed by asynchronous breeding through the summer. Texas egg records extend from March 3 to July 31.

Breeding Biology: Pair bonds are probably fairly permanent in this relatively sedentary species, and toward the end of winter males show such signs of the reawakening of sexual behavior as tolerating females on the same feeding platform and sometimes directly feeding them. Territorial display begins early; the males sing in prominent locations and chase other males away. Thus nonbreeding assemblages gradually break up, and females begin to seek out suitable nest sites. The female usually builds the nest alone, over a period of 3 to 9 days, and lays the first egg within a week of its completion. Normally the female incubates alone, although at times males have been observed sitting on the nest. Both parents feed the nestlings, which leave the nest in 7–11 days and are able to fly well by the time they are 19 days old. By about 45 days the young are completely independent of their parents, which by then have usually begun another brood. At least three broods are commonly raised in a season, and in some areas as many as four are fairly common.

Suggested Reading: Laskey 1944; Bent 1968.

Rose-breasted Grosbeak
Pheucticus ludovicianus

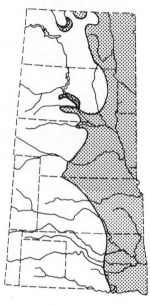

Breeding Status: Breeds in western Minnesota and eastern North Dakota to the Souris River and rarely the Missouri River (*Wilson Bulletin* 85:230–36), southward through eastern South Dakota west at least to Todd County (*Wilson Bulletin* 85:1–11), eastern Nebraska west to Garfield and Buffalo counties (*Nebraska Bird Review* 29:19), eastern Kansas west locally to Rawlins County (Rising 1974), and eastern Oklahoma (west locally at least to Kay and Cleveland counties). Hybridization with the black-headed grosbeak makes the western limits of this species rather difficult to define (*Auk* 79:399–424).

Breeding Habitat: The species is associated with relatively open deciduous forests on floodplains, slopes, and bluffs. The relative development of the understory is apparently not particularly important in this species (*Wilson Bulletin* 86:7).

Nest Location: In a South Dakota study, most grosbeak nests (both species) were 10–19 feet above the ground, and box elders were favored nesting trees. More generally the forks and crotches of various deciduous trees are used, and the nest is poorly constructed of twigs and grasses, lined with fine twigs and rootlets.

Clutch Size and Incubation Period: From 3 to 5 eggs (6 North Dakota nests averaged 3.8). The eggs are pale grayish to bluish, with dark brownish spots and blotches around the larger end, often forming a cap or wreath. The incubation period is 12–14 days. Usually single-brooded; double-brooding is reported in semicaptive birds.

Time of Breeding: North Dakota egg dates are from May 31 to June 27. In Kansas, eggs have been reported from May 11 to July 11, with a probable peak of laying in early June.

Breeding Biology: Immediately after the males return to their breeding areas in spring, they establish territories and begin to announce them with a warbled song and aggressive encounters with other males. Females arrive a few days later and are initially chased aggressively by males. Soon the male stops chasing the female, and she may attack him instead. Courtship feeding of the female is apparently uncommon in this species. The female builds the nest with the help of the male, and the male regularly partici-

448

pates in incubation. Both sexes care for the young, and at least in two cases the males have been known to take over the care of young birds while the female began a second nesting. The young remain in the nest 9–12 days.

Suggested Reading: Dunham 1966; Bent 1968.

Black-headed Grosbeak
Pheucticus melanocephalus

Breeding Status: Breeds in western North Dakota east to the Missouri River and rarely to the Souris River, in western South Dakota east at least to Charles Mix County, and occurs as hybrids to Clay County (*Wilson Bulletin* 86:5), in Nebraska eastward locally to at least Rock, Garfield, and Hall counties (*Nebraska Bird Review* 29:19), in northwestern Kansas (east locally or sporadically at least to Cloud and perhaps Sedgwick counties, but mostly confined to the northwestern counties), and northeastern Colorado (at least the Platte Valley). The only other breeding record for our region seems to be Quay County, New Mexico. Eastern limits of breeding are confused by frequent hybridization with the rose-breasted grosbeak (*Auk* 79:399-424; *Wilson Bulletin* 85:230-36).

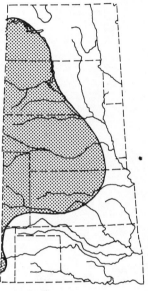

Breeding Habitat: The species is associated primarily with relatively open stands of deciduous forest in floodplains or uplands, especially those with well-developed understories. Also found secondarily in orchards, brushy woodlands or chaparral, and parks or suburbs with many trees.

Nest Location: The nest location and nest structure of this species seems to be identical to that of the rose-breasted grosbreak. Nests average about 10 feet above the ground and are usually in crotches or forks of horizontal or vertical branches of deciduous trees. The nest is a bulky structure of slender twigs, stems, and rootlets, lined with rootlets and fine stems.

Clutch Size and Incubation Period: From 2 to 5 eggs, usually 3 or 4. The eggs are not distinguishable from those of the rose-breasted grosbeak. The incubation period is 12 days. Probably single-brooded.

Time of Breeding: The probable breeding season in North Dakota extends from late May to late July. In Kansas the egg records are from May 11 to July 10, with a peak of egg-laying in early June.

Breeding Biology: So far as is known, the breeding biology of this species is essentially identical to that of the rose-breasted grosbeak. Studies of these two closely related forms in North Dakota indicate that the courtship behavior of the two is very similar, and

449

thus the color differences among the males are likely to be important in avoiding more widespread hybridization than occurs. Males apparently do not discriminate between the songs of their own and the other species, but do make visual discriminations when confronted with mounted males placed in their territories.

Suggested Reading: Weston 1947; Kroodsma 1970.

Blue Grosbeak
Guiraca caerulea (*Passerina caerulea*)

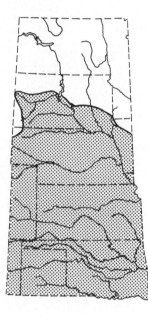

Breeding Status: Breeds from central South Dakota (north at least to Haakon and Dewey counties, *Nebraska Bird Review* 29:11) southward through virtually all of Nebraska (local in east), western Iowa (uncommonly), northwestern Missouri (occasionally), Kansas (increasingly common to west), eastern Colorado (uncommonly), Oklahoma (throughout), northern Texas and northeastern New Mexico.

Breeding Habitat: The species is associated with weedy pastures, old fields growing up to saplings, forest edges, streamside thickets, and hedgerows.

Nest Location: Nests are placed 3–12 feet high in a variety of shrubs, small trees, vines, and so forth, usually at the edge of a clearing. The nest is compactly constructed of grasses, bark strips, and stems, frequently incorporating snakeskins, and is lined with grasses and rootlets.

Clutch Size and Incubation Period: From 3 to 5 eggs, usually 4. The eggs are pure white to a pale blue that fades to white. The incubation period is 12 days, with about 2 days required to complete hatching in one case (Bailey and Niedrach 1965). Double-brooded.

Time of Breeding: Kansas egg records are from May 21 to June 30, with a peak of egg-laying in late May or early June. In Oklahoma, egg records are from May 26 to August 6, with fledglings just out of the nest seen as late as August 28.

Breeding Biology: Although relatively little has been written on the breeding biology of this species, it is probably much like that of the indigo and lazuli buntings. The nests are typically built quite close to the ground, and incubation is by the female alone. The young are fed actively by both parents and remain in the nest for 9–13 days. In one Colorado nesting, a female finished laying her second clutch in a nest 58 feet away from the first nest site almost exactly a month after she completed her first clutch. She would thus have had to begin the second nest very shortly after fledging the first brood.

Suggested Reading: Stabler 1959; Bent 1968.

Indigo Bunting
Passerina cyanea

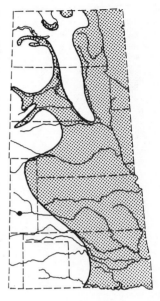

Breeding Status: Breeds in North Dakota primarily east of the James River, but common in the lower Missouri and rarely extending west into its tributaries, through South Dakota west locally to the Black Hills and the Wyoming border, Nebraska west to the Pine Ridge area and in the South Platte Valley to the Colorado border, in Kansas west at least to Ford and Clark counties, in Oklahoma west regularly to Grady County and occasionally to Alfalfa County. The western limits of this species are greatly confused by hybridization with the lazuli bunting in North Dakota (Kroodsma 1970), South Dakota and Nebraska (*Auk* 76: 433–63), western Kansas (Johnston 1964), and Oklahoma (Sutton 1967).

Breeding Habitat: The species is associated with relatively open hardwood forests on river floodplains or in uplands. Although it occasionally breeds inside forests, it is most often associated with open, drier woodland, favoring sites with decreased forest canopy and sapling density, and increasing shrub density. Thus, orchards, weedy fields, forest edges, second-growth, and similar successional habitats are all widely used.

Nest Location: Nests are in the crotches of shrubs, in vine tangles, or in low trees, at heights of 2–12 feet above the ground. The nest is built of grasses, twigs, bark strips, and weeds, lined with grasses and other soft materials.

Clutch Size and Incubation Period: From 2 to 4 eggs (17 Kansas clutches averaged 3.1). The eggs are white to pale bluish white. The incubation period is 12–13 days. Frequently double-brooded.

Time of Breeding: Minnesota egg dates are from June 6 to late June, and fledglings have been seen as early as July 4. In Kansas, egg records are from May 11 to August 20, with a peak of egg-laying in mid-June. Oklahoma egg dates range from May 13 to July 11, and dependent young have been seen as late as September 14.

Breeding Biology: In this species it is apparently the female that not only selects the specific nest site but also does all the actual construction. Sometimes nesting concentrations are very dense, such as 14 nests found in a 3-acre cotton patch in Mississippi. Both sexes help feed the young, though only the female broods them, and sometimes the male may take charge of older nestlings so the female can begin a second nesting. Usually a new nest is built for the second clutch, but sometimes the same one is used twice.

Suggested Reading: Bent 1968; Sibley and Short 1959.

451

Lazuli Bunting
Passerina amoena

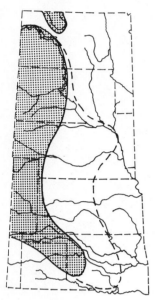

Breeding Status: Breeds in North Dakota, primarily in the Missouri and Souris valleys and their tributaries, in South Dakota primarily west of the Missouri River, in western Nebraska (common in the Pine Ridge area, occasionally east to Cherry County; the Platte Valley population is largely hybrid), eastern Colorado, southwestern Kansas (common in Morton and Hamilton counties), western Oklahoma (panhandle and east at least to Ellis and Roger Mills counties; hybrids seen east to Marshall County). Presumably the Texas panhandle is part of the breeding range, but there have been no recent nesting records, and likewise there are no definite breeding records for northeastern New Mexico. The eastern limits of this form's range are greatly obscured by hybridization (*Auk* 76:443-63), and there may have been some recent range retractions as well (*Wilson Bulletin* 87:145-77).

Breeding Habitat: Seemingly, this species' habitat needs are identical to those of the indigo bunting—rather diverse habitats having an abundance of shrubs, low trees, and herbaceous vegetation. In much of the arid west it is associated with rivers or streams, which are often the only areas that support shrubs and trees.

Nest Location: Nests are usually 1-4 feet above the ground, in shrubby growth, and rarely as high as 10 feet. The nest cannot be distinguished from that of the indigo bunting, nor can the eggs.

Clutch Size and Incubation Period: From 3 to 5 eggs, usually 4. The eggs are very pale bluish white. The incubation period is 12 days. Double-brooded.

Time of Breeding: In North Dakota, the probable breeding season is from early June to late August. In the Black Hills of South Dakota the nesting season is from late May through July, with nestlings seen as early as June 10. In Kansas, eggs are laid in June and July, and Oklahoma eggs have been found as early as May 26.

Breeding Biology: The breeding biology of this species can be considered identical to that of the indigo bunting. In the Great Plains these two species overlap appreciably, and the two forms occasionally hybridize (about 6-7 percent in one study). Playbacks of songs indicate that in some areas males respond only to the song of their own species and ignore that of the other, but in one area of sympatry (Chadron, Nebraska) males responded to both song types. This suggests that learning may be involved in song recognition. Mixed matings in areas of sympatry are infrequent and seem to exhibit delayed breeding characteristics compared with nonmixed pairs.

Suggested Reading: Emlen, Rising, and Thompson 1975; Bent 1968.

Painted Bunting
Passerina ciris

Breeding Status: Breeding is mostly restricted to the eastern half of Oklahoma, westward to Alfalfa, Blaine, Caddo, and Jefferson counties, but locally west to the Texas border and to northeastern Beaver County. Breeding also occurs infrequently in southeastern Kansas, west to Barber and north to Shawnee and Douglas counties (*Bulletin of the Kansas Ornithological Society* 21:7–8).

Breeding Habitat: Habitats used in Oklahoma include thickets, shelterbelts, wooded ravines, and forest edges. In some parts of its range the species nests regularly in cities, and it also is frequently found in shrubby river bottoms.

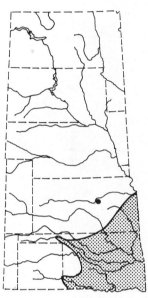

Nest Location: Nests are usually in low bushes or vine tangles at heights of 3-6 feet, rarely as high as 25 feet. The nest is a shallow cup constructed of grasses, leaves, and weed stems, lined with grasses, rootlets, and hair.

Clutch Size and Incubation Period: From 3 to 5 eggs, usually 4. The eggs are pale bluish or grayish white with reddish brown spotting. The incubation period is 11–12 days. Regularly double-brooded, sometimes triple-brooded in the south.

Time of Breeding: In Kansas, eggs are laid in June and July. Oklahoma nesting dates are from May 23 (eggs) to July 9 (brood fledging). Texas egg dates range from April 27 to August 19.

Breeding Biology: Male painted buntings are highly territorial and frequently engage in extended fights while they are establishing territories and while females are arriving. Apparently most of the obligations of reproduction, including nest-building, incubation, and brooding, are undertaken by the female alone. It has been suggested that the male does not even participate in feeding the young while they are in the nest, though he occasionally may feed them after they have fledged. This seems rather unlikely, particularly inasmuch as at least two breedings are common in this species, and thus it would be advantageous for the male to participate fully in care of at least the first brood.

Suggested Reading: Bent 1968; Parmelee 1959.

453

Dickcissel
Spiza americana

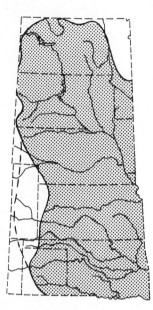

Breeding Status: Breeds in suitable habitats throughout North Dakota except the extreme northwestern corner, in southwestern Minnesota, and western Iowa, essentially all of South Dakota, all of Nebraska except the southwestern counties, throughout Kansas (local in the extreme west) throughout Oklahoma (local in the panhandle), and the Texas panhandle (common only in northern counties). There are no breeding records for northeastern New Mexico and no definite breeding records for eastern Colorado, where it probably nests in the Platte Valley at least to Logan County (*Nebraska Bird Review* 29:11). This species' range has increased considerably in this century (*Auk* 93:112-15).

Breeding Habitat: The species is associated with grasslands having tall grasses, forbs, or shrubs, or with croplands planted to crops such as timothy, alfalfa, and clover.

Nest Location: Nests are from ground level (usually) to about 12 feet (rarely) above the ground in grasses, weeds, or sometimes trees or shrubbery. They are often in natural depressions, well hidden by vegetation. The nest is built of weeds, grass stems, and leaves, lined with fine grasses, rootlets, and hair. Of 108 Oklahoma nests, only 26 were on the ground, and most were in various tree species, but generally throughout the range ground-nesting seems prevalent.

Clutch Size and Incubation Period: From 3 to 5 eggs (14 Kansas clutches averaged 4.1). The eggs are pale blue and unmarked. The incubation period is 11-12 days. Possibly double-brooded, at least in southern areas.

Time of Breeding: In North Dakota the probable breeding season is from early June to mid-August. Kansas egg records are from May 1 to July 10, with a peak of egg-laying in early May. Oklahoma egg records are from May 26 to August 8, and Texas egg records are from April 16 to July 30.

Breeding Biology: This is one of the rather few species of North American passerine birds that regularly practices polygyny; a Kansas study indicated that 18 percent of the males had more than one mate, 40 percent were monogamous, and 42 percent were unmated. The variable success of males in attracting females seems to be related to the nest sites available in their individual territories. The females build the nests and perform all the incubation; the male also feeds the young little if at all. Males usually obtain second mates during the laying or incubation phases of the first nesting cycle. Young birds remain in the nest 8-10 days and tend to be very silent until within a day or two of leaving it.

Suggested Reading: Harmeson 1974; Zimmerman 1966.

Evening Grosbeak
Hesperiphona verspertina

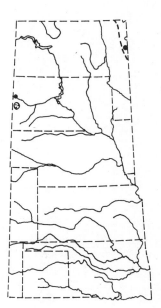

Breeding Status: Rare breeder. It is reported to breed at Itasca State Park, Clearwater County, Minnesota, and breeding has been inferred in Beltrami County. The only other area of known breeding is western South Dakota, where it regularly nests in Spearfish Canyon in the Black Hills, and has bred near Belle Fourche, Butte County.

Breeding Habitat: The species is associated with mature coniferous forests for the most part, but nests have also been found in willows growing beside rivers and even in city gardens near roads or sidewalks.

Nest Location: Nests are 15–125 feet above the ground, usually between 20 and 60 feet, on the forks of horizontal limbs of conifers. The nest is rather loosely constructed, consisting of a foundation of twigs on which are placed mosses, lichens, and grass fibers or rootlets, and is lined with finer materials.

Clutch Size and Incubation Period: From 2 to 5 eggs, usually 3–4. The eggs are blue to bluish-green, with dark spots, blotches, and sometimes fine lines. The incubation period is 11–14 days. Apparently double-brooded.

Time of Breeding: No specific egg dates are available for Minnesota, but Manitoba and Michigan egg records are concentrated between June 18 and June 24. Dependent fledglings have been seen in Minnesota as early as July 18. A female that had apparently laid 3 eggs was collected on June 11 in Spearfish Canyon, northern Black Hills.

Breeding Biology: Surprisingly little is known of the breeding biology of this handsome species. In winter the birds are gregarious and are strongly attracted to box elder trees, where they feed on the hanging seeds, as well as to various species of maples. Courtship displays seem to consist of the male's crouching, then spreading and quivering his wings while fluffing his plumage. Females solicit courtship feeding by bobbing their heads and swaying their bodies in front of males while fluttering their wings; this or a similar display with tail-raising precedes copulation. Males apparently accompany their mates while the females gather nesting materials, but presumably the female does all the nest-building. The female also incubates, but the male feeds her both off and on the nest. The fledging period is still unreported, but in one case the first egg of a second clutch appeared when the single nestling of the first cycle was only 11 days old and still in the nest.

Suggested Reading: Bent 1958.

Purple Finch
Carpodacus purpureus

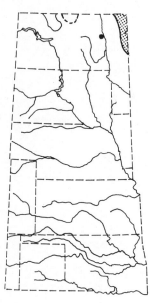

Breeding Status: Known breeding is limited to north-central Minnesota (Clearwater County); probably the more northerly wooded areas of Minnesota and the Turtle Mountains of North Dakota are also part of the breeding range. The only definite recent nesting records for North Dakota are from Grand Forks County, in 1971 and 1972.

Breeding Habitat: The species frequents coniferous forests, mixed forests, and plantings of conifers.

Nest Location: Nests are usually in coniferous trees, especially spruces, but sometimes are in deciduous trees or shrubs. Nests are near the treetops, up to about 40 feet high, on horizontal branches. They are built of grasses and roots, lined with fine grasses and hair.

Clutch Size and Incubation Period: From 4 to 6 eggs, usually 4 or 5. The eggs are greenish to bluish, with brown specks and spots, often concentrated at the larger end. The incubation period is 13 days. Single-brooded.

Time of Breeding: Minnesota breeding records extend from May 22 (nest-building) to July 4 (young just out of nest).

Breeding Biology: As part of his display the male does a "dance" in front of the female, often with a bit of nesting material in the mouth. Crest-raising, tail-cocking, and wing-drooping are all part of this sequence, which apparently serves as a premating display in some instances. Both sexes apparently assist in nest-building, and the male sometimes has been reported on the nest, though it is doubtful that he actually incubates. The young are fed by both parents and leave the nest in about 14 days.

Suggested Reading: Bent 1968.

House Finch
Carpodacus mexicanus

Breeding Status: Breeds from the Scottsbluff area of Nebraska (one record from Lincoln County; *Nebraska Bird Review* 21:38) southward through eastern Colorado, northwestern Kansas (*Bulletin of the Kansas Ornithological Society* 28:9), the Oklahoma panhandle (Cimarron County), northeastern New Mexico (Union and Quay counties), and the western panhandle of Texas (Oldham County).

Breeding Habitat: The species breeds in open woods, river-bottom thickets, scrubby or desert vegetation, ranchlands, and suburbs or towns.

Nest Location: Most nesting in trees occurs in the open interior area, often in the fork of an upper limb, and usually only about 5–7 feet above the ground. Old nests of other species, including woodpeckers, swallows, and even hawks, are often used, but in cities most nests are placed on ledges, in cavities, or on other supports provided by buildings. Nests vary greatly but often are composed of dry stems and leaves with a soft lining.

Clutch Size and Incubation Period: From 2 to 6 eggs, usually 4–5. The eggs are bluish white with spots, specks, and streaks of brown or black. The incubation period is 12–14 days. Double-brooding has frequently been noted, and as many as three broods have been reported in Colorado.

Time of Breeding: Nesting in Colorado begins in mid-April; second nestings occur in middle to late May, and third broods may extend into July.

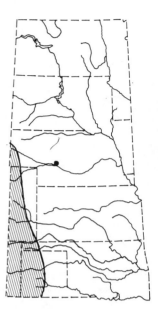

Breeding Biology: This species has a courtship display similar to that of the purple finch, with the male approaching the female with his tail spread and cocked and his wings lowered, uttering chirps and trills. Courtship feeding of the female also occurs during pair-formation and frequently during incubation. Both sexes help in nest-building, which requires from 2 to 11 days, with males helping mainly in the early stages. The female incubates and broods alone, but both sexes bring in food. The young remained in the nests for an average of 15 days in a California study and nearly 18 days in a Hawaiian study. In Colorado it has been reported that females frequently begin to gather nesting materials for their second brood while followed by begging young that are still partially covered with down.

Suggested Reading: Van Riper 1976; Evenden 1957.

Pine Siskin
Carduelis pinus (Spinus pinus)

Breeding Status: Breeds regularly only in north-central and northern Minnesota (probably at least Clearwater County), in the Black Hills of South Dakota (abundant), and in the Pine Ridge area of northwestern Nebraska (common to abundant). Elsewhere breeding is erratic, as in North Dakota, where sporadic breeding has occurred throughout the state, and also in South Dakota (*South Dakota Bird Notes* 27:4). Nesting has occurred in

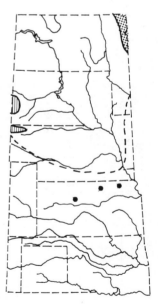

457

a number of eastern and southeastern Nebraska counties follow-ing cold springs (*Wilson Bulletin* 41:77), and in Ellis, Cloud, and Pottawatomie counties, Kansas. There is also occasional nesting in eastern Colorado (Lincoln County). The only breeding record for Oklahoma is from Cimarron County.

Breeding Habitat: In the Black Hills and Pine Ridge area the species breeds in pine forests or isolated stands and also in spruces as well as in deciduous trees in canyons, hollows, or gulches. Evergreen plantings are also used in some areas.

Nest Location: Nests are usually in conifers, 6–25 feet above the ground, on horizontal branches well out from the trunk. The nest is a large but shallow cup of twigs, grasses, bark strips, and rootlets, lined with rootlets, mosses, and other soft materials.

Clutch Size and Incubation Period: From 2 to 6 eggs, usually 3–4. The eggs are pale greenish blue with blackish specks or spots. The incubation period is 13 days. Single-brooded.

Time of Breeding: In North Dakota the estimated breeding sea-son is from mid-April to mid-June, and in South Dakota it is probably from May through July. In Colorado siskins are said to nest in the plains from May 15 to June 1, and Nebraska egg records are from March 29 to May 13.

Breeding Biology: During the nonbreeding season, siskins are highly gregarious; in late winter courtship begins in large flocks, when singing and chasing begins. Courtship feeding of the female is frequent, as are song flights by the male around a particular female. Frequently, nesting occurs in rather loose colonies, with the birds alternating between nest-building and social flocking. The female chooses the nest site and carries in the necessary materials; the male accompanies her and performs courtship feeding during this period as well as during incubation. Gregar-ious tendencies persist through the incubation period, and thus there is little territorial exclusion. Only the female incubates, but both sexes feed the young, which leave the nest in 14–15 days.

Suggested Reading: Bent 1968; Weaver and West 1943.

American Goldfinch
Carduelis tristis (*Spinus tristis*)

Breeding Status: Breeds in suitable habitats over virtually the entire region, becoming less common southwardly and reaching its usual southern limits in central Oklahoma. It probably breeds south to Grady, Caddo, and Roger Mills counties in Oklahoma, it occurs in the Texas panhandle during summer, though there are no breeding records, and there are likewise no breeding records for northeastern New Mexico, although birds have been seen in the Cimarron Valley during the summer.

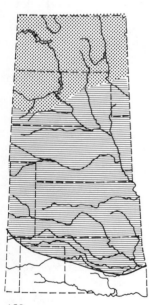

Breeding Habitat: The species frequents open grazing country, farmyards, swamps, weedy fields, and other habitats where thistles or cattails are likely to occur.

Nest Location: Nests are usually 5–15 feet high in a cluster of upright branches or the fork of a horizontal limb of a tree, rarely between 1 and 35 feet high. Nests are tightly woven of plant fibers, lined with the down of thistles and cattails, frequently bound around the rim with spider webbing.

Clutch Size and Incubation Period: From 3 to 6 eggs (8 Kansas clutches averaged 4.4, and 5 North Dakota clutches averaged 4.8). The eggs are pale bluish white and unmarked. The incubation period is 12–14 days. Occasionally double-brooded.

Time of Breeding: North Dakota egg dates are from July 2 to August 15. In Kansas, egg records extend from June 20 to September 10, with an early August peak. In Oklahoma eggs have been found from June 1 to August 27.

Breeding Biology: These gregarious birds remain in flocks well into late spring, and pair-formation begins among flocking birds in May or possibly earlier. It is achieved by courtship singing, courtship flights by a female and varying numbers of males, a hovering song flight by males, and true song resembling that of a canary. Pair bonds are maintained by courtship feeding, which occurs from egg-laying through the nestling period. Nesting is delayed until there is an abundant supply of composite seeds to feed the young. Nest-building and incubation are by the female alone, but both sexes feed the young by regurgitation. The nestlings fledge at 10–16 days, at which time the male takes over most of the feeding. This frees the female to begin a new nest, which sometimes happens within 3 days of fledging.

Suggested Reading: Stokes 1950; Bent 1968.

Lesser Goldfinch
Carduelis psaltria (Spinus psaltria)

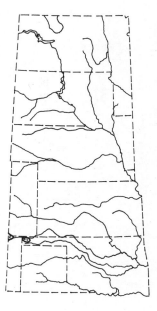

Breeding Status: Breeding in our region is apparently limited to northwestern Cimarron County, Oklahoma, and probably Union County, New Mexico (common summer resident at Capulin Mountain National Monument).

Breeding Habitat: The species is associated with mixed scrub oaks and ponderosa pine in Colorado, with relatively open oak or oak-pine woodlands in general, and also with chaparral and streamside thickets.

Nest Location: Nestings are often loosely colonial, with the nests placed 1–25 feet above the ground, generally in low trees or bushes. The nest is neatly constructed of grasses, weed stems, and similar materials, lined with wool or feathers.

Clutch Size and Incubation Period: From 3 to 6 white eggs, usually 4–5. The incubation period is 12 days (Bailey and Niedrach 1965). Single-brooded, but known to renest.

Time of Breeding: Egg records for Oklahoma are in the first half of June. In Colorado the nesting period is from late May to early July.

Breeding Biology: On the basis of studies in California, there appears to be a rather rapid development of pair bonds in lesser goldfinches, occurring within 2 weeks after spring flocks arrive. Increasing aggression breaks up the flocks, and a series of vocalizations and flights similar to those of the common goldfinch are associated with pair-formation. As the members of a pair become more tolerant of each other, they perform a "billing" display, which soon develops into true courtship feeding of the female. Males accompany their mates as they search for nestsites, but only occasionally do the males help carry materials to the nest. It requires 4–8 days for the female to complete the nest, and territories are not defended until the nest site is established. Nesting tends to occur in loose colonies; thus the territories serve primarily to isolate nesting females. As in other goldfinches, the young are fed regurgitated food, mostly seeds, and they fledge in 12–15 days.

Suggested Reading: Coutlee 1968; Linsdale 1957.

Red Crossbill
Loxia curvirostra

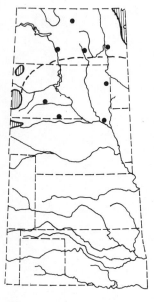

Breeding Status: Regular breeding is probably confined to northern Minnesota (at least including Clearwater County), but there is also a breeding record for Clay County. In North Dakota breeding is erratic (Towner, Burleigh, and Stutsman counties). There is occasional breeding in the Black Hills of South Dakota and also in the Pine Ridge area of Nebraska (*Nebraska Bird Review* 40:71). There is a single old nesting record for Kansas, from Shawnee County.

Breeding Habitat: The species is associated primarily with spruce and pine forests in our region. It is also rarely found breeding in deciduous woods but essentially is adapted to opening the cones of spruces and, to a lesser extent, pines.

Nest Location: Nests are normally in conifers, at heights of 10–40 feet, on horizontal branches well away from the trunk. They are rather bulky structures, of twigs, plant stems, needles, and *Usnea* lichens, lined with grass and other soft materials.

Clutch Size and Incubation Period: From 3 to 5 eggs, usually 4. The eggs are pale bluish to greenish white, with dark spots and dots, especially at the larger end. The incubation period is 13–16 days. Sometimes double-brooded.

Time of Breeding: Nesting in the Black Hills probably extends from January through July and sometimes lasts longer. In other areas such as Colorado nesting seems to begin extremely early, even in December, and eggs have also been found there as late as mid-September. Breeding is apparently timed to coincide with the period of maximum seed availability from conifers; thus the time of breeding depends on the species of trees in the area, since different conifers shed their seeds at different times. Breeding can occur while the birds are in full molt, or even in birds still in juvenile plumage.

Breeding Biology: Observations of this species in Colorado west of Denver indicate colonial nesting; about 24 pairs were found within a square mile of forest, but few were found elsewhere. Hatched young were found there as early as January 16, as well as nests in progress. The females did the building, which required about 5 days, and another 4 days elapsed before the first egg was laid. Four more days were spent in egg-laying and 14 in incubation, and 20 days were needed for fledging. Fledging times of 16–25 days have been reported; these variations probably depend on food supplies. Some pairs raise two broods in rapid succession. The nesting cycles of these birds are highly irregular; not only may two widely spaced breedings occur in a single year, but young birds may breed the same year that they are hatched.

Suggested Reading: Bailey, Niedrach, and Bailey 1953; Nethersole-Thompson 1975.

White-winged Crossbill
Loxia leucoptera

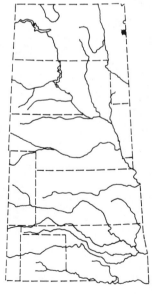

Breeding Status: A very rare or accidental breeder in the region. Nesting has been inferred at Itasca State Park, Clearwater County, Minnesota, where dependent fledglings have been seen. There is also an unsubstantiated report of nesting in Stutsman County, North Dakota, for several years starting in 1894. No other records seem to exist for the region.

Breeding Habitat: The species is generally associated with tamarack forests, more rarely breeding in spruce or pine forests. This species has a smaller bill than the red crossbill and is less well adapted to opening large cones such as pine cones.

Nest Location: Nests are elevated 5–70 feet above the ground, on horizontal limbs of conifers. The nest is a deep cup of twigs, rootlets, weed stalks, and bark strips, lined with fine grasses,

461

feathers, and so forth, and not distinguishable from the nest of the red crossbill.

Clutch Size and Incubation Period: From 2 to 5 eggs, usually 3–4. The eggs are greenish or bluish white with dark spotting or blotching. The incubation period is unreported but is probably 13–16 days.

Time of Breeding: Not definitely established for our region. In Canada nesting has been reported in every month but occurs mainly in spring and fall.

Breeding Biology: Little has been written of the courtship of this species, but part of it seems to consist of the males flying above the females in broad circles, while singing continuously. Like other crossbills, the male probably crouches and lowers his wings to exhibit the rump feathers, which are erected. Courtship feeding is almost certainly a part of it too, since it is known that males regularly feed their incubating mates. There has been at least one observation of a male helping in nest construction, but he only added the lining. Both sexes feed the young by regurgitation of semidigested seeds. The age of fledging has not been reported but is likely to be about 3 weeks. At least in other crossbills, the adults continue to feed their young for about a month after fledging, which might be related to their highly specialized feeding behavior.

Suggested Reading: Bent 1968; Newman 1972.

Green-tailed Towhee
Pipilo chlorura (Chlorura chlorura)

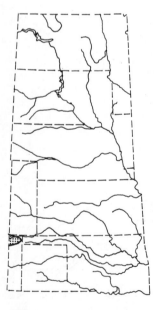

Breeding Status: In our region apparently limited as a breeding species to Union County, New Mexico, where it has bred near Clayton and is common during summer at Capulin Mountain National Monument. There are no breeding records from Oklahoma or adjacent parts of Colorado.

Breeding Habitat: The species is associated with relatively arid and brushy foothills, sagebrush, open pine forests, and chaparral. In Colorado and New Mexico it occurs from foothills into the lower mountain zone from about 6,000 to 9,000 feet, inhabiting sagebrush slopes, river valleys, and scrub oaks.

Nest Location: Nests are on or near the ground in dense shrubbery, averaging about 16 inches above the ground. They are often in sage and less frequently in low trees such as scrub oaks. The nest is constructed of twigs and stems, grasses, and bark, lined with grasses and frequently also horsehair or similar hair.

Clutch Size and Incubation Period: From 3 to 5 eggs, usually 4. The eggs are white with extensive brownish spotting and speckling. The incubation period has not yet been reported. Probably double-brooded, or at least prone to renest.

Time of Breeding: Colorado egg dates are from June 10 to June 24, and New Mexico egg records are from May 20 to July 10.

Breeding Biology: Like the other towhees and the fox sparrow, this species is a ground forager, but it is the most arid-adapted of the North American towhees. Like the related collared towhee of Mexico, it favors dense and brushy vegetation, and like the rufous-sided towhee it often utters catlike mewing call notes. In spite of this bird's widespread occurrence through the arid parts of western United States, very little is known of its breeding biology. In spite of its desert adaptations, however, it is known to be unable to drink salt water.

Suggested Reading: Bent 1968.

Rufous-sided Towhee
Pipilo erythropthalmus

Breeding Status: Breeds in suitable habitats through much of western and northern North Dakota and northern Minnesota, southward along the Missouri Valley through South Dakota (west to the Black Hills and north along the James River Valley), most of Nebraska (west to Sioux, Thomas, and Deuel counties), northeastern Colorado (probably to Logan County, *Nebraska Bird Review* 29:21), eastern Kansas (west to Rawlins, Marion, and Comanche counties), and extreme northern Oklahoma (one 1917 record from Washington County, and a 1977 record from near Colcord (*American Birds* 31:1157).

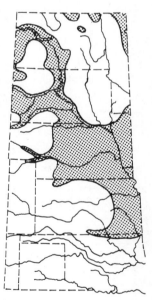

Breeding Habitat: The species is associated with brushy fields, thickets, woodland edges or openings, second-growth forests, and city parks or suburbs with trees and tall shrubbery.

Nest Location: Nests are usually on the ground but may also be in shrubs or vine tangles, as high as 12 feet (rarely) above the ground. The nest is a rather bulky structure built of leaves, weed stems, grasses, bark strips, and fine grasses and other soft materials. Ground nests are often placed under vegetation or brush piles, concealing them from above, and in tree nests the canopy is also usually very dense above the nest.

Clutch Size and Incubation Period: From 2 to 6 eggs, usually 3–4 (14 Kansas clutches averaged 4.0). The eggs are white with reddish brown dots or spotting. The incubation period is 12–14 days. Sometimes double-brooded; possibly 3 broods in southern areas.

Time of Breeding: North Dakota egg dates are from June 6 to June 20, but Minnesota egg dates range from May 22 to July 23,

463

with fledglings seen as early as June 13. Kansas egg dates are from April 21 to August 10, with a peak in early May.

Breeding Biology: At least in coastal California, territories and pair bonds are established very early, about 2 months before nesting starts. Pair formation is achieved by males singing persistently from a variety of locations in their territory. As pair bonds form the rate of singing drops off, and the two birds forage within the male's territory. The female builds the nest with little or no help from the male, although he sometimes carries about small twigs. The female incubates, but both sexes feed the young. The young leave the nest in 9–11 days.

Suggested Reading: Davis 1960; Baumann 1959.

Brown Towhee
Pipilo fuscus

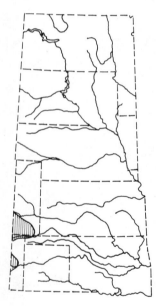

Breeding Status: Limited as a breeding species in the Great Plains to northeastern New Mexico (Quay and probably Union counties), Cimarron County, Oklahoma, and southeastern Colorado (Baca and perhaps Las Animas counties).

Breeding Habitat: In southeastern Colorado this species is found in piñon-juniper woodland, as well as in cholla cactus and sage desert areas. In Oklahoma it is limited to mesa areas, where it is found in valleys, slopes, or mesa tops among cactus, juniper, and piñon pines. Generally it is found in chaparral, brushland, woodland, and open habitats such as lawns and gardens.

Nest Location: In our region nests are usually a few feet above ground in junipers, sometimes up to 7 feet high and occasionally placed on beams in barns. The nest is a deep cup of grasses and plant stems, lined with fine grasses and often horsehair.

Clutch Size and Incubation Period: Usually 3 eggs, sometimes 4. The eggs are very pale gray or blue with spots and blotches of brownish and blackish tones. The incubation period is 11 days. Known to be double-brooded in some areas; up to three broods have been reported in California.

Time of Breeding: Colorado egg dates are from May 12 to June 6. In Oklahoma complete clutches have been found as early as May 4 and as late as June 6, but a recently completed nest was also reported in early September, suggesting possible fall breeding after a very dry summer.

Breeding Biology: Territories in this species, as studied in California, are rather uniform in size and average 1–2 acres. One male that was banded as a juvenile in Pasadena maintained his territory in the same yard for nearly 5 years, during which time he had at least three different mates. The female probably builds the nest, and she certainly performs all the incubation. The young remain in the nest for 8 days, and if there is no second nesting the fledglings may remain with the adults 4–6 weeks. When there is a second nesting the adults drive the young birds from their territory at the time the next clutch hatches.

Suggested Reading: Bent 1968; Marshall 1960.

Lark Bunting
Calamospiza melanocorys

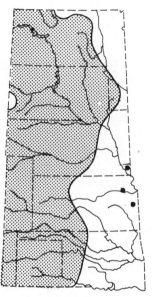

Breeding Status: Breeds in suitable habitats throughout central and western North Dakota, southwestern Minnesota (north to Traverse County), South Dakota (except the high parts of the Black Hills), western and central Nebraska (east occasionally to York, Clay and Lancaster counties), rarely in Missouri (*Wilson Bulletin* 82:465), western Kansas (east rarely as far as Franklin and Shawnee counties), eastern Colorado, western Oklahoma (east to Grant and Tillman counties), and probably most of the Texas panhandle (rare and irregular). In New Mexico it breeds at least in Union County and probably elsewhere.

Breeding Habitat: The natural habitat of this species consists of mixed short grasses and sagebrush, but areas of taller grasses with scattered shrubs are also used, as are disturbed grasslands such as weedy roadsides, retired croplands, and fields of alfalfa or clover.

Nest Location: Nests are placed in the ground, often at the base of a shrub or coarse forb, and frequently are sunk flush with the ground level. All of 83 Kansas nests were situated at the bases of plants in such a way as to provide clear visibility in at least two directions. The nest is constructed of grasses, weed stems, and fine roots, lined with various soft materials. It closely resembles that of the dickcissel, which also has pale bluish eggs.

Clutch Size and Incubation Period: From 3 to 6 eggs, usually 5 (20 North Dakota clutches averaged 4.3, and 45 of 70 Kansas clutches contained 5 eggs). The eggs are light greenish blue, rarely with brownish spotting. The incubation period is about 12 days. Apparently single-brooded, but known to renest.

Time of Breeding: North Dakota egg dates are from June 3 to July 21, and nestlings have been seen as late as August 2. Kansas records indicate that clutches are initiated as early as May 11 and completed as late as June 20.

Breeding Biology: Lark buntings arrive on their breeding areas in flocks, within which courtship begins, and thus dispersal occurs gradually. There seems to be relatively little territorial develop-

465

ment, since nests are often placed only 10–15 yards apart, and males sometimes sing from adjacent fenceposts. Both sexes incubate, but females evidently do the most. At least through the incubation period, to about the middle of July, the males continue to sing and perform song-flight displays. The abundance and local distribution of nesting birds seem to vary considerably from year to year; areas with dense populations one year may be virtually deserted the next. By late August the males have lost their distinctive nuptial plumage, and the fall migration begins soon afterward.

Suggested Reading: Bent 1968; Butterfield 1969.

Savannah Sparrow
Passerculus sandwichensis

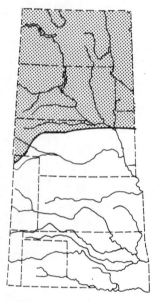

Breeding Status: Breeds in suitable habitats throughout North Dakota, western Minnesota, South Dakota, except the wooded areas of the Black Hills, western and northern Nebraska (south to Garden and southern Cherry counties), probably northwestern Iowa (no specific nesting records), and eastern Colorado (locally around large reservoirs). There are no breeding records for Kansas or for northwestern Missouri (where it summers uncommonly at Squaw Creek National Wildlife Refuge).

Breeding Habitat: In our region the species is associated with tallgrass and mixed-grass prairies, in the wet meadow zones of ponds, lakes, and streams. In mountainous areas it also nests in the moist open areas of mountain parks. Along coastlines it commonly nests in saltwater marshes.

Nest Location: Nests are on the ground, in thick herbaceous cover, almost always in a natural hollow or scraped-out depression, and are generally hidden from above by overhanging vegetation. The nest is built of coarse grass stems with a lining of finer grasses.

Clutch Size and Incubation Period: From 3 to 6 eggs (5 North Dakota nests averaged 5.0). The eggs are pale greenish or bluish white, with highly variable brown markings. The incubation period is 10–12 days. Regularly double-brooded.

Time of Breeding: In North Dakota, egg records are from May 28 to July 9. Minnesota egg records extend from May 30 to July 21; nestlings have been seen as early as June 10, and fledglings as early as June 23.

Breeding Biology: Males regularly arrive on their breeding areas before females and establish territories almost immediately, frequently in the same areas as in the previous year. The territories tend to be relatively small, and in one study they averaged only about ¼ acre, although birds with larger territories were most successful in attracting mates. In a Nova Scotia study it was found that of 13 territory-holders, 9 were monogamous, 3 were bigamous, and 1 failed to obtain a mate. Males do not participate in nest-building or in incubation, but they usually do help feed the nestlings. Fledging occurs at 9–10 days, and the young may be fed for about 2 more weeks by one or both of the parents.

Suggested Reading: Welsh 1975; Potter 1972.

Grasshopper Sparrow
Ammodramus savannarum

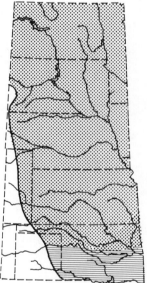

Breeding Status: Breeds in suitable habitats virtually throughout North Dakota, western Minnesota, South Dakota (excepting the Black Hills, where it is sporadic), Nebraska (irregular in extreme west), western Iowa, northwestern Missouri, eastern Colorado (irregular), Kansas, Oklahoma (except the extreme western panhandle), and the eastern panhandle of Texas (Lipscomb to Wheeler counties). Rare or occasional in northeastern New Mexico, with no breeding records.

Breeding Habitat: The species is primarily associated with mixed-grass prairies but is also found in short-grass and tall-grass prairie, sage prairie, and disturbed grassland habitats such as retired cropland, hayfields, and stubble fields. Areas that have grown up to shrubs are avoided, but scattered trees provide acceptable habitat and are used as song perches.

Nest Location: Nests are on the ground, in dense herbaceous vegetation, and are often at the bases of grass clumps, concealed from above. The nest is in a slight depression and is built mostly of grasses, with finer grasses used as lining.

Clutch Size and Incubation Period: From 3 to 6 eggs (6 North Dakota clutches and 5 Kansas clutches averaged 4.8). The eggs are white, with specks and spots of brown, often near the larger end. The incubation period is probably 11–12 days. Sometimes double-brooded: reported rarely triple-brooded in Florida.

Time of Breeding: North Dakota egg dates are from May 30 to July 21, with a peak in breeding from early June to late July. Kansas egg records are from May 1 to June 30, with a peak in egg-laying about May 21. Oklahoma egg records are from May 1 to June 28, but recently fledged young have been seen as late as August 18.

Breeding Biology: The grasshopper sparrow, like the lark bunting, is slightly colonial, and in some areas there are marked

467

year-to-year variations in breeding densities. Territories are immediately established after males return in spring and are advertised by the familiar grasshopperlike "song" and by wing-flicking. Although semicolonial, territories often tend to be rather large, 2–3 acres, and are strongly defended through the incubation period. The female alone incubates and broods the young, and the male may sing as late as the time of hatching. Besides the well-known grasshopperlike song, males also utter a more sustained song that apparently serves to attract and maintain a mate. Both sexes feed and tend the young, which remain in the nest about 9 days.

Suggested Reading: Smith 1963; Bent 1968.

Baird Sparrow
Ammodramus bairdii

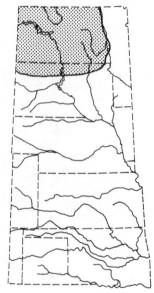

Breeding Status: Breeds in suitable habitats virtually throughout North Dakota except the Agassiz Lake basin, but only locally and sporadically present at the eastern edge of this basin in Clay County, Minnesota. Also breeds in north-central and northwestern South Dakota; the southern range limits are rather uncertain.

Breeding Habitat: In North Dakota the species is associated with large areas of ungrazed or lightly grazed mixed-grass prairies, wet meadows, or tall-grass prairie habitats associated with wetlands, and with various disturbance habitats such as retired croplands, stubble fields, or hayfields.

Nest Location: Nests are on the ground in dense herbaceous vegetation, usually tall grass that is held up by a nearby shrub. The nest is usually in a small depression and is constructed mostly of grasses, which are also used for lining.

Clutch Size and Incubation Period: From 3 to 6 eggs (15 North Dakota clutches averaged 4.7). The eggs are grayish white with extensive spotting or blotching of brown tones. The incubation period is 11–12 days. Single-brooded.

Time of Breeding: North Dakota egg dates are from June 5 to July 12, with dependent fledglings seen as early as June 30 and as late as August 18.

Breeding Biology: This attractive prairie-adapted sparrow remains relatively inconspicuous until it arrives on its breeding grounds, when males suddenly become territorial and begin actively singing their bell-like songs and displaying aggressively. However, females become extremely inconspicuous as soon as they begin nesting, and relatively few nests have been found. The female does all the incubating, while the male continues to sing and forage. She also broods and feeds the young for the first few days after hatching, when the male may finally begin to assist in bringing food. The young leave the nest when 8–10 days old, and by about 13 days they are able to fly a few yards.

Suggested Reading: Cartwright, Shortt, and Harris 1937; Bent 1968.

Henslow Sparrow
Ammodramus henslowii (*Passerherbulus henslowii*)

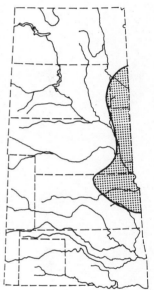

Breeding Status: Breeds locally in southwestern Minnesota (at least Lac qui Parle, Pipestone, and Jackson counties), extremely eastern South Dakota (Moody County), probably western Iowa (apparently rare), southeastern Nebraska (Lancaster and Washington counties), and eastern Kansas (Cloud, Shawnee, Douglas, Morris, and Anderson counties). Territorial males have been seen in Washington County, Oklahoma, but there are no breeding records for the state.

Breeding Habitat: The species is primarily associated with weedy prairies and meadows, neglected grassy fields, and pasturelands, especially those that are rather low-lying and damp. Scattered low bushes are often present.

Nest Location: Nests are on or near the ground and sometimes are in depressions in the ground. The nest is usually placed at the base of a thick clump of grass, with the nest bottom a few inches above the ground. It is constructed of grass and weed leaves and lined with finer grasses and sometimes also hair.

Clutch Size and Incubation Period: From 3 to 5 eggs. The eggs are white, with dots, spots, and sometimes blotches of brown, mostly at the larger end. The incubation period is about 11 days. Probably double-brooded.

Time of Breeding: Not many specific dates are available, but at least in Kansas the eggs are evidently laid in May and June.

Breeding Biology: Like the related grasshopper sparrow, this species tends to be a localized and semicolonial nester, and as many as ten pairs have been reported breeding on a half-acre field in Iowa. Territories are established within such aggregations, but most territorial disputes are limited to rather formal "songfests" rather than physical encounters. The intensity of singing varies greatly and is very high during territorial establishment and nest-building. Nest-building requires 5–6 days and is done mostly or entirely by the female. Territories averaged about an acre each in one Michigan study area and gradually increased in size through

469

the summer. Only the female incubates and broods the young, which remain in the nest for 9–10 days.

Suggested Reading: Robins 1971; Hyde 1939.

Le Conte Sparrow
Ammospiza leconteii (*Passerherbulus caudacutus*)

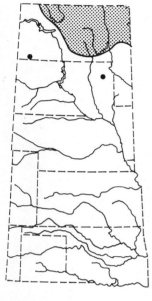

Breeding Status: Breeds in western Minnesota and the eastern portions of North Dakota (west uncommonly to the Missouri Slope and rarely to Adams County), southward rarely to South Dakota (probably bred in Day County in 1955, possibly bred in Beadle and Perkins counties). Not known to breed in Iowa or Nebraska.

Breeding Habitat: The prime habitat of this species are hummocky alkaline wetlands (fens), but the species also occurs less commonly in tall-grass prairie, the wet meadow zone of prairie ponds or lakes, and domestic hayfields or retired croplands.

Nest Location: Nests are placed on the ground in dense herbaceous vegetation, usually in the drier border areas of wetlands where vegetation is luxuriant. Nests in North Dakota are often built in cordgrass (*Spartina*). The nest is built of grasses woven among standing plant stems and thus often is elevated slightly above the ground, and it is lined with fine grasses.

Clutch Size and Incubation Period: From 3 to 5 eggs, usually 4. The eggs are white, with spots, dots, and blotches of brown, rather evenly distributed. The incubation period is 11–13 days. Probably single-brooded but known to renest.

Time of Breeding: In North Dakota the probable breeding season is from late May to mid-August, with a peak in early June to late July. Egg dates are from May 30 to July 21.

Breeding Biology: This elusive marshland-adapted species has been described as "mouselike" in behavior, and its territorial advertisement song sounds more like that of a grasshopper than a bird. Like the grasshopper sparrow, it also has a more prolonged and repeated song that is less frequently heard, and it sometimes sings in flight. Territories are established, but they rarely overlap, and territorial interactions are rarely seen. Incubation and brooding are by the female alone; presumably the male helps feed the young.

Suggested Reading: Murray 1969; Bent 1968.

470

Sharp-tailed Sparrow
Ammospiza caudacuta

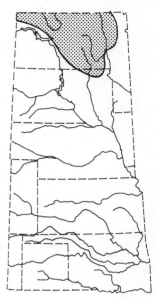

Breeding Status: Breeds in northwestern Minnesota (Kittson, Marshall, Mahnomen, and Clay counties), in eastern North Dakota west to the Missouri Slope, and locally or occasionally in northeastern South Dakota, with the extreme southern limits rather uncertain (possibly Day County, from which there are summer records).

Breeding Habitat: In our region this species is associated primarily with alkaline, hummocky bogs (fens) and the marshy zones of prairie lakes and ponds during years when water levels are low. Less often, wet meadow zones are used. In coastal areas, brackish or salt marshes are prime habitat. In northern Minnesota the species uses swampy lakes with tamarack borders.

Nest Location: Nests are on the ground, usually sunken to the ground level but sometimes built among upright stems and thus elevated in thick clumps of grass. The nests are constructed of dry grasses and lined with finer grasses.

Clutch Size and Incubation Period: From 3 to 6 eggs. The eggs are pale greenish white with brown dots and spots. The incubation period is 11 days. Frequently double-brooded.

Time of Breeding: The probable breeding season in North Dakota is from early June to late August, with a peak from mid-June to early August. Egg dates range from June 12 to July 12.

Breeding Biology: In contrast to the related Le Conte sparrow, males of this species are nonterritorial and simply occupy home breeding ranges. Furthermore, the birds tend to be somewhat colonial and are evidently promiscuous. The nonterritorial and semicolonial nature of this species may enable females to locate males, since the males' songs are weak and uttered relatively infrequently. The songs, which include a flight song, probably serve as an index of sexual excitement, since they do not advertise territories. Males take no role in nesting activities. The young remain in the nest for 9–10 days and continue to be fed for about 20 days after leaving the nest.

Suggested Reading: Murray 1969; Woolfenden 1956.

Vesper Sparrow
Pooecetes gramineus

Breeding Status: Breeds in suitable habitats virtually throughout the Dakotas, western Minnesota and Iowa, most of Nebraska (probably south to about the Platte River), northwestern Missouri (occasionally), eastern Colorado, and perhaps extreme

471

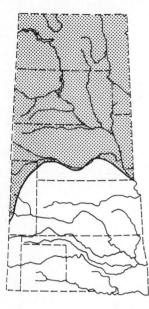

northeastern New Mexico (no specific records). There are no breeding records for Kansas, Oklahoma, or the Texas panhandle.

Breeding Habitat: The species frequents overgrown fields, prairie edges, and similar habitats where grasslands join or are mixed with shrubs and scattered small trees. Habitats providing song perches at least 25 feet high or higher are favored, but shrubs may be used if only they are available.

Nest Location: Nests are on the ground, in sparse vegetation or at times even beside a clod of dirt. The nest is built of grasses and weed stalks, with a lining of finer grasses and sometimes hair.

Clutch Size and Incubation Period: From 3 to 6 eggs (9 North Dakota clutches averaged 4.0). The eggs are white to greenish white with varying amounts of spots, blotches, and scrawls of brown. The incubation period is 12–13 days. Regularly double-brooded.

Time of Breeding: In North Dakota the breeding season is from mid-May to early August, with a peak from late May to mid-July, and egg dates range from July 2 to July 22. Colorado egg dates are from May 11 to July 4.

Breeding Biology: Vesper sparrows occupy a considerably larger home range than do many prairie-adapted sparrows and frequently defend territories of about 2 acres. Most singing is done from fairly high perches, but, rarely, song flights are also performed. It is believed that the female does most of the incubating, but males have been seen covering eggs, and they sometimes also brood the young. On the average, the young remain in the nest for 9 days, but they remain semidependent on their parents until they are 30–35 days old. In one Michigan study, a pair hatched a second brood 29 days after the hatching of the first, and among another group of 29 pairs, 15 pairs raised a single brood, 13 raised two broods, and 1 raised three broods in a single season.

Suggested Reading: Bent 1968.

Lark Sparrow
Chondestes grammacus

Breeding Status: Breeds in suitable habitats throughout the southern half of the region, northward through western Iowa, Nebraska (common in the Sandhills area and to the west), most of South Dakota (more common west of the Missouri River), North Dakota (primarily west of the Missouri, uncommon to rare and local in north-central and eastern areas), and western Minnesota (local; breeding records for Marshall and Norman counties).

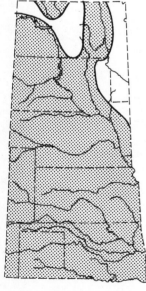

472

Breeding Habitat: Associated with natural grasslands or weedy fields that adjoin or contain scattered trees, shrubs, and coarse forbs.

Nest Location: Nests are usually on the ground but rarely may be up to about 25 feet above the ground in trees. (Of 91 Oklahoma nests, 39 were on the ground and 52 were from a few inches to 15 feet above ground). Ground nests are usually in depressions, often in shaded, bare sites, and are made of grasses and rootlets. Tree nests are more bulky but are made of the same materials.

Clutch Size and Incubation Period: From 3 to 6 eggs, usually 4–5. The eggs are white, with spots and scrawls of blackish concentrated around the larger end. The incubation period is about 12 days. Possibly double-brooded in some southern areas, otherwise probably single-brooded.

Time of Breeding: North Dakota egg dates are from June 20 to July 11. In Kansas, egg dates range from May 1 to July 20, with a possible peak in late May. Oklahoma records are from April 14 to July 14.

Breeding Biology: Lark sparrows are strongly territorial early in the breeding season, and the male may sing while on the ground, perched, or flying; in the last situation tail-spreading is also a conspicuous part of the display. An interesting aspect of display occurs in association with copulation. The female crouches while the male holds a twig in his beak, and during mating the twig is passed to the female, which then flies off with the twig, presumably to the nest site. Males may also drop twigs at a potential nest site, but apparently they never actually assist in the construction of the nest. Nest-building takes 2–3 days, and late nests require less time than early ones. The female does all the incubating and brooding, but males actively participate in feeding the young. They remain in the nest 9–10 days, by which time they are able to fly short distances. Although double-brooding has not been conclusively proved, one adult was seen feeding a juvenile while also actively nest-building, which strongly suggests that it occurs.

Suggested Reading: Newman 1970; Bent 1968.

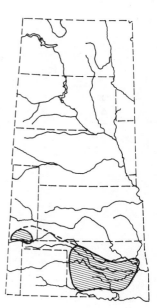

Rufous-crowned Sparrow
Aimophila ruficeps

Breeding Status: Breeds locally in central and western Oklahoma (Cimarron County, the Wichita, Quartz, and Arbuckle mountains, and locally elsewhere in gypsum hills or on rocky slopes, ridges, or exposures east to Cherokee County and south to Johnston and Atoka counties). Also probably breeds locally in the Texas panhandle (no actual breeding records), in southeastern

473

Colorado (Baca and Las Animas counties, but no actual nest records), and northeastern New Mexico (no actual records for region). There are no breeding records for Kansas.

Breeding Habitat: The species is primarily associated with dry and desertlike habitats with extensive bare spaces, often dominated by scrub oaks. Rocky slopes with large boulders, small cedars, and stunted oaks are the primary Oklahoma habitat.

Nest Location: Nests are usually on the ground, often flush with the surface, but sometimes are placed in shrubs. They are often on hillsides, sometimes oriented to face the morning sun. The nest is composed of grasses and weed stems and often has animal hair in the lining.

Clutch Size and Incubation Period: From 3 to 5 white eggs, often 4. The incubation period is unreported. Probably sometimes double-brooded, judging from the apparent length of the breeding season.

Time of Breeding: Oklahoma breeding records are from mid-April (adults carrying food) to early June (nestlings). Texas egg records are from April 4 to July 25, with most between May 1 and May 31.

Breeding Biology: These sparrows apparently do not flock during the nonbreeding season, except as family groups, and there is some indication that adults may remain paired through the winter. During the breeding season the birds are distinctly territorial, with the males singing advertising songs from various song posts or infrequently while in flight. However, regular flight songs are typical of this species. The egg-laying period may be related to the timing of the wet season, and in Oklahoma males have been heard singing as late as September. Incubation is by the female, and the young are tended by both parents.

Suggested Reading: Wolf 1977; Bent 1968.

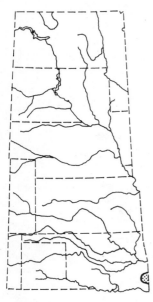

Bachman Sparrow (Pine Woods Sparrow)
Aimophila aestivalis

Breeding Status: Restricted as a breeder in our region to eastern Oklahoma (at least McCurtain and Pontotoc counties, possibly also Pittsburg, Okmulgee, and Woods counties).

Breeding Habitat: Habitat in Oklahoma consists of broom-sedge grassland, with scattered young pines and blackberry thickets. In general, the species seems to prefer brushy hillsides or wooded borders in the northern part of its range, and open pine stands with grasses and scattered shrubs, oaks, or other hardwoods in more southern areas.

Nest Location: Nests are on the ground, often at the outer edges of grass clumps in slight depressions, with a clear view in front of the nest. Nests in Texas and elsewhere have been reported to face away from the midday sun, often toward the west. They are constructed mostly of grasses, with a lining of finer grasses.

Clutch Size and Incubation Period: From 3 to 5 white eggs. The incubation period is 13–14 days. At least two broods, possibly three in the southern areas.

Time of Breeding: In Oklahoma, birds have been observed carrying nesting material, and a female almost ready to lay was collected in late April. No other specific information on nesting is available for our region.

Breeding Biology: Observations of wintering birds in Louisiana suggest that adults of this species may remain permanently paired, as has been suggested for the rufous-crowned sparrow. On the breeding areas, males establish territories that are well separated; the unusually loud songs of this species may be related to this aspect of its breeding biology. Nests are extremely hard to find, but apparently the female does all the nest-building and also all the incubation. Sometimes the male accompanies his mate on material-gathering sorties, but generally he continues to advertise the territory during this phase. Both sexes participate in caring for the young, and the nestling period lasts about 10 days.

Suggested Reading: Bent 1968; Wolf 1977.

Cassin Sparrow
Aimophila cassinii

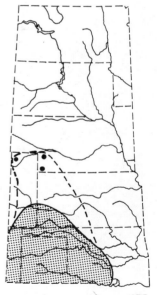

Breeding Status: Breeds in northeastern New Mexico, the Texas panhandle, western Oklahoma (east to about Cleveland and Love counties), southeastern Colorado (Baca and Prowers counties), and southwestern Kansas (regularly to Hamilton, Finney, and Comanche counties). Less commonly it extends to Cheyenne County, Kansas, and to northeastern Colorado; additionally, there is a single breeding record for southwestern Nebraska (*American Birds* 28:922). Details on the New Mexico distribution have recently appeared (*American Birds* 31: 933).

Breeding Habitat: Typically this species occurs in grassland habitats, within which there are shrubs or small trees. The habitats vary from nearly uniform but rather arid grassland to rather dense mesquite woodland; in Oklahoma sandy prairies with scattered sage, yucca, cactus, mesquite, and shinnery oaks are preferred.

Nest Location: Nests are either on the ground or elevated a few inches in grass or sage. Typically they are on the ground, concealed in weeds or at the base of small bushes. The nest is constructed of grasses, weeds, bark, and plant fibers, with finer grasses for lining.

Clutch Size and Incubation Period: From 3 to 5 white eggs, usually 4 (4 Kansas clutches averaged 3.8). The incubation period is unknown. Probably double-brooded, or at least a renester.

Time of Breeding: In Kansas the eggs are laid between mid-May and mid-July, and in Oklahoma the egg dates are from May 26 to July 22. In Texas, egg records extend from March 1 to August 1, suggesting multiple brooding.

Breeding Biology: Although a flocking species during the non-breeding season, this sparrow is apparently territorial, with breeding groups often occurring in loose colonies. Individual territorial males may be spaced 50–100 yards apart, and they advertise their territories by singing from perches or by a song flight, which is well developed in this species. The flight song may also be an important part of pair-formation. Some birds migrate westward in early summer while still in breeding condition and may even breed afterward; in this respect the species seems unique among North American sparrows. Very little is known of the incubation and brooding phases of this species' biology, partly because it is so intolerant of disturbance. Both sexes have been seen feeding nestlings. This species is highly arid-adapted and apparently can breed where no drinking water is locally available, but current information suggests that an arid spring in the southern plains may cause birds to leave and attempt to breed elsewhere where food supplies are more favorable.

Suggested Reading: Wolf 1977; Bent 1968.

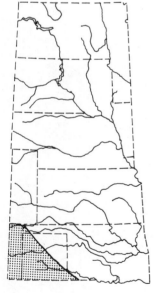

Black-throated Sparrow
Amphispiza bilineata

Breeding Status: Breeding in our area is restricted to northeastern New Mexico (locally common), the Texas panhandle (scarce and local, breeding record for Armstrong County), Cimarron County in Oklahoma, and probably southeastern Colorado (no specific breeding records).

Breeding Habitat: In our area this species occupies thinly grassed pastureland with scattered mesquite, yucca, prickly pear cactus, and cholla cactus. Generally, arid uplands are favored, but it also extends into the depths of Death Valley.

Nest Location: Nests are in small bushes, or sometimes in cholla cactus. The nest is composed of twigs, grass, and leaf fibers, lined with hair or other soft materials.

Clutch Size and Incubation Period: From 3 to 4 light blue eggs, which fade on prolonged exposure to light. The incubation period is unknown. Probably multiple-brooded.

Time of Breeding: Texas egg records are from March 10 to September 8, and in New Mexico eggs have been noted from May 20 to July 30. Oklahoma egg dates are from May 12 to June 12, with nestlings seen as late as June 22.

Breeding Biology: This desert-adapted sparrow is notable for its capacity to survive in areas lacking water; it obtains fluids by eating green vegetation or insects and is able to concentrate its urine wastes as well as to utilize saline water more effectively than other North American seed-eating songbirds. Males also hold relatively large territories, which they advertise and defend with a repertory of song phrases that varies remarkably both between and within individuals. Little has been written on the incubation and brooding phases of breeding biology, but in southern California singing and territorial defense begin as early as February, and dependent young have been seen as late as late August, suggesting that multiple brooding may be typical in favorable years.

Suggested Reading: Heckenlively 1970; Bent 1968.

Dark-eyed Junco (Slate-colored and White-winged Juncos)
Junco hyemalis

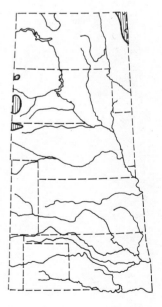

Breeding Status: Breeds in northwestern Minnesota (Clearwater and Marshall counties), in the Black Hills of South Dakota, and in the Pine Ridge area of Nebraska (Sioux and Dawes counties).

Breeding Habitat: In South Dakota, breeding habitats include pine or spruce forests, aspen groves, and deciduous woods in hollows, canyons, and gulches. Minnesota habitats are primarily coniferous forests.

Nest Location: Nests are either on the ground (commonly), or in trees as high as 8 feet above the ground. Ground nests are usually well concealed under weeds or grasses or may be under tree roots or fallen trees. Often they are on a steep slope, under a rock ledge, or in a rock crevice; of 29 Black Hills nests, almost half were under logs, tree roots, or overhanging ledges. The nest is built of grasses, rootlets, twigs, and so on, with a lining of finer grasses, rootlets, and hair.

Clutch Size and Incubation Period: From 3 to 6 eggs, usually 4–5. The eggs are grayish to pale bluish white, with dark markings concentrated around the larger end. The incubation period is 12–13 days. Double-brooded.

477

Time of Breeding: Minnesota egg records are from May 18 to July 18. In South Dakota the breeding season is from late May to late July; egg dates are from May 20 to July 22.

Breeding Biology: Juncos are notable for their sociable winter flocking behavior, which persists until the birds return to their breeding areas. In the Black Hills, territorial singing sometimes can be heard in early March. When a female enters a male's territory he follows her with tail lifted and fanned and wings drooping. Several days are spent in establishing and strengthening the pair bond, during which the birds remain close together and the male continues to display frequently by wing-drooping and tail-fanning. The female builds the nest over a period of several days, and she apparently does all the incubation and brooding. Both parents bring food to the young, which fledge in 10–13 days. The young continue to be semidependent on their parents for about 3 weeks after leaving the nest, and juveniles have been seen with their father as late as 46 days after fledging. Presumably by that time the female might be incubating a second clutch, if not feeding young.

Suggested Reading: Bent 1968; Hostetter 1961.

Chipping Sparrow
Spizella passerina

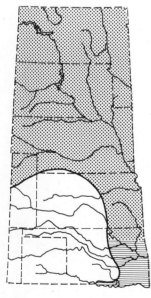

Breeding Status: Breeds in suitable habitats throughout the northern half of our region, extending south into Colorado (possibly widespread, but specific records are lacking), eastern Kansas (west at least to Barber and Shawnee counties), and eastern Oklahoma (west regularly to Osage and Pushmataha counties, sometimes to Comanche County). Uncommon to scarce in the Texas panhandle; no breeding records. Formerly also bred in Cimarron County, Oklahoma.

Breeding Habitat: This is a forest-edge species, found in the margins of deciduous forests, in parks, gardens, or residential areas, and in farmsteads, orchards, and grassland habitats with scattered trees. Generally, trees surrounded by an open area with only herb stratum vegetation and some open ground for foraging provide the best environmental combination.

Nest Location: Nests are usually elevated from 1 to 10 feet in shrubs or trees, rarely to 30 feet, and are generally near the trunk and top of smaller trees, or lower in the branches and farther from the trunk in larger open-grown trees. Nest placement seems to be determined primarily by accessibility, concealment, and a low density of overhead vegetation. The nest is built of grasses, forbs, and rootlets, lined with fine grasses and often hair.

Clutch Size and Incubation Period: From 2 to 5 eggs, usually 4. The eggs are pale greenish blue, with brown to blackish spotting, streaking, or blotches near the larger end. The incubation period is 11–14 days. Double-brooded.

Time of Breeding: North Dakota egg dates are from May 25 to June 20, and dependent young have been seen as late as August 24. In Kansas, egg records are only for early May, but laying probably extends into June or July. Oklahoma breeding dates are from April 25 (nest-building) to early July (fledglings).

Breeding Biology: Territorial establishment begins almost immediately after the males return to their breeding grounds in spring, and the males spend a good deal of time each day in singing and chasing intruders. Territories average about an acre in area but sometimes are as small as half an acre. The female gathers all the nesting material and constructs the nest. Usually she also does all the incubating, but an exceptional instance of male incubation has been reported. The female broods the young, but both sexes feed them, and they fledge in about 10 days, with an observed range of 8–12 days among 52 broods. By the time they are 14 days old the young are able to fly several feet.

Suggested Reading: Walkinshaw 1944; Tate 1973.

Clay-colored Sparrow
Spizella pallida

Breeding Status: Breeds in suitable habitats virtually throughout North Dakota and western Minnesota, extending southward through northeastern South Dakota (one nesting record from the Black Hills). There is one breeding record for south-central Nebraska (Hall County) and also a probable breeding record for southwestern Kansas (Morton County), but no other indications of breeding in the southern half of the region seem to exist.

Breeding Habitat: The optimum habitat consists of mixed-grass prairie with scattered low thickets of shrubs such as wolfberry or silverberry, but grasslands with taller shrubs or small trees are also utilized. Brushy woodland margins or successional stages of forests following fires or logging are used, as are disturbance habitats such as retired croplands and shelterbelts on agricultural land.

Nest Location: Nests are on or near the ground, often in low shrubs at heights of up to about 3 feet. The nest is usually well hidden, either on a shrub branch or in a clump of dead grass, and is usually concealed by leaves. In North Dakota, 34 nests averaged only 11 inches above ground, and none was more than 30 inches. The nest is built of grasses, weeds, and rootlets, lined with

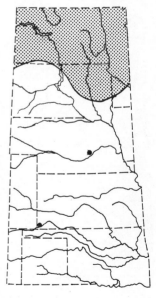

479

finer grasses, rootlets, and sometimes hair. The nest and eggs resemble those of chipping sparrows, but in this species hair is less frequently used and the nest is somewhat more bulky.

Clutch Size and Incubation Period: From 3 to 5 eggs (20 North Dakota clutches averaged 3.8). The eggs are pale bluish green, with blackish spots or scrawls near the larger end. The incubation period is 10 to 11½ days. Normally single-brooded, though sometimes two broods are raised; a persistent renester.

Time of Breeding: North Dakota egg dates are from May 29 to July 22, with a peak of records during the first half of June. Minnesota records are from May 22 to June 13, with a majority between May 31 and June 6.

Breeding Biology: These plain-colored prairie-adapted sparrows begin to establish territories a week or two after they arrive in spring, selecting grassy areas containing at least one clump of shrubbery, and defending areas varying from ¼ acre to 1 acre in extent. Field sparrows and sometimes other sparrows also are threatened by territorial males. The female constructs the nest; and the male often accompanies her on her material-gathering trips. Incubation usually begins with the laying of the third egg, and reportedly both sexes participate. Both sexes feed the young, which leave the nest at a surprisingly early age of 7–8 days. However, they are unable to fly until they are about 2 weeks old. Up to four nesting attempts are sometimes made, and a few cases of definite double-brooding have been reported.

Suggested Reading: Salt 1966; Bent 1968.

Brewer Sparrow
Spizella breweri

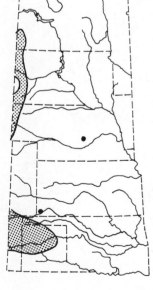

Breeding Status: Breeds locally in southwestern North Dakota (Bowman, Slope, Golden Valley, and Billings counties), the western edge of South Dakota (excluding the Black Hills), northwestern Nebraska (breeds in Sioux County and probably elsewhere, and there is one record from Howard County), possibly northeastern Colorado (nesting record for Hereford, Weld County, is just beyond this book's limits), probably northeastern New Mexico (no specific breeding records), Cimarron County, Oklahoma (irregularly), and the Texas panhandle (locally; breeding records for Randall, Armstrong, and Briscoe counties). Nesting also occurs in Morton County, Kansas, where numerous territorial males were observed in 1978.

Breeding Habitat: In our region this species is especially characteristic of short-grass prairies with sage or other semiarid shrubs such as rabbitbrush; in New Mexico the species' range is almost coextensive with those of big sagebrush and rabbitbrush. Else-where, the species also occurs in timberline areas of Canada.

Nest Location: Nests are in sage or similar shrubs, almost always less than 4 feet above the ground and averaging about 6 inches. Rarely, nests have been found on the ground, in slight depressions. The nest is built of grass and plant fibers, lined with finer grass and sometimes hair. Concealment above the nest seems to be an important criterion for site selection.

Clutch Size and Incubation Period: From 3 to 5 eggs (20 North Dakota clutches averaged 3.8). The eggs are bluish, with dots, spots, and blotches of dark brown, mostly near the larger end. The incubation period is 12–13 days.

Time of Breeding: In North Dakota, eggs have been found in early June. In Colorado, eggs have been found between May 28 and July 21, and in New Mexico the egg records extend from May 20 to July 10.

Breeding Biology: This species occurs in two widely different climatic zones—the arid sage-dominated western states and the arctic timberline of northern Canada—and in both is dependent upon open, shrub-dominated habitat. There is little information on breeding biology, and yet in eastern Washington as many as 47 pairs may occur in 100 acres of favorable habitat. In a Montana study, spray-killing all the sagebrush on a study area reduced the Brewer sparrow population by about half. The sparrows will nest in dead sagebrush, but it provides considerably less concealment than do live plants.

Suggested Reading: Bent 1968; Best 1972.

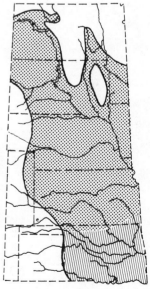

Field Sparrow
Spizella pusilla

Breeding Status: Breeding occurs locally in western and south-eastern North Dakota (mostly Missouri and Little Missouri drain-ages, also the southern Lake Agassiz basin), southwestern Minne-sota (Pope County, probably elsewhere), most of South Dakota excepting the Black Hills, western Iowa, and Nebraska (rare in the panhandle), eastern Colorado (probably a local resident, but no specific records), virtually all of Kansas, Oklahoma except for the panhandle, and probably the Texas panhandle (scarce and local; nearest breeding record is for Wilbarger County).

Breeding Habitat: Brushy, open woodland, forest edge, brushy ravines or draws, sagebrush flats, abandoned hayfields, forest clearings, and similar habitats offering a combination of low, grassy areas and shrubs or low trees are utilized by this species. It

481

uses habitats rather similar to those of the chipping sparrow but occurs at an earlier successional stage than does that species.

Nest Location: Nests are on the ground (early spring nests) or in thick shrubs (later nests) to as high as about 3 feet. Nests are built of dead grass stems and leaves, with a lining of rootlets and hair.

Clutch Size and Incubation Period: From 2 to 5 eggs, usually 3–4 (21 Kansas clutches averaged 4.1). The eggs vary from creamy to pale greenish or bluish white, with dots and spots of brown usually concentrated around the larger end. The incubation period is 11–12 days. Often double-brooded, sometimes three broods.

Time of Breeding: North Dakota egg dates are for June, but young have been seen in Minnesota as early as May 21, and nest-building has been seen as late as July 28. Iowa egg records are from May 12 to July 30, and those from Kansas are from April 21 to September 10. Oklahoma egg dates are from April 20 to July 31.

Breeding Biology: Males establish territories as soon as they return to their breeding areas; these average about 3 acres but range from less than 2 to about 6 acres. Older males are the first to return and usually reoccupy exactly the same territories they held the previous year. Females return somewhat later than males, and if a female's former partner has already found a new mate she may settle on a nearby territory with a new mate. Singing drops off sharply as soon as mates have been found, and nest-building soon begins. The female chooses the site and does the building, often closely accompanied by the male. Early nests may take 3–7 days to construct, compared with 2–3 days for later ones. Incubation is entirely by the female, who is sometimes fed on the nest by her mate. Both sexes tend the young, which leave the nest in 7–8 days and 5 days later are able to fly short distances. The young become independent at 26–34 days of age, by which time the adults are likely to have begun a second nesting.

Suggested Reading: Best 1977; Crooks and Hendrickson 1953.

White-throated Sparrow
Zonotrichia albicollis

Breeding Status: Breeding in our region is limited to north-central and northwestern Minnesota (Clearwater and Marshall counties), and to the Turtle Mountains and Pembina Hills of northeastern North Dakota.

Breeding Habitat: In our region the species is associated with various semiopen wooded habitats, such as coniferous forests with well-developed woody undergrowth, aspen groves with a

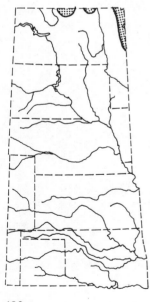

shrubby understory, willow-bordered marshes, and sometimes planted conifer groves.

Nest Location: Nests are typically on the ground, in areas of small trees or clumps of shrubs, with extensive ground cover. The nest is usually at the edge of a clearing, well concealed by the ground vegetation, but, rarely, may be off the ground in a bush, a brush heap, or under tree roots. Blueberries frequently serve for nest cover, and there is usually also a large object such as a tree or stump nearby that apparently serves as a lookout perch. The nest is constructed of diverse materials such as grasses, rootlets, needles, and mosses, lined with finer grasses, rootlets, and hair.

Clutch Size and Incubation Period: From 3–6 eggs, usually 4. The eggs are grayish, heavily dotted or spotted with brown. The incubation period is about 12–14 days. Usually single-brooded, but sometimes two broods.

Time of Breeding: Minnesota egg records are from May 30 to August 4, and dependent fledglings have been seen as early as June 30 and as late as August 21.

Breeding Biology: Almost as soon as they return to their breeding areas, males establish and defend territories, which they advertise by singing from a few favored singing posts, usually 20–40 feet above the ground in spruces. Territories range from less than an acre to more than 2 acres, the size apparently varying with the habitat. The female builds the nest without help from the male, and only the female incubates. Both sexes care for the young, which leave the nest in 8–9 days and are able to fly a few days later. Second broods are evidently rare, but up to three nesting attempts have been observed.

Suggested Reading: Bent 1968.

Swamp Sparrow
Melospiza georgiana

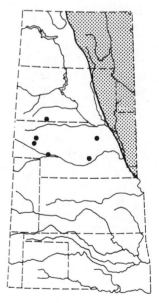

Breeding Status: Breeds throughout most of western Minnesota and locally in eastern North Dakota (Turtle Mountains, northern Agassiz Lake basin, Drift Plain, and Missouri Coteau), as well as southward through eastern South Dakota, western Iowa, and northwestern Missouri. Breeding in Nebraska is apparently highly local but includes the vicinity of Neligh in Antelope County, Smith Lake in Sheridan County (Rosche 1977), the Loup Valley in Howard County (*Nebraska Bird Review* 38:18), and Crescent Lake N.W.R., Garden County.

Breeding Habitat: In North Dakota the species generally frequents alkaline bogs (fens), especially those having cattails, phragmites, and shrubs or small trees. It also breeds in wet meadows, along swampy shorelines of lakes or streams, and to a limited degree in coastal meadows.

Nest Location: Nests are rarely on the ground but instead are built about a foot above the substrate, often in water up to 24 inches deep. They are constructed among the stalks of cattails or in bushes and are rather bulky structures of grass with a finer grass lining.

Clutch Size and Incubation Period: From 3 to 6 eggs, usually 4–5. The eggs are pale green with dots, spots, and blotches of brown. The incubation period is 12–13 days. Normally single-brooded, sometimes two broods.

Time of Breeding: Egg records in Minnesota are from May 18 to June 17, and in North Dakota adults carrying food have been seen as late as August 8.

Breeding Biology: Probably because it is restricted to wet and inaccessible habitats, rather little is known of the breeding biology of this relatively insectivorous sparrow. A breeding density of 2 pairs in 9½ acres of Maryland bog has been reported, but territory size and other aspects of breeding remain essentially unstudied. Apparently the female incubates alone, although the male has been observed feeding his brooding mate. The nestling period was judged to be about 12–13 days by one early observer, and about 9 days by a more recent one, which seems more realistic.

Suggested Reading: Bent 1968.

Song Sparrow
Melospiza melodea

Breeding Status: Breeds in suitable habitats virtually throughout the Dakotas, western Minnesota, and western Iowa, but becomes relatively local and uncommon in Nebraska, and apparently rarely breeds south of the Platte River (recent records for Hall and Lancaster counties, early records to Webster and Nemaha counties, *Nebraska Bird Review* 43:3). Although common at Squaw Creek National Wildlife Refuge in northwestern Missouri, it is apparently only a local breeder in northeastern Kansas, and there are no breeding records for more southerly areas of our region. However, there are several breeding records for eastern Colorado (Kingery and Graul, 1978).

Breeding Habitat: This is a typical edge species, breeding in thickets of shrubs and trees among grasslands, brushy margins or openings of forests, brushy edges of ponds or lakes, in shrub swamps, shelterbelts, farmsteads, and sometimes in parks or suburbs.

Nest Location: Nests are elevated from a few inches to about 10 feet in shrubs or saplings or may be on the ground under bushes, clumps of grass, or brush piles. The nest is built of grasses, weeds, and bark fibers, lined with fine grasses, rootlets, and sometimes hair.

Clutch Size and Incubation Period: From 3 to 5 eggs (11 North Dakota clutches averaged 4.4). The eggs are greenish white with extensive spotting or blotching of reddish brown to purple tones. The incubation period is 12–13 days. Regularly double-brooded, sometimes three broods.

Time of Breeding: North Dakota egg dates are from May 25 to July 11, with dependent young seen as late as September 17. In Minnesota, eggs have been reported from April 27 to August 14.

Breeding Biology: The song sparrow is one of America's best-studied songbirds, thanks to the classic banding efforts of M. M. Nice. Males are highly territorial and often maintain the same territories year after year. Females return to their old territories about half the time but only infrequently (8 of 30 cases in one study) remate with their previous partners. They often settle into adjacent territories if their males have found new mates or may move as far as a mile from their place of hatching. Usually the female builds the nest alone, but there are cases of unmated males building nests, and of helping mates in nest construction. About 3–4 days are needed for nest-building, and the female incubates alone. Both sexes feed the young, which remain in the nest for about 10 days and become independent when 28–30 days old. Periods between the fledging of two successful broods range from 30 to 41 days.

Suggested Reading: Bent 1968; Nice 1943.

McCown Longspur
Calcarius mccownii

Breeding Status: Breeding is now largely limited to the western quarter of North Dakota, east to Renville and eastern Bowman counties, rarely to McLean County. Breeding also once occurred in northwestern South Dakota, but apparently there have been no authenticated nesting records since 1949. Formerly, breeding occurred as far east as Pipestone County, Minnesota, and as far south as western Nebraska. Recently, birds have been found

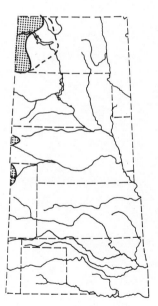

485

breeding in southern Sioux County, northwestern Nebraska (*Nebraska Bird Review* 34:75). Breeding also occurs in Weld County, Colorado.

Breeding Habitat: The species is associated with short-grass prairies and grazed mixed-grass prairies, as well as stubble fields or fields with newly sprouting grains in North Dakota.

Nest Location: Nests are placed in small depressions in the ground, usually amid sparse plant cover. Of 40 Wyoming nests, 19 were beside grass clumps and the rest were associated with rabbitbrush or horsebrush cover. The nests are usually constucted entirely of grasses, with a lining of finer grasses and sometimes wool where it is available.

Clutch Size and Incubation Period: From 3 to 6 eggs, usually 3–4. The eggs are white to greenish white with dark brown to black lines, streaks, spots, and dots. The incubation period is 12 days. Probably single-brooded.

Time of Breeding: North Dakota egg dates are May 17 to July 22, with most between May 27 and June 10. Dates of clutches from eastern Wyoming are from May 17 to June 29, and Colorado egg dates range from May 23 to July 11, with fledged young seen as early as May 27.

Breeding Biology: Male longspurs arrive on their breeding grounds of eastern Wyoming in late April and soon begin to select territories. These are marked by flight songs as well as by singing from shrubs or rocks. As competition increases, territories gradually decrease in size to an area about 250 feet in diameter. The courtship display is remarkable; the male moves around the female in a narrow circle, holding the nearer wing erect and thus exposing the white lining. The female gathers the nesting material and makes any nest excavation that may be necessary. She also performs all the incubation and does most of the brooding, though she is occasionally relieved by the male during the later brooding stages. The young leave the nest at 10 days and 2 days later are able to fly for short distances.

Suggested Reading: Mickey 1943; Bent 1968.

Chestnut-collared Longspur
Calcarius ornatus

Breeding Status: Breeds in suitable habitats virtually throughout North Dakota (rare in southern Agassiz Lake basin) and in adjacent western Minnesota (recent records for Wilkins and Clay counties, an old record for Polk County), extending southward through much of South Dakota except the Black Hills and southeastern South Dakota, and through the northwestern corner of

Nebraska (Sioux, Dawes, Sheridan, and Box Butte counties). Also breeds in Weld County of northern Colorado. Breeding probably formerly extended to western Kansas, in the vicinity of Ellis, Trego, and Logan counties.

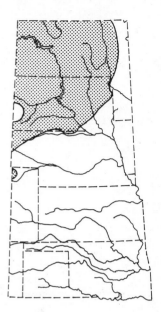

Breeding Habitat: In North Dakota, the prime habitat consists of grazed or hayed mixed-grass prairies, but the birds also use transitional areas between short-grass and mixed-grass habitats, meadow zones of salt grass around ponds and lakes, and disturbance habitats such as mowed hayfields, heavily grazed pastures, and other similar low-stature grassy areas.

Nest Location: Nests are on the ground, in depressions under sparse vegetative cover and often under grass tufts. The nests are built of grasses, with a lining of fine grasses, rootlets, and hair or feathers when these are available.

Clutch Size and Incubation Period: From 3 to 6 eggs (45 North Dakota clutches averaged 4.2). The eggs are creamy white with spots, scrawls, and blotches of dark brown or black tones. The incubation period is 12-13 days. At least occasionally double-brooded.

Time of Breeding: North Dakota egg dates are from May 6 to July 25, and those from South Dakota range from May 20 to June 17, with most before June 5. Minnesota records are from May 19 to June 16.

Breeding Biology: Males establish territories shortly after their spring arrival; they prefer grassy plains that have sparse vegetation and at least one large rock or fencepost to serve as a singing post. Such singing points are often a central part of the territory; the nest is usually within 25 feet, and the total territory is about 100 feet in diameter. In some marginal areas, however, territories up to 10 acres have been estimated. Although flight songs are used in territorial advertisement, they are not as frequent or as formalized as in the McCown longspur. Both species gradually gain altitude with rapid wingbeats, but the McCown longspur sails downward quickly with wings upstretched, whereas the chestnut-collared longspur circles and undulates while singing more softly before gradually descending. The female builds the nest alone and also does the incubation. Both sexes feed the young, which leave the nest in 9-11 days or rarely as late as the 14th day. By the 14th day the young can fly very well, and by about 26 days they are independent of their parents.

Suggested Reading: Moriarty 1965; Bent 1968.

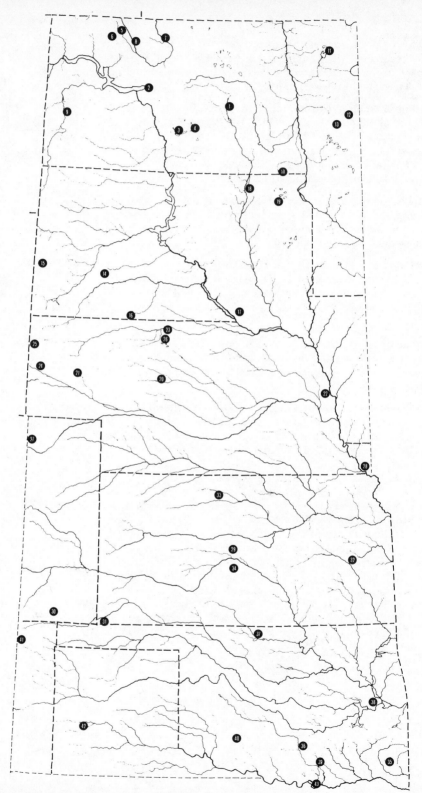

Fig. 8. Bird-watching localities in the Great Plains states.

A Guide to Bird-watching Localities
in the Great Plains

Space does not allow for an extended description of all the excellent birding localities within this region; O. S. Pettingill's *Guide to Bird Finding West of the Mississippi* (Oxford, 1953), long out of print but being revised, is still the best such guide. A more recent and very useful guide is Jessie Kitching's *Birdwatcher's Guide to Wildlife Sanctuaries* (New York: Arco, 1976), which includes information on many federal, state, and local sanctuaries.

Partly on the basis of the availability of relatively complete bird lists available for the areas, I have included here forty-four state and federal parks, refuges, and other sanctuaries located within the region covered by this book. Nearly all the national wildlife refuges are included, except a number of smaller refuges for which bird lists are not available. Further, a group of four North Dakota national wildlife refuges in the Souris Valley is collectively considered in a "Souris Loop" category.

In the discussion that follows and the associated list of summer birds, the localities are grouped by states, and the states are organized in a general north-to-south sequence. The 236 species included in the associated bird list (Appendix B) are only those that are known to breed or to have bred in the region. Accidental species, introduced "pest species," and species not known to breed in the region though occasionally summering within it are excluded from the list.

North Dakota

1. Arrowwood National Wildlife Refuge. Situated about 14 miles north of Jamestown, North Dakota. Contains nearly 16,000 acres of lakes, marshes, grasslands, wooded areas, and fields. Bird checklist of 250 species, including 106 breeding species, available from refuge manager, Pingree, North Dakota 58476.

2. Audubon National Wildlife Refuge. Situated at the east end of Lake Sakakawea, North Dakota, between Minot and Bismarck. Contains about 13,500 acres administered by the federal government and 11,200 acres supervised by the state. Mostly consists of short-grass prairie and reservoir shoreline, as well as prairie potholes and marshes. Bird checklist of 260 species, including 75 breeders and 14

489

accidentals, is available from the refuge manager, R.R. 1, Coleharbor, North Dakota 58531.

3. Long Lake National Wildlife Refuge. Situated about four miles southeast of Moffitt, North Dakota. Contains more than 22,000 acres, mostly prairie grasslands, ravines, fields, trees and shrub plants, and marsh or lake areas. Bird checklist of 199 species, including 75 breeders and 6 accidentals, available from the refuge manager, Moffitt, North Dakota 58560.

4. Slade National Wildlife Refuge. Situated between Jamestown and Bismarck, North Dakota. Consists of 3,000 acres of prairie pothole habitat, with many marshes and small lakes. Bird checklist of 202 species, including 88 breeders and 5 accidentals, available from the refuge manager, Dawson, North Dakota 58428.

5–8. Souris Loop National Wildlife Refuges. These include four national wildlife refuges: Des Lac, Box 578, Kenmore, North Dakota 58746; Lostwood, Lostwood, North Dakota 58724; J. Clark Salyer (formerly Lower Souris), Upham, North Dakota 58799; and Upper Souris, R.R. 1, Foxholm, North Dakota 58738. Collectively they take in more than 100,000 acres of mixed prairie, marshland, and river-bottom habitat. A collective bird checklist of 262 species, including 138 breeders, is available from the manager of any of the four refuges. The breeders include all five species of grebes, as well as such grassland sparrows as Baird and Le Conte.

9. Theodore Roosevelt National Memorial Park. Situated near Medora, North Dakota, and administered by the U.S. National Park Service. Adjacent to the Missouri River National Grasslands; the area consists mostly of short-grass vegetation on uplands, brushy flats, and ravines, and wooded river valleys. A bird checklist titled "Birds of the Grasslands" and containing 188 species is available at small cost from the Theodore Roosevelt Nature and History Association. Medora, North Dakota 58645.

10. Tewaukon National Wildlife Refuge. Situated in southeastern North Dakota near Cayuga, and consisting of nearly 8,000 acres of prairie grassland, marshes, and larger water areas. A bird checklist of 235 species, including 88 breeders, is available from the refuge manager, Cayuga, North Dakota 58103.

MINNESOTA

11. Agassiz National Wildlife Refuge. Situated 11 miles east of Holt, in Marshall County, Minnesota. This area, once a part of glacial Lake

490

Agassiz, contains 61,500 acres of grasslands with hardwood groves, potholes, and lakes. A bird checklist of 245 species reported on the refuge is available from the refuge manager, Middle River, Minnesota 56737. An annotated checklist has also been published (*Flicker* 27: 138–47).

12. Itasca State Park. Situated 22 miles north of Park Rapids, Minnesota, encompassing Lake Itasca and the headwaters of the Mississippi River. It includes more than 32,000 acres, with many dense forests of conifers and hardwood trees. A checklist of early summer birds (*Loon* 37:27–39) includes 111 probable breeding species and 30 additional probable nonbreeders. A recent complete checklist of 208 bird species in the park and surrounding areas has also been published (*Loon* 49:81–95), including 94 known breeders. Many boreal species, such as winter wrens, several warblers, and boreal chickadees are more likely to be found here than anywhere else in the region covered by this book.

13. Tamarac National Wildlife Refuge. Situated 18 miles northeast of Detroit Lakes in Becker County, Minnesota. It consists of 42,000 acres, in the transition area between conifers and hardwood forests, and has 24 lakes within the refuge boundaries. A bird checklist of 221 species is available from the refuge manager, Rochert, Minnesota 56578.

South Dakota

14. Badlands National Monument. Situated 62 miles east of Rapid City, South Dakota. This area of more than 100,000 acres comprises mostly short-grass plains, buttes, ridges, and cliffs, in the White River drainage. There is no published checklist, but a mimeographed list of 208 species seen in the monument and its vicinity is available from the manager, P.O. Box 72, Interior, South Dakota 57750.

15. Black Hills. This general area of more than a million acres includes Wind Cave National Park, Black Hills National Forest, Custer State Park, and the adjacent 600,000-acre Buffalo Gap National Grassland. The comprehensive reference for the entire area is the booklet by Pettingill and Whitney (see South Dakota references). Additionally, there is a list of 186 species for the Black Hills National Forest (including Wind Cave and Custer State Park) and a separate list of 198 species for the Buffalo Gap National Grassland available from the district ranger, USFS, Hot Springs, South Dakota 57747. Last, a checklist of birds of Wind Cave National Park (28,000 acres) is available from park headquarters, Hot Springs, South Dakota 57747, and a paper on summer bird watching at Wind Cave was published in 1969 in *South Dakota Bird Notes* 22:53–60.

16. Lacreek National Wildlife Refuge. Situated about 15 miles southeast of Martin, South Dakota. Consists of extensive marshes and shallow lakes in the valley of the South Fork of the White River, just north of the Nebraska Sandhills. A checklist of 235 species seen on the refuge since 1936 (including 91 breeding species) is available from the refuge manager, Martin, South Dakota 57551. The breeding trumpeter swans are of special interest and may be easily seen during most seasons.

17. Lake Andes National Wildlife Refuge. Situated north of Fort Randall Dam in southeastern South Dakota. It consists of 5,450 acres, around the Lake Andes marsh. There is an outdated bird checklist of 200 species (70 breeders), and a new one is in preparation. Nearby is the recently established Karl E. Mundt National Wildlife Refuge below Fort Randall dam. It consists of less than 1,000 acres but is an important wintering area for bald eagles. For information, contact the manager, Box 391, Lake Andes, South Dakota 57356.

18. Sand Lake National Wildlife Refuge. Situated 25 miles northeast of Aberdeen South Dakota, in the James River Valley, originally part of the shoreline of glacial Lake Dakota. In consists of more than 21,000 acres of grasslands, marshes, and shallow impoundments, as well as shelterbelts and fields. A checklist of 241 species, including 104 breeders and 15 accidentals, is available from the refuge manager, Columbia, South Dakota 57433.

19. Waubay National Wildlife Refuge. Situated 8 miles north of Waubay, South Dakota, in the glaciated Coteau Hills of northeastern South Dakota. It contains nearly 5,000 acres of marshlands, lakes, grasslands, brushy areas, and oak timber. A checklist of 246 species, including 103 breeding species and 14 accidentals, is available from the refuge manager, R.R. 1, Waubay, South Dakota 57273.

NEBRASKA

20. Bessey Division, Nebraska National Forest. This human-planted forest of various pine species, and the adjoining areas of sandhills, grasslands, and riverine woodlands, administered by the U.S. Forest Service, is situated near Thedford, Thomas County. A checklist of 95 summer bird species, of which 36 are known to breed within the forest, is available from the superintendent, Halsey, Nebraska 69142.

21. Crescent Lake National Wildlife Refuge. Situated 28 miles north of Oshkosh, Nebraska, in Garden County. This refuge consists of

46,000 acres of relatively pristine sandhills grasslands and contains numerous shallow lakes and marshes. A checklist of 218 species (64 breeders) observed on the refuge during the 1960s is available from the refuge manager, Ellsworth, Nebraska 69340. A survey of the nesting birds of the refuge was published in 1966 in the *Nebraska Bird Review* 34:31–35.

22. De Soto National Wildlife Refuge. Situated on the Nebraska-Iowa border, between Blair, Nebraska, and Missouri Valley, Iowa. The refuge consists of an oxbow cutoff (De Soto Lake) of the Missouri River and encompasses 7,800 acres of floodplain land including river-bottom forest and adjacent grasslands, marshes, and cultivated lands. A check list of 195 species, including 97 species known to have nested in the refuge, is available from the refuge manager, R.R. 1-B, Missouri Valley, Iowa 51555.

23. Fort Niobrara National Wildlife Refuge. Situated 5 miles east of Valentine, Cherry County, Nebraska. This refuge contains 19,100 acres, and is largely concerned with the management of bison and other large mammals. About two-thirds of the refuge consists of sandhills prairie, and the rest is mostly of mixed hardwoods along the Niobrara river. A checklist of 201 species (76 breeders) is available from the refuge manager, Hidden Timber Star Route, Valentine, Nebraska 69201.

24–25. Scotts Bluff and Agate Fossil Beds National Monuments. Scotts Bluff National Monument is 5 miles southwest of the town of Scottsbluff, on the old Oregon Trail. It consists of short-grass plains, eroded rock slopes and cliffs, and a small amount of wooded vegetation. The address of the superintendent of Scotts Bluff and Agate Fossil Beds monuments is Box 427, Gering, Nebraska 69341. Agate Fossil Beds National Monument is in central Sioux County 20 miles south of Harrison and includes nearly 2,000 acres of short-grass plains and fossil-bearing rock outcrops. No checklists are available for these specific areas, but an annotated checklist of 266 species, covering most of northwestern Nebraska, has been privately printed by Richard Rosche, P.O. Box 482, Crawford, Nebraska 69339, and is available from him ($2).

26. Valentine National Wildlife Refuge. This refuge is in the center of the Nebraska Sandhills, in Cherry County, about 20 miles south of Valentine. It covers more than 71,000 acres and includes 36 lakes plus numerous marshes, surrounded by sand dunes from 40 to 200 feet high. A checklist of 221 species (93 breeders) recorded on the refuge is available from the refuge manager, Hidden Timber Star Route, Valentine, Nebraska 69201. Many typical grassland species, such as long-billed curlews and sharp-tailed grouse, are abundant on this enormous refuge.

COLORADO

27. Pawnee National Grassland. Consists of 775,000 acres of grasslands in northeastern Colorado. Descriptions of the birdlife of this and other eastern Colorado areas can be found in *A Birder's Guide to Eastern Colorado*, available ($3.25) from L. & P. Press, Box 19401, Denver 80219. A checklist of the birds of the Pawnee Grassland is also available from the U.S. Forest Service, Bldg. 85, Denver Federal Center, Denver, Colorado 80225.

MISSOURI

28. Squaw Creek National Wildlife Refuge. This refuge is 5 miles south of Mound City, Holt County, in extreme northwestern Missouri. It consists of nearly 7,000 acres of marshes, Missouri River bottomlands, wooded bluffs, and farmlands. The checklist of 263 species includes 104 known breeders, and there are 27 additional accidentals reported only once or twice. It is available from the refuge manager, Box 101, Mound City, Missouri 64470.

29. Cheyenne Bottoms Waterfowl Management Area. This state-controlled area is northeast of Great Bend, Barton County, Kansas. It consists of about 18,000 acres of marshlands, as well as adjacent bottomlands associated with the Arkansas River. A checklist of about 320 species, including 102 that have bred in the area, is available from the area manager, Route 1, Great Bend, Kansas 67530.

30–31. Comanche and Cimarron National Grasslands. These areas of Great Plains grasslands are in southwestern Kansas and adjacent southeastern Colorado. They include many semiarid grassland and sagebrush-dominated habitats, as well as some lakes, marshes, and ponds. A checklist of 235 bird species is available from the district ranger's office, U.S. Forest Service, Elkart, Kansas 67950; Springfield, Colorado 81073; or La Junta, Colorado 81050.

32. Flint Hills National Wildlife Refuge. This refuge is on the John Redmond Reservoir of the Neosho River in Coffey County, Kansas. It includes 18,500 acres and is managed primarily for waterfowl. A checklist of 189 bird species (50 breeders) seen within the refuge area since it was established in 1963 is available from the refuge manager, P.O. Box 1306, Burlington, Kansas 66839.

33. Kirwin National Wildlife Refuge. This refuge is about 10 miles southeast of Phillipsburg, in Phillips County, Kansas. It consists of

10,800 acres, mostly marshes, grasslands, croplands, and reservoir acreage impounded by the North Fork of the Solomon River. A checklist of 186 species (38 breeders) is available from the refuge manager, Kirwin, Kansas 67644.

34. Quivira National Wildlife Refuge. This refuge is 12 miles northeast of Stafford, in south-central Kansas. It consists of 21,800 acres, including 13,000 acres of grassland and 4,700 acres of marsh, as well as farmlands and some low sandhills. A checklist of 245 bird species (56 breeders) is available from the refuge manager, P.O. Box G, Stafford, Kansas 67578.

OKLAHOMA

35. McCurtain Game Preserve. This sanctuary is administered by the Oklahoma Department of Wildlife Conservation and consists of about 15,220 acres of virgin oak-pine forest in north-central McCurtain County. An unpublished Ph.D. dissertation by W. A. Carter ("Ecology of the summer nesting birds of the McCurtain Game Preserve," Oklahoma State University, Stillwater) lists 63 species of summer birds, including 56 known or probable nesters. Several southeastern species, such as the brown-headed nuthatch, Kentucky warbler, and pine warbler, are found only in this part of Oklahoma.

36. Platte National Park. (Chickasaw National Recreation Area). This small (900 acre) national park and associated recreation area around the Lake of the Arbuckles consists of hardwood forests along the stream valleys and prairie on hilltops and upper slopes. A mimeographed checklist of birds containing 137 species is available from the superintendent, P.O. Box 201, Sulphur, Oklahoma 73086.

37. Salt Plains National Wildlife Refuge. This refuge is 3 miles southeast of Jet, Oklahoma, in Alfalfa County. It consists of nearly 20,000 acres associated with the Salt Plains Reservoir on the Salt Fork of the Arkansas River. Most of the area is covered by the Salt Plains Reservoir, but there are also extensive salt flats that provide a unique habitat as well as upland, forest, and rangeland. A checklist of 256 species, plus 18 very rare or accidental species, is available from the refuge manager, Jet, Oklahoma 73749.

38. Sequoyah National Wildlife Refuge. This refuge is in east-central Oklahoma, around the western part of the Robert S. Kerr Reservoir. It was established in 1970 and includes 20,800 acres, about half of which is water. Most of the rest is steep shoreline or river bottomland, with many ponds and sloughs. A bird checklist of 245 species seen since 1970, including 93 species known to breed locally, is available from the refuge manager, Box 398, Sallisaw, Oklahoma 74955.

39. Tishomingo National Wildlife Refuge. This refuge is 6 miles southeast of Tishomingo, on Lake Texoma in eastern Oklahoma. It consists of 16,600 acres, including about 4,000 acres of reservoir, as well as marshes, cropland, and grassland. A checklist of 252 species, including 76 species known to breed locally, is available from the refuge manager, P.O. Box 248, Tishomingo, Oklahoma 73460.

40. Wichita Mountains Wildlife Refuge. This refuge is 12 miles north of Cache, Oklahoma, in Comanche County. It consists of 59,000 acres of rugged topography, with many valleys, lakes, and ponds, and a predominance of oak vegetation. A bird checklist of 241 species, including 52 that have been known to breed locally, is available from the refuge manager, P.O. Box 448, Cache, Oklahoma 73527. A more complete analysis of the birds of this area appeared in the *Great Plains Journal* 16 (1977):135–62.

New Mexico

41. Capulin Mountain National Monument. This national monument is 3 miles north of Capulin, Union County, New Mexico. It consists of 775 acres, centered on an old volcanic crater that rises 8,215 feet above sea level. The area around the crater consists of grassland and various forest communities, such as pines, junipers, oaks, and scrub. A mimeographed checklist of birds, containing 104 species, is available from the superintendent, Capulin, New Mexico 88414.

Texas

42. Buffalo Lake National Wildlife Refuge. Situated 30 miles southwest of Amarillo, Texas. This 7,700-acre refuge includes about 1,000 acres of surface water resulting from the impoundment of Tierra Blanca Creek, as well as adjoining grasslands. The bird checklist contains 275 species, of which 41 are known to have bred locally, and is available from the refuge manager, P.O. Box 228, Umbarger, Texas 79091.

43. Hagerman National Wildlife Refuge. Situated 15 miles northwest of Sherman, Texas, around Lake Texoma, on the Texas-Oklahoma border. This 11,320-acre area is associated with Lake Texoma and includes habitats similar to those in the Tishomingo National Wildlife Refuge in Oklahoma. A bird checklist containing 265 species, including 62 known to nest on the refuge, is available from the refuge manager, Route 3, Box 123, Sherman, Texas 75090.

44. Muleshoe National Wildlife Refuge. Situated 20 miles south of Muleshoe, slightly outside the limits of this book's coverage. The refuge contains 5,800 acres of lakes, marshes, short-grass plains, and other minor habitats. The sink-type lakes provide the most important wintering habitat in North America for lesser sandhill cranes, which sometimes number over 100,000. The bird checklist includes 263 species, of which 45 are reported to have nested on the refuge, and is available from the refuge manager, P.O. Box 549, Muleshoe, Texas 79347.

Abundance and Breeding Status of Birds at Selected Parks and Refuges in the Great Plains

Species	North Dakota							Minnesota			South Dakota					Nebraska				Kansas				Missouri	Oklahoma				Texas	
	Arrowwood N.W.R.	Audubon N.W.R.	Long Lake N.W.R.	Slade N.W.R.	Souris Loop Refuges	Theodore Roosevelt N.R.	Teewaukon N.W.R.	Agassiz N.W.R.	Itasca State Park	Tamarac N.W.R.*	Badlands N.M.*	Sand Lake N.W.R.	Lacreek N.W.R.	Waubay N.W.R.	Black Hills	Crescent Lake N.W.R.	Niobrara N.W.R.	Valentine N.W.R.	De Soto N.W.R.*	Cheyenne Bottoms W.M.A.	Flint Hills N.W.R.	Kirwin N.W.R.	Quivira N.W.R.	Squaw Creek N.W.R.	Sequoyah N.W.R.	Salt Plains N.W.R.	Tishomingo N.W.R.	Wichita N.W.R.	Buffalo Lake N.W.R.	Muleshoe N.W.R.
Common Loon	R	o	o	o	r		r	o	C	C					c															
Red-necked grebe	U	o	o	o	u	u	r	C	C	r																				
Horned grebe	C	C	O	O	U	r	U	C	r	r		A	C	C	O	r						o			o	R			R	
Eared grebe	C	U	U	U	U	U	D	U	C	u	c	A	C	C		A	u	C		r		o								
Western grebe	C	C	C	C	C	D	U	C	r	u	c	A	C	C		C		C		r										
Pied-billed grebe	C	C	C	C	C	c	A	C	U	U	u	A	C	C		C	r	C		R	o	o	c	c		U		U		
White pelican	c	a	c	c	c		u	c		o	r	C	A	A		u	u	c		c	u	u	c	u	o		o	o	o	o
Double-crested cormorant	c	A	u	u	U		C	C		r	r	C	A	A		C	o	a		U	u	C	c	r	o	R		o	o	o
Anhinga																								R						

Species
Great blue heron
Green heron
Little blue heron
Cattle egret
Great egret
Snowy egret
Black-crowned night heron
Yellow-crowned night heron
Least bittern
American bittern
White-faced ibis
Trumpeter swan
Canada goose
Common mallard
Black duck
Gadwall
Northern pintail
Green-winged teal
Blue-winged teal
Cinnamon teal
American wigeon
Northern shoveler
Wood duck
Redhead
Ring-necked duck

*Breeding status is not indicated for Tamarac, Badlands, and Salt Plains; only known breeders are listed for De Soto.

Note: Letters indicate relative abundance: a, abundant; c, common; o, occasional; u, uncommon; r, rare. Upper-case letters signify known breeding at the specified location. Species that are accidental, introduced "pests", or not known to breed in the region are not listed.

Species	North Dakota							Minnesota			South Dakota					Nebraska				Kansas				Missouri	Oklahoma				Texas	
	Arrowwood N.W.R.	Audubon N.W.R.	Long Lake N.W.R.	Slade N.W.R.	Souris Loop Refuges	Theodore Roosevelt N.R.	Teewaukon N.W.R.	Agassiz N.W.R.	Itasca State Park	Tamarac N.W.R.*	Badlands N.M.*	Sand Lake N.W.R.	Lacreek N.W.R.	Waubay N.W.R.	Black Hills	Crescent Lake N.W.R.	Niobrara N.W.R.	Valentine N.W.R.	De Soto N.W.R.*	Cheyenne Bottoms W.M.A.	Flint Hills N.W.R.	Kirwin N.W.R.	Quivira N.W.R.	Squaw Creek N.W.R.	Sequoyah N.W.R.	Salt Plains N.W.R.	Tishomingo N.W.R.	Wichita N.W.R.	Buffalo Lake N.W.R.	Muleshoe N.W.R.
Canvasback	U	R	O	U	U		U	U		r		U	U	U		o	o	O		U	U	r	u	r			r			o
Lesser scaup	R	r	O	U	U		o	U				U	r	U		O														
Common goldeneye		r							u	o				o		r	u			u	u	o								
Bufflehead					o				u					o	o							r								
White-winged scoter						o		r		o				o	r	o														o
Ruddy duck	C	C	C		O	O	C	C	r			C	C	C		C		C		U	u		o	u					o	O
Hooded merganser			u		R	O	C	u	C							r											r		o	
Turkey vulture					O	U	u		U	o	c	u	u	u	C	o	u	o		o	A		A	o	C	C	C	C		O
Black vulture					O			r		u		u	u	u						r		A	c	c	C	C	A	C		r
Mississippi kite																														
Goshawk					U			u	U					r	R					o		r				u		U		
Sharp-shinned hawk								u	u	r		U	o	U	r	o		o		u	U	o		U	o	o	o	r	u	u
Cooper hawk		r						C	u	u		U	U	U	U	C	o	r		C	A	o	C	O	u	u	o	o	u	r
Red-tailed hawk	C	C		r	C	C			C	o		C	C	C	U		C	C		U		C	u	u	C	C	R	R	u	u

Red-shouldered hawk								r		o			r											O	O		O			
Broad-winged hawk								o	c	o				o										u	U		R			
Swainson hawk	U	o		O	U	R	u			r		U	o	U	r	U	u	U		C		C	A		U		o	u		U
Ferruginous hawk					O	O					u	o	o			r						r								U
Harris hawk																														o
Golden eagle		r								r				o	U	o														
Bald eagle		r						o	u	u										o										
Marsh hawk	C	u	C	C	C	C	C	C	u	u	c	C	U	C	R	C	c	C		C	C	o		O	o	a	r	o		C
Osprey		o						r	c	u			o	r				r										r		
Prairie falcon	r	r	o			R		r			u		u	r	U	r						u							r	
Peregrine falcon										r			u	o	R	r				r		u							o	
Merlin		o				R				o			o		R	r														
American kestrel	R	o		O	O	C	U	C	c	u	c	U	U	u		r	c	C		C	C	o	u	u	O	c	o	R	O	o
Spruce grouse								r	r																					
Ruffed grouse								C	c	c																				
Pinnated grouse			R	r		r		r				R	r			r	U	U												r
Sharp-tailed grouse	C	C	C	C	C			C		r	c		O	R	C	C	C	A												
Bobwhite													o			r	U	r	A	C	A	C	A	U	A	A	A	c	C	C
Scaled quail																													O	C
Ring-necked pheasant	U	C	C	C	U		C	r		r	u	A	A	C	U	A	C	A	A	A		C	A	C		c			O	O
Gray partridge	C	C	U	C	C		U	u		r		U	r	U				r												
Wild turkey															C		C			o		r				a	o	C		
King rail	r									r			r							C			U	U	u	r	R			
Virginia rail	U	O	U	U	C	u	U	U	u	u		C	C	U	r	o	r	C		C		o	U	O	o	r			r	
Sora	C	C	C	U	C	U	C	c	c	u		C	A	C	r	O	r	U		C	c	o	U	u	o		r			
Yellow rail					r								r																	
Black rail																				R			r				r			
Common gallinule																r		r		U				o		r	r		r	
American coot	C	C	C	C	C	c	C	A	u	u	c	C	C	C	r	C	u	C		C	c	u	C	O	C	a	o	r	o	O
Piping plover	U	O		U	R		o	r				U	o	O				o	C	r			u	o			o			
Snowy plover																				C		o	C			a	C		c	C
Killdeer	U	A	C	A	C	C	C	C	c	c	c	C	C	C	C	A	C	C	C	C	A	c	C	C	A	a	C	C	c	C

State	Refuge	American woodcock	Common snipe	Long-billed curlew	Upland sandpiper	Spotted sandpiper	Willet	Marbled godwit	American avocet	Black-necked stilt	Wilson phalarope	California gull	Ring-billed gull	Franklin gull	Forster tern
North Dakota	Arrowwood N.W.R.	U	U	U	U	U	U	U	A	U	O	c	c	u	U
North Dakota	Audubon N.W.R.	r	A	U	U	C	C	C	C	C	A	o	c	c	O
North Dakota	Long Lake N.W.R.		C	C	U		C				C	O	C	C	C
North Dakota	Slade N.W.R.		C	C	C	C	O				U	O	C	C	O
North Dakota	Souris Loop Refuges		C	C	C	c	U	C			U	C	C	C	C
North Dakota	Theodore Roosevelt N.R.	u	R	R	O						u	u			r
North Dakota	Teewaukon N.W.R.	o	U	U	U	U	U	u		C	U	o	c	a	u
Minnesota	Agassiz N.W.R.		O	O	U	r	U	r			U	c	C	c	
Minnesota	Itasca State Park	u	u		C						u	c	r		
Minnesota	Tamarac N.W.R.*	u	u					r	u		u	u	o		r
South Dakota	Badlands N.M.*				c	c									
South Dakota	Sand Lake N.W.R.		C	C	C	C	C	U	C	A	c	C	C	C	C
South Dakota	Lacreek N.W.R.		U	C	C	C	C	U	C		U	o	C	C	C
South Dakota	Waubay N.W.R.		U	U	O	O	O	C			U	c	C	C	U
South Dakota	Black Hills	R	C												
Nebraska	Crescent Lake N.W.R.	R	C	A	o	o	r	C	A			o	c	C	
Nebraska	Niobrara N.W.R.	c	c	C	C	o	o	u	r	C		c	u	u	
Nebraska	Valentine N.W.R.	u	u	U	C	c	U	C		C	c	o	o		c
Nebraska	De Soto N.W.R.*				C										
Kansas	Cheyenne Bottoms W.M.A.		u	U	u	u	u	r	A	r	C			c	U
Kansas	Flint Hills N.W.R.		u	A	u			A	r		c	u	u		
Kansas	Kirwin N.W.R.		u	o	o	o			o			o	o		
Kansas	Quivira N.W.R.	u	u	C	c		u	u	C	r	u	c	c	u	
Missouri	Squaw Creek N.W.R.	o	r	R	C			r		o		o	o	o	o
Oklahoma	Sequoyah N.W.R.		c	u								o			o
Oklahoma	Salt Plains N.W.R.		o						a			o	o		
Oklahoma	Tishomingo N.W.R.	r	r					o	o			o	o		o
Oklahoma	Wichita N.W.R.	r	r				o	o	r			o	o		r
Texas	Buffalo Lake N.W.R.	o	o	u	u		o	r	C	r	c	o	o		
Texas	Muleshoe N.W.R.	o	u						A	u		u			r

Species	1	2	3	4	5	6	7	8	9	10	11	12	13	14	15	16	17	18
Common tern		A	U	O	U		U	u	u	u	u	r	C	u	C			o
Least tern							r					u	o					
Black tern	U	A	U	O	C	c	c	C	C	U	c		C	A	C			

Species	19	20	21	22	23	24	25	26	27	28	29	30	31	32	33	34	35
Common tern				r	C	U			u	C			c	o			r	u
Least tern			r	C	U					C			c	a	o			r	u
Black tern	A	u	A			C	a	c	C	C	o	a	o	o	u	c

Species	col
Mourning dove	C C C A C C C C u c c C C C A C A C A A A A A A A C a A C c A
Yellow-billed cuckoo	O o u o u O r r r o O C U A O C C C c C C U U
Black-billed cuckoo	U U O U u C u U o c U O U O C U A U o o U
Roadrunner	R o O U U U
Barn Owl	r R U O o R r u r r U
Screech owl	r r R o u U O u R U U U U U U u U O c O U
Great horned owl	U o U U U C U c C u c U U C C U C C C C A C C C A c c C C O C
Burrowing owl	r O r r R R U r c U C R O U O C O C u C A
Barred owl	u c u C A u r C A C u
Long-eared owl	R R R o r o u O R r u
Short-eared owl	U U O R r U o u U U u u O o c o r
Saw-whet owl	r u u r u
Chuck-will's-widow	o o C c O C
Whip-poor-will	o u r o R U C O
Poor-will	o r u r U
Common nighthawk	u o u U R C o u c u c U u c C C C C c U A o C o C c C c C u c C
Chimney swift	r o U u o U U U O c c C O
White-throated swift	c C
Ruby-throated hummingbird	r o r R o u u C u r U C o o r u u c U C C c C u
Belted kingfisher	u r r r R o u u C c c U C U C o U O R u o c U C c C u o u
Common flicker	C U u U C C C c C c c C O C C o C C C C A C A C C c O r
Pileated woodpecker	r C r O o
Red-bellied woodpecker	U U A C C a C C
Red-headed woodpecker	R r R R U U o u r c U U u O U o U o C A C C U O C U o
Lewis woodpecker	U U
Yellow-bellied sapsucker	C c u o o U C C o C u u
Hairy woodpecker	u u O R U U u C c c u O o C C C o C U u u U O c c C
Downy woodpecker	C u O U C U c u c u U U C C o U u C U A O U C O c C C

Species	Arrowwood N.W.R.	Audubon N.W.R.	Long Lake N.W.R.	Slade N.W.R.	Souris Loop Refuges	Theodore Roosevelt N.R.	Teewaukon N.W.R.	Agassiz N.W.R.	Itasca State Park	Tamarac N.W.R.*	Badlands N.M.*	Sand Lake N.W.R.	Lacreek N.W.R.	Waubay N.W.R.	Black Hills	Crescent Lake N.W.R.	Niobrara N.W.R.	Valentine N.W.R.	De Soto N.W.R.*	Cheyenne Bottoms W.M.A.	Flint Hills N.W.R.	Kirwin N.W.R.	Quivira N.W.R.	Squaw Creek N.W.R.	Sequoyah N.W.R.	Salt Plains N.W.R.	Tishomingo N.W.R.	Wichita N.W.R.	Buffalo Lake N.W.R.	Muleshoe N.W.R.
	North Dakota							**Minnesota**			**South Dakota**					**Nebraska**				**Kansas**				**Missouri**	**Oklahoma**				**Texas**	
Ladder-backed woodpecker																												r	U	U
Black-backed three-toed woodpecker									r						R															
Northern three-toed woodpecker															U															
Eastern kingbird	C	C	U	C	C	C	C	C	C	u	u	A	C	A	C	C	C	C	A	C	A	A	A	C	A	A	A	A	A	
Western kingbird	C	C	U	U	C	C	C	U	r	u	c	A	C	A	U	C	C	C	R	C	A	A	A	C	A	a	O	r	A	
Cassin kingbird															U	U	R	U												
Scissor-tailed flycatcher													u	u		r	o	o	U	C	A		A		A	C	A	C		C
Great crested flycatcher	O						O	o	u	u		o							U	U	A	o	o	C	A	c	O	A	A	
Ash-throated flycatcher															U	U			R						C	c	C	C		
Eastern phoebe	r				O	O	O	O	r	r	c	o	r	o	r	O	C	o	R	o	C	o	o	C	C	c	C	C	r	
Say phoebe	U	U		O	U	O	O	U			c				U	C				C	C	C	O		r	C	C	C	C	C

Species															
Acadian flycatcher	U			A		U		O	C		U				
Willow flycatcher	U		O	R		C	u	C	C	O	u			r	
Least flycatcher			C	R	r	C	c	R	O	O	u				
Dusky flycatcher				O	c	c	c	O	c						
Western flycatcher								C	C						
Eastern wood pewee	O		O		C	C	c	C			O	u			
Western wood pewee			o	o		u				O	u				
Olive-sided flycatcher	o			U	o			o		C					
Vermilion flycatcher									C						o
Horned lark	C	c	C	U	r	o	c	C	C	C	C	C	c	o	C A A
Violet-green swallow					C			C							
Tree swallow	O	U	U	O	U	u	c	O	r	O	U	U	o	o	O
Bank swallow	C	C	U	C	C	u	o	C	o	O	C	C	a	o	
Rough-winged swallow	U	U	O	C	U	r	o	C	r	C	u	C	a	O	O
Barn swallow	C	C	C	A	A	C	c	A	C	C	C	A	a	A	U u
Cliff swallow	o	C	A	U	A	U	o	C	C	C	A	O	a	U	u
Purple martin	U r	U	C	C	U	U	c	O	o	U	U	C	o	r	u
Gray jay	u	o	O	U	u	u		C	C	C	C	C			
Blue jay	U	r	U	C	C	c	r	C	C	C	C	C	a	C	o
Black-billed magpie				O	O			U				U			
White-necked raven	U	U	C	C	O	o		O	r					r	U
American crow	U	U	U	C	C	C	u	U	U	C	O	C	C	C	o
Fish crow						C						o			
Pinyon jay								C							
Black-capped chickadee	U		C	R	C	A	c	C	O	C	U	C		C	C
Carolina chickadee		U		C		o		C		R	C A		C	a	C C
Boreal chickadee			R												
Tufted titmouse					r	A		C A U				r A		C	C
White-breasted nuthatch	C	R	C	o	u	C		C A	o	C	c	C	C	C	c
Red-breasted nuthatch	O		C	C		o		U C	C		o	C	o	o	c
Pygmy nuthatch								U							
Brown creeper	U		C	o		U		U			C		O		

State	Refuge	Dipper	House wren	Bewick wren	Winter wren	Carolina wren	Long-billed marsh wren	Short-billed marsh wren	Canyon wren	Rock wren	Northern mockingbird	Gray catbird	Brown thrasher	Sage thrasher	American robin
Texas	Muleshoe N.W.R.		r	u					U		C		C		
Texas	Buffalo Lake N.W.R.		O	r			r	U	U	U	A	C	A	o	o
Oklahoma	Wichita N.W.R.		C	C		U					A	C	A		R
Oklahoma	Tishomingo N.W.R.		o	c		C					A	R	C		R
Oklahoma	Salt Plains N.W.R.		o	c		c					A	C	O		c
Oklahoma	Sequoyah N.W.R.		O	o		C	U				A	C	A		A
Missouri	Squaw Creek N.W.R.		u	r		C	U	U			C	C	C		A
Kansas	Quivira N.W.R.		U	u		u	U	u			C	u	C		C
Kansas	Kirwin N.W.R.		C	u		C	u	u			A	u	A		C
Kansas	Flint Hills N.W.R.		u	C		C					U	A	A		C
Kansas	Cheyenne Bottoms W.M.A.		u	u		r	u	O	r		U	U	U		U
Nebraska	De Soto N.W.R.*		A	C			R	R	R	r	U	C	C		C
Nebraska	Valentine N.W.R.		O	r				o	R	o	r	U	C		C
Nebraska	Niobrara N.W.R.		C	o			A	o	R	r	U	C	C		C
Nebraska	Crescent Lake N.W.R.		o	C			A	r		o	o	C	C		C
South Dakota	Black Hills	U	C	U					U	C	C	C	C	A	A
South Dakota	Waubay N.W.R.		O	C			C	U			r	C	C		C
South Dakota	Lacreek N.W.R.		O	U			C	O		o	o	C	C		U
South Dakota	Sand Lake N.W.R.		U	C			C	C			C	C	C		C
South Dakota	Badlands N.M.*		c	c		c			c	c	r	c	c		c
Minnesota	Tamarac N.W.R.*		o	C		o	R	o			c	u	u		a
Minnesota	Itasca State Park		C	C	C	R	c	c			C	u	u		C
Minnesota	Agassiz N.W.R.		U	C		C	C	o			U	u	u		C
North Dakota	Teewaukon N.W.R.		U	C		C	C				U	U	U		C
North Dakota	Theodore Roosevelt N.R.		C	O					C	C	U	C	C		C
North Dakota	Souris Loop Refuges		U	U		r	U	U			U	U	U	r	O
North Dakota	Slade N.W.R.		C	O		C	C	O			U	U	U		C
North Dakota	Long Lake N.W.R.		C	C		O	C	C			U	C	U		U
North Dakota	Audubon N.W.R.		C	O		C	C	C			O	C	C		U
North Dakota	Arrowwood N.W.R.		U	U		C	U	U			C	C	C		C

Wood thrush									r	u									U					U	O		o			
Hermit thrush								C	c															o						
Swainson thrush								u	u	u					C	o														
Veery						R	o	u	C	c					U										o					
Eastern bluebird	R			O	O	u	o		U	u		U		U	U	o	U	R	U		C	o	c	U	C	c	o	C	u	o
Mountain bluebird						C					c				C					r									u	
Townsend solitaire						o									C	r														
Blue-gray gnatcatcher									R												A		c	U	C	c	C	U	r	o
Golden-crowned kinglet								o	C	o					U															
Ruby-crowned kinglet									u	r					R	r										o				
Sprague pipit	U	o	u		C																									
Cedar waxwing	U	r	r		C		O	c	C	u		u	o	U	U	o			U					U						
Loggerhead shrike	R	o	O		U	C	o			r	c	u	O	u	U	o	C	C		C	C	o	c	u	C	c	C	o	o	C
Black-capped vireo																									r					
White-eyed vireo																								r	O	r	C	r		
Bell vireo											u		O		r	r	u	O	C	U	C	U	A	C	u	c	C			
Yellow-throated vireo					O			r	U	o				r								o			o					
Solitary vireo									C						C															
Red-eyed vireo	u			R	C	C	o	u	C	c			O		C		U	C	R				u	C	O	u	c	u		
Philadelphia vireo				r	O		r						r												o					
Warbling vireo	U				C	C	U	C	U	u		U	C	u	C			C	A	U	C		c	C	O	u	r		r	
Black-and-white warbler					O	C		u	C	o					r		u		U						U		o	u		
Prothonotary warbler																							u	R	U	o	C			
Swainson warbler																									o					
Worm-eating warbler																									u					
Golden-winged warbler									C	r																				
Blue-winged warbler																									r					
Tennessee warbler									R																u					
Nashville warbler								U	C	o																				
Northern parula									C	u															O					
Yellow warbler	C	C	C	O	C	C	C	C	U	c		C	C	C	C	C	C	C	A	O		c	c	C	C	c	o		o	
Magnolia warbler									U	r																				

Species	Arrowwood N.W.R.	Audubon N.W.R.	Long Lake N.W.R.	Slade N.W.R.	Souris Loop Refuges	Theodore Roosevelt N.R.	Teewaukon N.W.R.	Agassiz N.W.R.	Itasca State Park	Tamarac N.W.R.*	Badlands N.M.*	Sand Lake N.W.R.	Lacreek N.W.R.	Waubay N.W.R.	Black Hills	Crescent Lake N.W.R.	Niobrara N.W.R.	Valentine N.W.R.	De Soto N.W.R.*	Cheyenne Bottoms W.M.A.	Flint Hills N.W.R.	Kirwin N.W.R.	Quivira N.W.R.	Squaw Creek N.W.R.	Sequoyah N.W.R.	Salt Plains N.W.R.	Tishomingo N.W.R.	Wichita N.W.R.	Buffalo Lake N.W.R.	Muleshoe N.W.R.
Cape May warbler						c		r																						
Black-throated blue warbler								r	R																					
Yellow-rumped warbler						U		u	r	C					C										u					
Black-throated green warbler						o		r	C	o															u					
Cerulean warbler									C																					
Blackburnian warbler									C																u					
Yellow-throated warbler									r										U						u		o			
Chestnut-sided warbler									C																					
Bay-breasted warbler									r																					
Pine warbler									C																					
Palm warbler						o			c										R											
Ovenbird						U		o	u	C					C				R					U						

Species	1	2	3	4	5	6	7	8	9	10	11	12	13	14	15	16	17	18	19	20	21	22	23	24	25	26	27	28	29	30
Northern waterthrush									r																					
Louisiana waterthrush																									o		o			
Kentucky warbler																									o					
Mourning warbler									C	r															u					
MacGillivray warbler														U																
Common yellowthroat		c	o	o	C	C	C	C	C	u	u	U	A	O	C	U	C	C	A	C	A			U	C	O	u	O		
Yellow-breasted chat				O	C								O	o	C	o	C	o		o				o	C	C	o	A		
Hooded warbler									r																					
Canada warbler								u	U	r					r															
American redstart	R				U	U	o		C	C	c			r	o	C		U	O	A				U	U					
Bobolink	C	C	C	C	C	o	C	C	C	r			O	C	C	r	C		C	R	u		U	U						
Eastern meadowlark								u	u				C			A	U	C	C	U	A	O	C	U	C	c	A	C		o
Western meadowlark	C	A	A	A	A	C	C	C	u	r	c	C	A	C	A	A	A	C	A	A		A	A	O			c	r	r A	A
Yellow-headed blackbird	C	C	C	C	A	C	A	C	u	r	c	A	A	C		A	U	C	R	A		C	U	U			r		o	o
Red-winged blackbird	C	A	A	A	A	C	A	C	C	a	c	A	A	C	C	A	C	A	A	A	A	A	C	C	A	a	A	U	A	C
Orchard oriole	U	U		C	O	R	U			r	c	U	C	O	U	C	O	C	C	C	C	O	C	C	C	c	C	r	C	u
Northern oriole	U	U	C	O	C	O	U	C	C	u		U		C	o	U	U	A	C	C	C	A	C	O	c	o	O	r	C	C
Brewer blackbird	U	C	U		U	C	U	u	u	u		C	U	U	C		U				C		O				r			
Great-tailed grackle																								o		o				
Common grackle	C	C	C	C	C	C	C	u	r	c		C	c	C	C	A	C	C	C	A	o	c	C	c	o	O	r			
Brown-headed cowbird	C	A	C	U	C	C	C	C	C	c	c	C	C	C	R	C	C	A	C	A	c	C	C	O	c	C	C	U	u	
Western tanager													C	o	u	o														
Scarlet tanager						r			r	C	u			r			u					O	o							
Summer tanager																				u	U	C				u				
Cardinal								r					r	u		C	U	A	C	C	A	A	a	A	C	o				
Rose-breasted grosbeak	r				O	R		u	C	u		o		u	o		o	A	o		U	o								
Black-headed grosbeak					O							U	o	C			U													
Blue grosbeak							u		r		r	o	U	O		U			o	c	O	U	c	O	U	u	c			
Indigo bunting					r	C	o		o	r	r	r			C	U	A		C	O	o	A					u			
Lazuli bunting				O	U			r		o	r	C	r		O			o	r						u					
Painted bunting																R			C		u		O	c	A	c				
Dickcissel	U	O	o	o	O	R		C		u	u	C	U	u	U	u	C	C	A	C	C	O	A	C	C	a	A	U	u	

Species	Arrowwood N.W.R.	Audubon N.W.R.	Long Lake N.W.R.	Slade N.W.R.	Souris Loop Refuges	Theodore Roosevelt N.R.	Teewaukon N.W.R.	Agassiz N.W.R.	Itasca State Park	Tamarac N.W.R.*	Badlands N.M.*	Sand Lake N.W.R.	Lacreek N.W.R.	Waubay N.W.R.	Black Hills	Crescent Lake N.W.R.	Niobrara N.W.R.	Valentine N.W.R.	De Soto N.W.R.*	Cheyenne Bottoms W.M.A.	Flint Hills N.W.R.	Kirwin N.W.R.	Quivira N.W.R.	Squaw Creek N.W.R.	Sequoyah N.W.R.	Salt Plains N.W.R.	Tishomingo N.W.R.	Wichita N.W.R.	Buffalo Lake N.W.R.	Muleshoe N.W.R.
Evening grosbeak	U	U	C	C	r	C	C	u	C	u																			o	o
Purple finch	U	C	C	C	C	C	C	C	C	u																				
House finch	U	U	C	U		C	U		u	r					A								o						O	
Pine siskin	C	U	U	O	C	C	C	C	C	c	u	C	C	C	C	C	C	U	U		o		c	C		c				
American goldfinch	r	C	A	O	A	A	O	r	C	r		O	A	A	C	C	A	C	A	A	o	o	o	o	C	O	A	A	r	
Lesser goldfinch	U	U	U	O	U	U	U		u	r		O	O	U	U	C	C	C	C				r	C			r	r	r	
Red crossbill	C	U	U	U	C	C	U	C	r	r		O	O	U	C	U	U	U	C				c	C						
White-winged crossbill	r	u		U	C	C	C	r	C	r		O	o	U		C	A	C	C				u	U	a				u	
Rufous-sided towhee	r	O		O	r	r	r	r	r	u	c	c	o	o	U	o	C	U	U	O	r	O	u	U	A	c	c	u	u	A
Lark bunting										o				o		C	U	U	o	u		o	c					r	u	
Savannah sparrow														U		C	A	C	C	C	C	o	u	C	O	O	r	u	R	u
Grasshopper sparrow																A	C	U	U	C	c	c	c	u	A	A	u	u	u	u
Baird sparrow											c											r		c						A
Le Conte sparrow									u															u						

Species																				
Henslow sparrow	R																	u		
Sharp-tailed sparrow	U	C	C	U			U	C	C	U		c	c	c	C	U		c	c	U
Vesper sparrow	U	C	C	U		r	C	C	A	C	U		o	U	C	C	A	C	c	r
Lark sparrow	o	o	C	U	o	c	U	U	r	C	C	A	C	U		c	U	C	c	
Rufous-crowned sparrow																				c
Cassin sparrow			u						C											
Dark-eyed junco	U	O	U	o	o	R	C		o	U	A	C		c	U	C		r		
Chipping sparrow	C	O	U	u	c	u	U	C	o	c	C	U	c	U	c		r	o	
Clay-colored sparrow	C	A	C	C	c	c	U	U	r	o	u		o						
Brewer sparrow	U		U										o						
Field sparrow	o	o	o		U	r	o	r	U	O	A		U	o	C		r		
White-throated sparrow	r																		
Swamp sparrow			C	o	C	o	C	C	C		O	C	O	C					
Song sparrow	C	C	C	c	u	c	U	C	U	O	U	O	U	u	c	o			
McCown longspur	O	O	O																
Chestnut-collared longspur	C	C	C	U	c		U	O	o	o		c	C		o				

References

REGIONAL AND STATE REFERENCES

The Great Plains

Allen, D. L. 1967. *The life of prairies and plains.* New York: McGraw-Hill.
American Ornithologists' Union. 1957. *Check-list of North American birds.* 5th ed. Baltimore: Lord Baltimore Press. (Supplement published in *Auk* [1973] 90:411-19.
Hubbard, J. P. 1974. Avian evolution in the aridlands of North America. *Living Bird* 12:155-96
Johnsgard, P. A. 1976. The grassy heartland. In *Our continent: A natural history of North America.* Washington, D. C.: National Geographic Society.
————. 1978. The ornithogeography of the Great Plains. *Prairie Naturalist* 10:97-112.
Küchler, A. W. 1964. *Potential natural vegetation of the conterminous United States.* American Geographical Society Special Publication no. 36. Washington, D. C.: American Geographical Society.
Mengel, R. E. 1970. The North American central plains as an isolating agent in bird speciation. In *Pleistocene and Recent environments of the central Great Plains.* Special Publication no. 3. Lawrence: University of Kansas Department of Geology.
Peterson, R. T. 1963. *The birds.* New York: Time, Inc.
Udvardy, M. D. F. 1958. Ecological and distributional analysis of North American birds. *Condor* 60:50-66

Colorado

Bailey, A. M., and Niedrach, R. J. 1965. *Birds of Colorado.* 2 vols. Denver: Denver Museum of Natural History.
C.F.O. Journal (formerly *Colorado Field Ornithologist*). Published three times a year by the Colorado Field Ornithologists, 31 issues through 1977.
Kingery, H. E., and Graul, W. D. 1978. *Colorado bird distribution latilong study.* Denver: Colorado Field Ornithologists and Colorado Division of Wildlife.

Iowa

Brown, W. H. 1971. An annotated list of the birds of Iowa. *Iowa State Journal of Sciences* 45:387-469.
Iowa Bird Life. Published quarterly by the Iowa Ornithologists' Union. 47 vols. through 1977.

Kansas

Johnston, R. F. 1964. The breeding birds of Kansas. *University of Kansas Museum of Natural History Publications* 12:575–655.

_____. 1965. Directory to the birds of Kansas. *University of Kansas Museum of Natural History Miscellaneous Publications* 41:1–67.

Kansas Ornithological Society Bulletin. Published quarterly by the Kansas Ornithological Society. 28 vols. through 1977.

Rising, J. D. 1974. The status and faunal affinities of the summer birds of western Kansas. *University of Kansas Science Bulletin* 50:347–88.

Tordoff, H.B. 1956. Check-list of the birds of Kansas. *University of Kansas Museum of Natural History Publications* 8:30–59.

Minnesota

Green, J. C., and Janssen, R. B. 1975. *Minnesota birds: Where, when, and how many.* Minneapolis: University of Minnesota Press.

Kellehur, K. 1967. Distribution of breeding birds in deciduous forests at the prairie-hardwood forest ecotone in northwestern Minnesota. Ph.D. diss., University of Minnesota, Minneapolis.

The Loon (formerly *The Flicker*). Published quarterly by the Minnesota Ornithologists' Union. 49 vols. through 1977.

Roberts, T. S. 1932. *The birds of Minnesota.* 2 vols. Minneapolis: University of Minnesota Press (rev. ed. 1936).

Missouri

Bennet, R. 1932. Check-list of the birds of Missouri. *University of Missouri Studies* 7:1–81.

The Bluebird. Published quarterly by the Audubon Society of Missouri. 44 vols. through 1977.

Nebraska

Johnsgard, P.A. 1979. The breeding birds of Nebraska. *Nebraska Bird Review* 47:3–14.

Nebraska Bird Review. Published quarterly by the Nebraska Ornithologists' Union. 45 vols through 1977.

Rapp, W. F.; Rapp, J. L. C.; Baumgarten, H. E.; and Moser, R. A. 1970. *Revised check-list of Nebraska birds, with supplement through 1970.* Occasional Papers of the Nebraska Ornithologists' Union no. 5a. Crawford, Neb.

Rosche, R. C. 1977. *Check-list of birds of northwestern Nebraska and southwestern South Dakota.* Crawford, Neb.: R. C. Rosche.

New Mexico

Bailey, F. M. 1828. *Birds of New Mexico.* Santa Fe: New Mexico Department of Fish and Game.

Hubbard, J. P. 1978. *Revised check-list of the birds of New Mexico.* New Mexico Ornithological Society Publication no. 6. Albuquerque, N.M.

Ligon, J. S. 1961. *New Mexico birds and where to find them.* Albuquerque: University of New Mexico Press.

The Southwestern Naturalist. Published quarterly by the Southwestern Association of Naturalists. 22 vols. through 1977.

New Mexico Ornithological Society Bulletin. Published 3-4 times per year. 8 vols. through 1978.

North Dakota

The Prairie Naturalist. Published quarterly by the North Dakota Natural Science Society. 9 vols through 1977.

Stewart, R. E. 1975. *Breeding birds of North Dakota.* Fargo, N. D.: Tri-College Center for Environmental Studies.

Oklahoma

Bulletin of the Oklahoma Ornithological Society. Published quarterly by the Oklahoma Ornithological Society. 10 vols. through 1977.

Sutton, G. M. 1967. *Oklahoma birds: Their ecology and distribution, with comments on the avifauna of the southern Great Plains.* Norman: University of Oklahoma Press.

_____.1974. *A check-list of Oklahoma birds.* Norman: Stovall Museum of Science and History, University of Oklahoma.

South Dakota

Harrell, B. E., ed. 1978. *The birds of South Dakota: An annotated check-list.* Vermilion: South Dakota Ornithologists' Union and W. H. Over Dakota Museum.

Pettingill, O. S., Jr., and Whitney, N. R., Jr. 1965. *Birds of the Black Hills.* Special Publication no. 1. Ithaca, N.Y.: Cornell Laboratory of Ornithology.

South Dakota Bird Notes. Published quarterly by South Dakota Ornithologists' Union. 29 vols. through 1977.

Whitney, N. R., Jr. 1965. Check-list of South Dakota birds. *South Dakota Bird Notes* 17:80–83.

Texas

Bulletin of the Texas Ornithological Society. Published twice a year by the Texas Ornithological Society. 10 vols. through 1977.

Hamilton, T. H. 1962. The habitats of the avifauna of the mesquite plains of Texas. *American Midland Naturalist* 67:85–105.

Oberholser, H. C. 1974. *The bird life of Texas.* Ed. E. B. Kincaid. 2 vols. Austin: University of Texas Press.

Addicott, A. R. 1938. Behavior of the bush-tit in the breeding season. *Condor* 40:49-62.

Adkisson, C. S. 1966. The nesting and behavior of mockingbirds in northern Lower Michigan. *Jack-Pine Warbler* 44:102-16.

Allen, A. A. 1924. A contribution to the life history and economic status of the screech owl (*Otus asio*). *Auk* 41:1-16.

Allen, R. A., and Nice, M. M. 1952. A study of the breeding biology of the purple martin (*Progne subis*). *American Midland Naturalist* 47:606-65.

Allen, R. P. 1952. *The whooping crane*. Research Report no. 2, New York: National Audubon Society.

Allen, T. T. 1961. Notes on the breeding behavior of the anhinga. *Wilson Bulletin* 73:115-25.

Ambrose, J. E., Jr. 1963. The breeding ecology of *Toxostoma curvirostre* and *T. bendirei* in the vicinity of Tucson, Arizona. M.S. thesis, University of Arizona, Tucson.

Angell, T. 1969. A study of the ferruginous hawk: Adult and brood behavior. *Living Bird* 8:225-41.

Armstrong, E. A. 1955. *The wren*. London: Collins.

Armstrong, W. H. 1958. Nesting and food habits of the long-eared owl in Michigan. *Michigan State University Museum Publications, Biological Series* 1 (2):63-96.

Austin, G. R. 1964. *The world of the red-tailed hawk*. Philadelphia and New York: Lippincott.

Austin, G. T. 1976. Sexual and season differences in foraging of ladder-backed woodpeckers. *Condor* 78:317-23.

Bailey, A. M.; Niedrach, R. J.; and Bailey, A. L. 1953. The red crossbills of Colorado. *Denver Museum of Natural History, Museum Pictorial* 9:1-64.

Bailey, P. F. 1977. The breeding biology of the black tern (*Chlidonias niger surinamensis* Gmelin). M.S. thesis, State University of Wisconsin, Oshkosh.

Baird, P. A. 1976. Comparative ecology of California and ring-billed gulls (*Larus californicus* and *L. delawarensis*). Ph.D. diss., University of Montana.

Bakus, G. J. 1959. Observations on the life history of the dipper in Montana. *Auk* 76:190-207.

Balda, R. P., and Bateman, G. C. 1973. The breeding biology of the pinon jay. *Living Bird* 11:5-42.

Balgooyen, T. G. 1976. Behavior and ecology of the American kestrel (*Falco sparverius* L). in the Sierra Nevada of California. *University of California Publications in Zoology* 103:1-83.

Banko, W. 1960. The trumpeter swan. U.S. Fish and Wildlife Service, *North American Fauna* 63:1-214.

Barclay, R. 1977. Solitary vireo breeding behavior. *Blue Jay* 35:33-37.

Barlow, J. C. 1962. Natural history of the Bell vireo, *Vireo bellii* Audubon. *University of Kansas Museum Publications* 12:241-96.

Barlow, J. C., and Rice, J. C. 1977. Aspects of the comparative behavior of red-eyed and Philadelphia vireos. *Canadian Journal of Zoology* 55:528-42.

Barlow, J. C., and Rising, J. D. 1965. The summer status of wood pewees in southwestern Kansas. *Bulletin of the Kansas Ornithological Society* 16:14–16.

Baskett, T. S. 1947. Nesting and production of the ring-necked pheasant in north-central Iowa. *Ecological Monographs* 17:1–30.

Baumann, S. A. 1959. The breeding cycle of the rufous-sided towhee *Pipilo erythropthalmus* (Linnaeus) in central California. *Wasmann Journal of Biology* 17:161–220.

Baxter, W. L., and Wolfe, C. W. 1973. *The ring-necked pheasant in Nebraska*. Lincoln: Nebraska Game and Parks Commission.

Beason, R. C., and Franks, E. C. 1974. Breeding behavior of the horned lark. *Auk* 91:65–74.

Beecham, J. J., and Kochert, M. N. 1975. Breeding biology of the golden eagle in southwestern Idaho. *Wilson Bulletin* 87:506–13.

Bent, A. C. 1907. The marbled godwit on its breeding grounds. *Auk* 24:160–67.

———. 1921. Life histories of North American gulls and terns. *United States National Museum Bulletin* 113:1–345.

———. 1926. Life histories of North American marsh birds. *United States National Museum Bulletin* 135:1–490.

———. 1937. Life histories of North American birds of prey. Part 1. *United States National Museum Bulletin* 167:1–409.

———. 1938. Life histories of North American birds of prey. Part 2. *United States National Museum Bulletin* 170:1–428.

———. 1939. Life histories of North American woodpeckers. *United States National Museum Bulletin* 174:1–322.

———. 1940. Life histories of North American cuckoos, goatsuckers, hummingbirds and their allies. *United States National Museum Bulletin* 176:1–506.

———. 1942. Life histories of North American flycatchers, larks, swallows and their allies. *United States National Museum Bulletin* 179:1–555.

———. 1946. Life histories of North American jays, crows, and titmice. *United States National Museum Bulletin* 191:1–495.

———. 1948. Life histories of North American nuthatches, wrens, thrashers, and their allies. *United States National Museum Bulletin* 195:1–475.

———. 1949. Life histories of North American thrushes, kinglets, and their allies. *United States National Museum Bulletin* 196:1–454.

———. 1950. Life histories of North American wagtails, shrikes, vireos, and their allies. *United States National Museum Bulletin* 197:1–411.

———. 1953. Life histories of North American wood warblers. *United States National Museum Bulletin* 203:1–734.

———. 1958. Life histories of North American blackbirds, orioles, tanagers and allies. *United States National Museum Bulletin* 211:1–549.

———. 1968. Life histories of North American cardinals, grosbeaks, buntings, towhees, finches, sparrows, and allies. In three parts. *United States National Museum Bulletin* 237:1–1889.

Bergman, R. D.; Swain, P.; and Weller, M. W. 1970. A comparative study of nesting Forster's and black terns. *Wilson Bulletin* 82:435–44.

Best, L. B. 1972. First-year effects of sagebrush control on two sparrows. *Journal of Wildlife Management* 36:534–44.

———. 1977. Territorial quality and mating success in the field sparrow (*Spizella pusilla*). *Condor* 79:192–203.

Bibbee, P. C. 1947. The Bewick's wren, *Thryomanes bewickii* (Audubon). Ph.D. diss., Cornell University.

Bicak, T.K. 1977. Some eco-ethological aspects of a breeding population of long-billed curlews (*Numenius americanus*) in Nebraska. M.A. thesis. University of Nebraska at Omaha.

Black, C. P. 1976. The ecology and bioenergetics of the northern black-throated blue warbler. Ph.D. diss., Dartmouth College.

Bock, C. E. 1970. The ecology and behavior of the Lewis woodpecker (*Asyndesmus lewis*). *University of California Publications in Zoology* 92:1–100.

Bowdish, B. S., and Philipp, P. B. 1916. The Tennessee warbler in New Brunswick. *Auk* 33:1–8.

Boyd, R. L. 1972. Breeding biology of the snowy plover at Cheyenne Bottoms Waterfowl Management Area, Barton County, Kansas. M.S. thesis, Kansas State Teachers College.

Braaten, D. J. 1975. Observations at three brown creeper nests in Itasca State Park. *Loon* 47:110–13.

Brackbill, H. 1958. Nesting behavior of the wood thrush. *Wilson Bulletin* 70:70–89.

———. 1970. Tufted titmouse breeding behavior. *Auk* 87:522–36.

Brakhage, G. K. 1965. Biology and behavior of tub-nesting Canada geese. *Journal of Wildlife Management* 29:751–71.

Brewer, R. 1963. Ecological and reproductive relationships of black-capped and Carolina chickadees. *Auk* 80:9–47.

Brown, J. L. 1964. The integration of agonistic behavior in the Steller's jay *Cyanocitta stelleri* (Gmelin). *University of California Publications in Zoology* 60:223–328.

Brown, L., and Amadon, D. 1968. *Eagles, hawks and falcons of the world.* 2 vols. New York: McGraw-Hill.

Bump, G.; Darrow, R.; Edminister, F., and Crissey, W. 1947. *The ruffed grouse: Life history, propagation, management.* Albany: York State Conservation Department.

Bunn, D. S., and Warburton, A. B. 1977. Observations on breeding barn owls. *British Birds* 70:246–56.

Bunni, M.K. 1959. The killdeer, *Charadrius v. vociferus* Linnaeus, in the breeding season: Ecology, behavior and the development of homiothermism. Ph.D. diss., University of Michigan.

Burger, J. 1974. Breeding adaptations of Franklin's gulls (*Larus pipixcan*) to a marsh habitat. *Animal Behaviour* 22:521–67.

Burger, J., and Miller, L. M. 1977. Colony and nest site selection in white-faced and glossy ibises. *Auk* 94:664–75.

Butterfield, J. D. 1969. Nest site requirements of the lark bunting in Colorado. M.S. thesis, Colorado State University.

Calder, W. A. 1973. Microhabitat selection during nesting of hummingbirds in the Rocky Mountains. *Ecology* 54:127–34.

Carter, B. C. 1958. The American goldeneye in central New Brunswick. Canadian Wildlife Service, *Wildlife Management Bulletin*, series 2, no. 9. pp. 1–47.

Cartwright, B. W.; Shortt, T. M.; and Harris, R. D. 1937. Baird's sparrow. *Transactions of the Royal Canadian Institute* 21:153–97.

Chamberlain, D. R., and Cornwell, G. W. 1971. Selected vocalizations of the common crow. *Auk* 88:613–34.

Chamberlain, M.L. 1977. Observations on the red-necked grebe nesting in Michigan. *Wilson Bulletin* 89:33–46.

Chapman, L. B. 1955. Studies of a tree swallow colony. *Bird-Banding* 26:45–70.

Clark, R. J. 1975. A field study of the short-eared owl *Otus flammeus* (Pontoppidan) in North America. *Wildlife Monographs*, vol. 47.

Coles, V. 1944. Nesting of the turkey vulture in Ohio caves. *Auk* 61:219–28.

Combellack, C.R.B. 1954. A nesting of violet-green swallows. *Auk* 71:435–42.

Cornwell, G. W. 1963. Observations on the breeding biology and behavior of a nesting population of belted kingfishers. *Condor* 65:426–31.

Coulter, M. W., and Miller, W. R. 1968. Nesting biology of black ducks and mallards in northern New England. *Vermont Fish and Game Department Bulletin*, no. 68-2, pp. 1–74.

Coutlee, E. 1968. Comparative behavior of lesser and Lawrence's goldfinches. *Condor* 70:228–42.

Cox, C. W. 1960. A life history of the mourning warbler. *Wilson Bulletin* 72:5–28.

Craig, W. 1943. The song of the wood pewee. *New York State Museum Bulletin* 334:1–186.

Crawford, R. D. 1977. Polygynous breeding of short-billed marsh wrens. *Auk* 94:359–62.

Cripps, B. J., Jr. 1966. The nesting cycle of the chestnut-sided warbler. *Raven* 37:43–48.

Crooks, M. P., and Hendrickson, G. O. 1953. Field sparrow life history in central Iowa. *Iowa Bird Life* 23:10–13.

Dane, C. W. 1966. Some aspects of breeding biology of the blue-winged teal. *Auk* 83:389–402.

Davis, C. A., and Griffing, J. P. 1972. Nesting of the white-necked raven in southeastern New Mexico. *New Mexico State University Agricultural Experiment Station Research Report* 231:1–5.

Davis, D. E. 1959. Observations on territorial behavior of least flycatchers. *Wilson Bulletin* 71:73–85.

Davis, J. 1960. Nesting behavior of the rufous-sided towhee in coastal California. *Condor* 62:434–56.

Davis, J.; Fisher, G. F.; and Davis, B. S. 1963. The breeding biology of the western flycatcher. *Condor* 65:337–82.

Day, K. C. 1953. Home life of the veery. *Bird-Banding* 24:100–106.

De Garis, C. F. 1936. Notes on six nests of the Kentucky warbler (*Oporornis formosus*). *Auk* 53:418–28.

519

Dennis, J. V. 1948. Observations on the orchard oriole in the lower Mississippi delta. *Bird-Banding* 19:12-20.

Dilger, W. C. 1956. Hostile behavior and reproductive isolating mechanisms in the avian genera *Catharus* and *Hylocichla*. *Auk* 73:313-53.

Dixon, J. B.; Dixon, R. E.; and Dixon, E. 1957. Natural history of the white-tailed kite in San Diego County, California. *Condor* 59:155-65.

Dixon, K. L. 1949. Behavior of the plain titmouse. *Condor* 51:110-36.

———. 1955. An ecological analysis of the interbreeding of crested titmice in Texas. *University of California Publications in Zoology* 54:125-206.

———. 1956. Territoriality and survival in the plain titmouse. *Condor* 58: 169-82.

Dow, D. D. 1965. The role of saliva in food storage of the gray jay. *Auk* 82: 139-54.

Drewien, R. C. 1973. Ecology of Rocky Mountain greater sandhill cranes. Ph.D. diss., University of Idaho.

Dunham, D. W. 1964. Reproductive displays of the warbling vireo. *Wilson Bulletin* 76:170-73.

———. 1966. Territorial and sexual behavior in the rose-breasted grosbeak. *Zeitschrift für Tierpsychologie* 23:438-51.

Dunkle, S. W. 1977. Swainson's hawks on the Laramie Plains, Wyoming. *Auk* 94:65-71.

Dunstan, T. C. 1973. The biology of ospreys in Minnesota. *Loon* 45:108-13.

Dunstan, T. C.; Mathisen, J. E.; and Harper, J. G. 1975. The biology of bald eagles in Minnesota. *Loon* 47:5-10.

Dunstan, T. C., and Sample, S. D. 1972. The biology of barred owls in Minnesota. *Loon* 44:111-15.

Eaton, S. W. 1957. A life history study of *Seiurus noveboracensis*. St. Bonaventure University, *Science Studies* 19:7-36.

———. 1958. A life history study of the Louisiana waterthrush. *Wilson Bulletin* 70:211-36.

Eckert, A. W. 1974. *The owls of North America.* New York: Doubleday.

Eckhardt, R. C. 1976. Polygyny in the western wood pewee. *Condor* 78:561-62.

Edson, J. M. 1943. A study of the violet-green swallow. *Auk* 60:396-403.

Emlen, J. T., Jr. 1954. Territory, nest building and pair formation in the cliff swallow. *Auk* 71:16-35.

Emlen, S. T.; Rising, J. D.; and Thompson, W. L. 1975. A behavioral and morphological study of sympatry in the indigo and lazuli buntings of the Great Plains. *Wilson Bulletin* 87:145-79.

Enderson, J. H. 1964. A study of the prairie falcon in the central Rocky Mountain region. *Auk* 81:332-52.

Engeling, G. A. 1950. Nesting habits of the mottled duck (*Anas fulvigula maculosa*) in Colorado, Fort Bent and Brazoria counties, Texas. M.S. thesis, Texas A. & M. College.

Erpino, M. J. 1968. Nest related activities of the black-billed magpie. *Condor* 70:154-65.

Errington, P. L; Hamerstrom, F.; and Hamerstrom, F. N. 1940. The great horned owl and its prey in the north-central United States. *Iowa State*

College Research Bulletin 277:758–850.

Erskine, A. J. 1972. Buffleheads. Canadian Wildlife Service Monograph Series, no. 4. Ottawa: Information Canada.

Erwin, W. G. 1935. Some nesting habits of the brown thrasher. Journal of the Tennessee Academy of Science 10:179–204.

Evenden, F. G. 1957. Observations on nesting behavior of the house finch. Condor 59:112–17.

Eyer, L. E. 1963. Observations on golden-winged warblers at Itasca State Park, Minnesota. Jack-Pine Warbler 41:96–109.

Faaborg, J. 1976. Habitat selection and territorial behavior of the small grebes of North Dakota. Wilson Bulletin 88:390–99.

Fehon, J. H. 1955. Life-history of the blue-gray gnatcatcher (Polioptila caerulea caerulea). Ph.D. diss., Florida State University.

Ficken, M. S. 1962. Agonistic behavior and territory in the American redstart. Auk 79:607–32.

Ficken, M. S., and Ficken, R. W. 1966. Behavior of myrtle warblers in captivity. Bird-Banding 37:273–79.

———. 1968a. Territorial relationship of blue-winged warblers, golden-winged warblers, and their hybrids. Wilson Bulletin 80:442–51.

———. 1968b. Reproductive isolating mechanisms in the blue-winged warbler, golden-winged warbler complex. Evolution 22:166–79.

Ficken, R. W.; Ficken, M. S.; and Morse, D. H. 1968. Competition and character displacement in two sympatric pine-dwelling warblers (Dendroica, Parulidae). Evolution 22:307–14.

Fisher, R. 1958. The breeding biology of the chimney swift. New York State Museum and Science Service Bulletin 368:1–41.

Fitch, F. W., Jr. 1950. Life history and ecology of the scissor-tailed flycatcher. Auk 67:145–68.

Fitch, H. S. 1963. Observations on the Mississippi kite in southeastern Kansas. University of Kansas Museum Publications 12:503–19.

Fitch, H. S., and Fitch, V. R. 1955. Observations on the summer tanager in northeastern Kansas. Wilson Bulletin 67:45–54.

Fitzner, J. N. 1978. The ecology and behavior of the long-billed curlew (Numenius americanus) in southeastern Washington. Ph.D. diss., Washington State University.

Fjeldså, J. 1973. Antagonistic and heterosexual behaviour of the horned grebe, Podiceps auritus. Sterna 12:161–217.

Frederickson, L. H. 1970. Breeding biology of American coots in Iowa. Wilson Bulletin 82:445–57.

———. 1971. Common gallinule breeding biology and development. Auk 88:914–19.

Frydendall, M. J. 1967. Feeding ecology and territorial behavior of the yellow warbler, Ph.D. diss., Utah State University.

Ganier, A. F. 1964. The alleged transportation of its eggs or young by the chuck-will's-widow. Wilson Bulletin 79:19–27.

Gibbon, R. S. 1966. Observations on the behavior of nesting three-toed woodpeckers, Picoides tridactylus, in central New Brunswick. Canadian Field-Naturalist 80:223–26.

Gibson, F. 1971. The breeding biology of the American avocet (*Recurvirostra americana*) in central Oregon. *Condor* 73:444–54.

Girard, G. L. 1939. Notes on life history of the shoveller. *North American Wildlife Conference Transactions* 4:363–71.

———. 1941. The mallard, its management in western Montana. *Journal of Wildlife Management* 5:233–59.

Glover, F. A. 1953. Nesting ecology of the pied-billed grebe in northwestern Iowa. *Wilson Bulletin* 65:32–39.

Glue, D. G. 1977. Breeding biology of long-eared owls. *British Birds* 70:318–31.

Godfrey, R. S. 1975. Behavior and ecology of American woodcock on the breeding range in Minnesota. Ph.D. diss., University of Minneapolis.

Goforth, W. R., and Baskett, T. S. 1971. Social organization of penned mourning doves. *Auk* 88:528–42.

Goodwin, D. 1967. *Pigeons and doves of the world.* London: British Museum (Natural History).

———. 1976. *Crows of the world.* Ithaca: Cornell University Press.

Goodwin, R. A. 1960. A study of the ethology of the black tern, *Chlidonias niger surinamensis.* Ph.D. diss., Cornell University.

Graber, J. 1961. Distribution, habitat requirements and life history of the black-capped vireo. *Ecological Monographs* 31:313–36.

Graber, R., and Graber, J. 1951. Nesting of the parula warbler in Michigan. *Wilson Bulletin* 63:75–83.

Grant, R. A. 1965. The burrowing owl in Minnesota. *Loon* 37:1–17.

Graul, W. D. 1974. Adaptive aspects of the mountain plover social system. *Living Bird* 12:69–94.

Green, R. 1976. Breeding behaviour of ospreys *Pandion haliaetus* in Scotland. *Ibis* 118:475–90.

Grice, D., and Rogers, J. P. 1965. *The wood duck in Massachusetts.* Massachusetts Division of Fisheries and Game, Final Report, Project W-19-R.

Griscom, L., et al. 1957. *The warblers of America.* New York: Devin-Adair.

Gullion, G. W. 1954. The reproductive cycle of American coots in California. *Condor* 71:366–412.

Haecker, F. A. 1948. A nesting study of the mountain bluebird in Wyoming. *Condor* 50:216–19.

Hahn, H. W. 1937. Life history of the ovenbird in southern Michigan. *Wilson Bulletin* 49:145–237.

———. 1950. Nesting behavior of the American dipper in Colorado. *Condor* 52:49–62.

Hamilton, R. C. 1975. Comparative behavior of the American avocet and the black-necked stilt (Recurvirostridae). *A.O.U. Monographs* 17:1–98.

Hanson, H. C., and Kossack, C. W. 1963. The mourning dove in Illinois. Illinois Department of Conservation Technical Bulletin no. 2, Urbana, Ill.

Harding, K. C. 1931. Nesting habits of the black-throated blue warbler. *Auk* 40:512–22.

Hardy, J. 1957. The least tern in the Mississippi Valley. *Michigan State University Museum Publications, Biological Series* 1(1):1–60.

———. 1961. Studies in behavior and phylogeny of certain New World jays (Garrulinae). *University of Kansas Science Bulletin* 62:13–149.

Harmeson, J. P. 1974. Breeding ecology of the dickcissel. *Auk* 91:348–59.

Hartshorne, J. M. 1962. Behavior of the eastern bluebird at the nest. *Living Bird* 1:131–49.

Hays, H. 1973. Polyandry in the spotted sandpiper. *Living Bird* 11:43–57.

Heckenlively, D. B. 1970. Song in a population of black-throated sparrows. *Condor* 72:24–36.

Hespenheide, H. A. 1964. Competition and the genus *Tyrannus*. *Wilson Bulletin* 76:265–81.

Hickey, J. J., ed. 1969. *Peregrine falcon populations: Their biology and decline*. Madison: University of Wisconsin Press.

Higgins, K. F., and Kirsch, L. M. 1975. Some aspects of the breeding biology of the upland sandpiper in North Dakota. *Wilson Bulletin* 87:96–102.

Hines, J. E. 1977. Nesting and brood ecology of lesser scaup at Waterhen Marsh, Saskatchewan. *Canadian Field-Naturalist* 91:248–55.

Hochbaum, A. H. 1944. *The canvasback on a prairie marsh*. Harrisburg: Stackpole.

Hofslund, P. B. 1959. A life history of the yellowthroat, *Geothlypis trichas*. *Proceedings of the Minnesota Academy of Science* 27:144–74.

Höhn, E. O. 1967. Observations on the breeding biology of Wilson's phalarope (*Steganopus tricolor*) in central Alberta. *Auk* 84:220–44.

Holcomb, L. C. 1972. Traill's flycatcher breeding biology. *Nebraska Bird Review* 40:50–67.

Horn, H. S. 1970. Social behavior of nesting Brewer's blackbirds. *Condor* 72:15–23.

Hostetter, D. R. 1961. Life history of the Carolina junco, *Junco hyemalis* Brewster. *Raven* 32:97–170.

Howell, J. C. 1942. Notes on the nesting habits of the American robin (*Turdus migratorius* L.). *American Midland Naturalist* 28:529–603.

Hoyt, S. 1957. The ecology of the pileated woodpecker. *Ecology* 38:246–56.

Hubbard, J. P. 1969. The relationships and evolution of the *Dendroica coronata* complex. *Auk* 86:393–432.

Hyde, A. S. 1939. The life history of the Henslow's sparrow. *University of Michigan Miscellaneous Publications* 41:1–72.

Jackson, A. 1976. A comparison of some aspects of the breeding ecology of red-headed and red-bellied woodpeckers in Kansas. *Condor* 78:67–76.

James, R. D. 1973. Ethological and ecological relationships of the yellow-throated and solitary vireos (Aves: Vireonidae) in Ontario. Ph.D. diss., University of Toronto.

Jenni, D. A. 1969. A study of the ecology of four species of herons during the breeding season at Lake Alice, Alachua County, Florida, *Ecological Monographs* 39:245–70.

Johnsgard, P. A. 1973. *Grouse and quails of North America*. Lincoln: University of Nebraska Press.

————. 1975. *Waterfowl of North America*. Bloomington: Indiana University Press.

Johnson, N. K. 1972. Breeding distribution and habitat preference of the gray vireo in Nevada. *California Birds* 3:73–78.

Johnston, R. F., and Hardy, J. W. 1962. Behavior of the purple martin. *Wilson Bulletin* 74:243–62.

Kangarise, C. M. 1979. Breeding biology of Wilson's phalarope in North Dakota. *Bird-banding* 50:12–22.

Kaufmann, G. W. 1971. Behavior and ecology of the sora, *Porzana carolina,* and Virginia rail. *Rallus limicola.* Ph.D. diss., University of Minnesota.

Kendeigh, S. C. 1941. Territorial and mating behavior of the house wren. *Illinois Biological Monographs* 18:1–120.

———. 1945. Nesting behavior of wood warblers. *Wilson Bulletin* 57:145–64.

Kent, F. W., and Vane, R. F. 1958. Nesting of the whip-poor-will in Iowa County. *Iowa Bird Notes* 28:71–79.

Kessel, B. 1957. A study of the breeding biology of the European starling (*Sturnus vulgaris* L.) in North America. *American Midland Naturalist* 58:257–331.

Kilham, L. 1959. Early reproductive behavior of flickers. *Wilson Bulletin* 71:323–36.

———. 1961. Reproductive behaviour of red-bellied woodpeckers. *Wilson Bulletin* 73:237–54.

———. 1962. Breeding behavior of yellow-bellied sapsuckers. *Auk* 79:31–43.

———. 1966. Reproductive behavior of hairy woodpeckers. 1. Pair formation and courtship. *Wilson Bulletin* 78:251–65.

———. 1968, 1972. Reproductive behavior in white-breasted nuthatches. *Auk* 85:477–92; 89:115–29.

———. 1973. Reproductive behavior in the red-breasted nuthatch. 1. Courtship. *Auk* 90:597–609.

———. 1974. Early breeding season behavior of downy woodpeckers. *Wilson Bulletin* 84:407–18.

———. 1977*a*. Early breeding season behavior of red-headed woodpeckers. *Auk* 94:231–39.

———. 1977*b*. Nesting behavior of yellow-bellied sapsuckers. *Wilson Bulletin* 89:310–24.

Killpack, M. L. 1970. Notes on sage thrasher nestlings in Colorado. *Condor* 72:486–88.

Kingsbury, E. W. 1933. The status and natural history of the bobolink *Dolichonyx oryzivorus.* Ph.D. diss., Cornell University.

Krause, H. 1965. Nesting of a pair of Canada warblers. *Living Bird* 4:5–11.

Kroodsma, R. L. 1970. North Dakota species pairs. I. Hybridization in buntings, grosbeaks and orioles. II. Species-recognition behavior of territorial male rose-breasted and black-headed grosbeaks (*Pheucticus*). Ph.D. diss., North Dakota State University.

Lancaster, D. A. 1970. Breeding behavior of the cattle egret in Colombia. *Living Bird* 9:167–93.

Lanyon, W. E. 1957. The comparative biology of the meadowlarks (*Sturnella*) in Wisconsin. *Publications of the Nuttall Ornithological Club*, no. 1. pp. 1–67.

———. 1961. Specific limits and distribution of ashy-throated and Nutting flycatchers. *Condor* 63:421–49.

Laskey, A. R. 1944. A study of the cardinal in Tennessee. *Wilson Bulletin* 56:24–44.

———. 1962. Breeding biology of mockingbirds. *Auk* 79:596–606.

Laun, C. H. 1957. A life history study of the mountain plover, *Eupoda montana* Townsend, in the Laramie Plains, Albany County, Wyoming. M.S. thesis, University of Wyoming.

Lawrence, L. de K. 1948. Comparative study of the nesting behavior of chestnut-sided and Nashville warblers. *Auk* 65:204-19.

———. 1949. Notes on nesting pigeon hawks at Pimisi Bay, Ontario. *Wilson Bulletin* 61:15-25.

———. 1953*a*. Nesting life and behavior of the red-eyed vireo. *Canadian Field-Naturalist* 67:47-87.

———. 1953*b*. Notes on the nesting behavior of the Blackburnian warbler. *Wilson Bulletin* 65:135-44.

———. 1967. A comparative life-history study of four species of woodpeckers. *A.O.U. Monographs* 5:1-156.

Lea, R. B. 1942. A study of the nesting habitats of the cedar waxwing. *Wilson Bulletin* 54:225-37.

Lederer, R. J. 1977. Winter feeding territories in the Townsend's solitaire. *Bird-Banding* 48:11-18.

Lewis, J. C. 1973. *The world of the wild turkey.* Philadelphia and New York: Lippincott.

Ligon, J. D. 1970. Behavior and breeding biology of the red-cockaded woodpecker. *Auk* 87:255-78.

Linsdale, J. M. 1957. Goldfinches on the Hastings Natural History Reservation. *American Midland Naturalist* 57:1-119.

Littlefield, C. D., and Ryder, R. A. 1968. Breeding biology of the greater sandhill crane on Malheur National Wildlife Refuge, Oregon. *Transactions of the North American Wildlife and Natural Resources Conference* 33:444-54.

Low, J. B. 1941. Nesting of the ruddy duck in Iowa. *Auk* 58:506-17.

———. 1945. Ecology and management of the redhead, *Nyroca americana,* in Iowa. *Ecological Monographs* 15:35-69.

Ludwig, J. P. 1965. Biology and structure of the Caspian tern (*Hydroprogne caspia*) population of the Great Lakes from 1896-1964. *Bird-Banding* 36: 217-33.

Lumsden, H. G. 1965. *Displays of the sharptail grouse.* Ontario Department of Lands and Forests Technical Series Research Report no. 66. Maple, Ont.

Lunk, W. A. 1962. The rough-winged swallow *Stelgidopteryx ruficollis* (Vieillot); A study based on its breeding biology in Michigan. *Publications of the Nuttall Ornithological Club,* no. 4, pp. 1-155.

McAllister, N. M. 1958. Courtship, hostile behavior, nest establishment, and egg-laying in the eared grebe. *Auk* 75:290-311.

McCabe, R. A., and Hawkins, A. S. 1946. The Hungarian partridge in Wisconsin. *American Midland Naturalist* 36:1-75.

McKinney, F. 1965. The displays of the American green-winged teal. *Wilson Bulletin* 77:112-21.

McLaren, M. A. 1975. Breeding biology of the boreal chickadee. *Wilson Bulletin* 87:344-54.

McNicholl, M. K. 1971. The breeding biology and ecology of Forster's tern (*Sterna forsteri*) at Delta, Manitoba, M.S. thesis, University of Manitoba.

Mader, W. J. 1975. Biology of Harris' hawk in southern Arizona. *Living Bird* 14:59–96.

Marshall, J. T., Jr. 1960. Interrelations of Abert and brown towhees. *Condor* 62:49–64.

Martin, D. J. 1973. Selective aspects of burrowing owl ecology and behavior. *Condor* 75:446–56.

Martin, S. G. 1971. Adaptations for polygynous breeding in the boblink. *American Zoologist* 14:109–19.

Matray, P. F. 1974. Broad-winged hawk nesting and ecology. *Auk* 91:307–24.

Maxwell, G. R., II. 1970. Pair formation, nest building and egg laying of the common grackle in northern Ohio. *Ohio Journal of Science* 70:284–91.

Maxwell, G. R., II, and Putnam, L. S. 1972. Incubation, care of young, and nest success of the common grackle (*Quiscalus quiscala*) in northern Ohio. *Auk* 89:349–59.

Mayfield, H. 1965. The brown-headed cowbird, with old and new hosts. *Living Bird* 4:13–27.

Meanley, B. 1955. A nesting study of the little blue heron in eastern Arkansas. *Wilson Bulletin* 67:84–99.

―――. 1963. Pre-nesting activity of the purple gallinule near Savannah, Georgia, *Auk* 80:545–47.

―――. 1969. *Natural history of the king rail.* U.S. Fish and Wildlife Service, North American Fauna, no. 67. Washington, D.C.

―――. 1971. *Natural history of the Swainson's warbler.* U.S. Fish & Wildlife Service, North American Fauna, no. 69.

Meanley, B., and Meanley, A. G. 1959. Observations on the fulvous tree duck in Louisiana. *Wilson Bulletin* 71:33–45.

Mendall, H.L. 1937. Nesting of the bay-breasted warbler. *Auk* 54:429–39.

―――. 1958. The ring-necked duck in the Northeast. *University of Maine Bulletin*, vol. 60, no. 16; and *University of Maine Studies,* 2d ser., no. 73:1–317.

Meng, H. 1951. The Cooper's hawk. Ph.D. diss., Cornell University.

Mengel, R.M., and Jenkinson, M. A. 1971. Vocalizations of the chuck-will's-widow and some related behavior. *Living Bird* 10:171–83.

Meyeriecks, A. J. 1960. Comparative behavior of four species of North American herons. Publications of the Nuttall Ornithological Club, no.2, pp. 1–158.

Mickey, F. W. 1943. Breeding habits of McCown's longspur. *Auk* 60:181–209.

Miller, E. V. 1941. Behavior of the Bewick wren. *Condor* 43:81–99.

Mitchell, R.M. 1977. Breeding biology of the double-crested cormorant at Utah Lake. *Great Basin Naturalist* 37:1–23.

Mock, D. M. 1976. Pair-formation displays of the great blue heron. *Wilson Bulletin* 88:185–230.

Morehouse, E. L., and Brewer, R. 1968. Feeding of nestling and fledgling eastern kingbirds. *Auk* 85:44–54.

Moriarty, L. J. 1965. A study of the breeding biology of the chestnut-collared longspur (*Calcarius ornatus*) in northeastern South Dakota. *South Dakota Bird Notes* 17:76–79.

Morse, T. E.; Jakabosky, J. L.; and McCrow, V. P. 1969. Some aspects of the breeding biology of the hooded merganser. *Journal of Wildlife Management* 33:596–604.

Mousley, H. 1934. A study of the home life of the northern crested flycatcher. *Auk* 51:207–16.

———. 1939. Home life of the American bittern. *Wilson Bulletin* 51:83–85.

Mumford, R. E. 1964. *The breeding biology of the Acadian flycatcher.* University of Michigan Museum Miscellaneous Publications, no. 125. Ann Arbor, Mich.

Murray, B. G., Jr. 1969. A comparative study of the Le Conte's and sharp-tailed sparrows. *Auk* 86:199–231.

Murray, B. G., Jr., and Gill, F. B. 1976. Behavioral interactions between blue-winged and golden-winged warblers. *Wilson Bulletin* 88:231–54.

Murton, R. K., and Carke, S. P. 1968. Breeding biology of rock doves. *British Birds* 61:429–48.

Nero, R. E. 1956. A behavior study of the red-winged blackbird. *Wilson Bulletin* 68:5–37, 129–50.

———. 1970. Great gray owls nesting near Roseau. *Loon* 42:88–93.

Nethersole-Thompson, D. 1975. *Pine crossbills: A Scottish contribution.* Berkhamstead: T. & A. D. Poyser.

Newman, G. A. 1970. Cowbird parasitism and nesting success of lark sparrows in southern Oklahoma. *Wilson Bulletin* 82:304–9.

Newman, I. 1972. *Finches.* Collins: London.

Nice, M. M. 1939. *The watcher at the nest.* New York: Macmillan

———. 1943. Studies in the life history of the song sparrow. II. The behavior of the song sparrow and other passerines. *Transactions of the Linnaean Society of New York* 6:1–238.

Nice, M. M., and Collias, N. E. 1961. A nesting of the least flycatcher. *Auk* 78:145–49.

Nice, M. M., and Nice, L. B. 1932. A study of two nests of the black-throated green warbler. *Bird-Banding* 3:95–105, 157–72.

Nice, M. M., and Thomas, R. H. 1948. A nesting of the Carolina wren. *Wilson Bulletin* 60:139–58.

Nickell, W. P. 1965. Habitats, territory and nesting of the catbird. *American Midland Naturalist* 73:433–78.

Noble, G. K.; Wurm, M.; and Schmidt, M. 1938. Social behavior of the black-crowned night heron. *Auk* 55:7–40.

Nolan, V., Jr. 1960. Breeding behavior of the Bell vireo in southern Indiana. *Condor* 62:225–44.

———. 1974. Notes on parental behavior and development of the young in the wood thrush. *Wilson Bulletin* 86:145–55.

———. 1978. Ecology and behavior of the prairie warbler. *A.O.U. Monographs*, vol. 26.

Norris, R. A. 1958. Comparative biosystematics and life history of the nuthatches *Sitta pygmaea* and *Sitta pusilla*. *University of California Publications in Zoology* 56:119–300.

Nowicki, T. 1973. A behavioral study of the marbled godwit in North Dakota. M.S. thesis, Central Michigan University.

Nuechterlain, G. 1975. Nesting ecology of western grebes on the Delta Marsh, Manitoba. M.S. thesis, Colorado State University.

Odum, E. P. 1941–42. Annual cycle of the black-capped chickadee. *Auk* 58:314–33, 518–35; 59:499–531.

Oeming, A. F. 1955. A preliminary study of the great gray owl (*Scotiaptex nebulosa nebulosa* Forster) in Alberta, with observations on some other species of owls. M.S. thesis, University of Alberta.

Offutt, G. C. 1965. Behavior of the tufted titmouse before and during the nesting season. *Wilson Bulletin* 77:382–87.

Ohlendorf, H. M. 1976. Comparative breeding ecology of phoebes in trans-Pecos Texas. *Wilson Bulletin* 88:255–71.

Ohlendorf, R. R. 1975. *Golden eagle country*. New York: Knopf.

Ohmart, R. D. 1973. Observations on the breeding adaptations of the road-runner. *Condor* 75:140–49.

Olson, S. T., and Marshall, W. H. 1952. The common loon in Minnesota. *Occasional Papers of the Minnesota Museum of Natural History*, no. 5, pp. 1–77.

Orians, G. H., and Christman, G. M. 1968. A comparative study of the behavior of red-winged, tricolored, and yellow-headed blackbirds. *University of California Publications in Zoology* 84:1–85.

Oring, L. W. 1969. Summer biology of the gadwall at Delta, Manitoba. *Wilson Bulletin* 81:44–54.

Oring, L. W., and Knudson, M. L. 1973. Monogamy and polyandry in the spotted sandpiper. *Living Bird* 11:59–73.

Palmer, R. S. 1941. A behavior study of the common tern. *Proceedings of the Boston Society of Natural History* 42:1–119.

———. 1962. *Handbook of North American birds*. Vol. 1. *Loons through Flamingos*. New Haven: Yale University press.

Parker, J. W. 1975. The breeding biology of the Mississippi kite in the Great Plains. Ph.D. diss., University of Kansas.

Parmelee, D. F. 1959. The breeding behavior of the painted bunting in southern Oklahoma. *Bird-Banding* 30:1–17.

Patterson, R. L. 1952. *The sage grouse in Wyoming*. Denver: Sage Books.

Peek, F. W. 1971. Seasonal change in the breeding behavior of the male red-winged blackbird. *Wilson Bulletin* 83:383–95.

Peterson, A. J. 1955. The breeding cycle in the bank swallow. *Wilson Bulletin* 67:235–86.

Phillips, R. S. 1972. Sexual and agonistic behavior in the killdeer (*Charadrius vociferus*). *Animal Behavior* 20:1–9.

Pickens, A. L. 1936. Notes on nesting ruby-throated hummingbirds. *Wilson Bulletin* 48:80–85.

Pitelka, F. A. 1940. Breeding behavior of the black-throated green warbler. *Wilson Bulletin* 52:2–18.

Platt, J. B. 1976. Sharp-shinned hawk nesting and nest site selection in Utah. *Condor* 78:102–3.

Pospichal, L. B., and Marshall, W. H. 1954. A field study of the sora rail and Virginia rail in central Minnesota. *Flicker* 26:2–32.

Porter, D. K.; Strong, M. S.; Giezentanner, J. B.; and Ryder, R. A. 1975. Nest ecology, productivity and growth of the loggerhead shrike on the shortgrass prairie. *Southwestern Naturalist* 19:429-36.

Poston, H. J. 1969. Home range and breeding biology of the shoveler. M.S. thesis, Utah State University, Logan, Utah.

Potter, E. F. 1973. Breeding behavior of the summer tanager. *Chat* 37:35-39.

Potter, P. E. 1972. Territorial behavior in Savannah sparrows in southeastern Michigan. *Wilson Bulletin* 72:48-59.

Power, H. W., III. 1966. Biology of the mountain bluebird in Montana. *Condor* 68:351-71.

Pratt, H. M. 1970. Breeding biology of great blue herons and common egrets in central California. *Condor* 72:407-16.

Preble, N. A. 1957. The nesting habits of the yellow-billed cuckoo. *American Midland Naturalist* 57:474-82.

Prescott, K.W. 1965. *The scarlet tanager.* New Jersey State Museum, Investigations, no. 2. Trenton, N.J.

Pruitt, J. 1975. The return of the great-tailed grackle. *American Birds* 29:985-92.

Purdue, J. R. 1976. Adaptation of the snowy plover on the Great Salt Plains, Oklahoma. *Southwestern Naturalist* 21:347-57.

Rawls, C. K., Jr. 1949. An investigation of life history of the white-winged scoter (*Melanitta fusca deglandi*). M.S. thesis, University of Minnesota.

Raynor, G. S. 1941. The nesting habits of the whip-poor-will. *Bird-Banding* 12:98-104.

Reese, J. G. 1972. A Chesapeake barn owl population. *Auk* 89:106-14.

Reller, A. W. 1972. Aspects of behavioral ecology of red-headed and red-bellied woodpeckers. *American Midland Naturalist* 88:207-90.

Robins, J. D. 1971. A study of Henslow's sparrow in Michigan. *Wilson Bulletin* 83:39-48.

Robinson, T. S. 1957. *The ecology of bobwhites in south-central Kansas.* University of Kansas Museum of Natural History and State Biological Survey Miscellaneous Publication no. 15. Lawrence, Kans.

Robinson, W. L., and Maxwell, D. E. 1968. Ecology study of the spruce grouse on the Yellow Dog Plains. *Jack-Pine Warbler* 46:75-83.

Rogers, J. A., Jr. 1977. Breeding displays of the Louisiana heron. *Wilson Bulletin* 89:266-85.

Rohwer, S. A. 1971. Systematics and evolution of Great Plains meadowlarks, genus *Sturnella.* Ph.D. diss., University of Kansas.

Root, R. B. 1969. The behavior and reproductive success of the blue-gray gnatcatcher. *Condor* 71:16-31.

Roseberry, J. L., and Klimstra, W. D. 1970. The nesting ecology and reproductive performance of the eastern meadowlark. *Wilson Bulletin* 82:243-67.

Rosene, W. 1969. *The bobwhite quail: Its life and management.* New Brunswick, N.J.: Rutgers University Press.

Salt, W. R. 1966. A nesting study of *Spizella pallida. Auk* 83:274-81.

Samuel, D. E. 1971. The breeding biology of barn and cliff swallows in West Virginia. *Wilson Bulletin* 83:284-301.

529

Santee, R., and Granfield, W. 1939. Behavior of the saw-whet owl on its nesting grounds. *Condor* 41:3–9.

Sappington, J.N. 1977. Breeding biology of house sparrows in north Mississippi. *Wilson Bulletin* 89:300–309.

Schaller, G. B. 1964. Breeding behavior of the white pelican at Yellowstone Lake, Wyoming. *Condor* 66:3–23.

Schemnitz, S. D. 1961. Ecology of the scaled quail in the Oklahoma panhandle. *Wildlife Monographs* 8:1–47.

Schnell, J. H. 1958. Nesting behavior and food habits of goshawks in the Sierra Nevada of California. *Condor* 60:377–403.

Schrantz, F. G. 1943. Nest life of the yellow warbler. *Auk* 60:367–87.

Schukman, J. N. 1974. Comparative nesting ecology of the eastern phoebe (*Sayornis phoebe*) and Say's phoebe (*Sayornis saya*) in west-central Kansas. M.S. thesis, Fort Hays State College.

Schwartz, C. W. 1945. The ecology of the prairie chicken in Missouri. *University of Missouri Studies* 20:1–99.

Sedgwick, J. A. 1975. A comparative study of the breeding biology of Hammond's and dusky flycatchers. M.S. thesis, University of Montana.

Selander, R. K., and Giller, D. R. 1959. Interspecific relations of woodpeckers in Texas. *Wilson Bulletin* 71:107–24.

Sheldon, W. G. 1967. *The book of the American woodcock.* Amherst: University of Massachusetts Press.

Sherrod, S. K.; White, C. M.; and Williamson, F. S. L. 1976. Biology of the bald eagle on Amchitka Island, Alaska. *Living Bird* 15:143–82.

Short, L. L., Jr. 1965. Hybridization in the flickers (*Colaptes*) of North America. *Bulletin of the American Museum of Natural History* 129:309–428.

———. 1971. Systematics and behavior of some North American woodpeckers, genus *Picoides* (Aves). *Bulletin of the American Museum of Natural History* 145:1–118.

Sibley, C. G., Short, L. L., Jr. 1959. Hybridization in the buntings (*Passerina*) of the Great Plains. *Auk* 76:443–63.

———. 1964. Hybridization in the orioles of the Great Plains. *Condor* 66:130–50.

Sibley, C. G., and West, D. A. 1959. Hybridization in the rufous-sided towhees of the Great Plains. *Auk* 76:326–38.

Sisson, L. 1976. *The sharp-tailed grouse in Nebraska.* Lincoln: Nebraska Game and Parks Commission.

Smith, R. L. 1963. Some ecological notes on the grasshopper sparrow. *Wilson Bulletin* 75:159–65.

Smith, S. T. 1972. Communication and other social behavior in *Parus carolinensis. Publications of the Nuttall Ornithological Club*, no. 11, pp. 1–125.

Smith, W. J. 1969. Displays of *Sayornis phoebe* (Aves, Tyrannidae). *Behaviour* 33:283–322.

———. 1970. Courtship and territorial displaying in the vermilion flycatcher, *Pyrocephalus rubinus. Condor* 72:488–91.

Smith, W. P. 1934. Observations on the nesting habits of the black-and-white warbler. *Bird-Banding* 5:31–36.

Snyder, N. F. R. 1974. Breeding biology of swallow-tailed kites in Florida. *Living Bird* 13:73-97.

Southern, W. E. 1958. Nesting of the red-eyed vireo in the Douglas Lake region, Michigan. *Jack-Pine Warbler* 36:105-30, 185-207.

Sowls, L. K. 1955. *Prairie ducks: A study of the behavior, ecology and management*. Washington, D. C.: Wildlife Management Institute; Harrisburg, Pa.: Stackpole Company (reprinted 1978, University of Nebraska Press).

Spencer, H. E., Jr. 1953. The cinnamon teal (*Anas cyanoptera* Vieillot): Its life history, ecology, and management. M.S. thesis, Utah State University.

Spencer, O. R. 1943. Nesting habits of the black-billed cuckoo. *Wilson Bulletin* 55:11-22.

Stabler, R. M. 1959. Nesting of the blue grosbeak in Colorado. *Condor* 61:46-48.

Stalheim, P. S. 1975. Breeding and behavior of captive yellow rails *Coturnicops noveboracensis*. *Avicultural Magazine* 81:133-41.

Steirly, C. C. 1957. Nesting ecology of the red-cockaded woodpecker in Virginia. *Atlantic Naturalist* 12:280-92.

Stewart, P. 1974. A nesting of black vultures. *Auk* 91:595-600.

Stewart, R. E. 1949. Ecology of a nesting red-shouldered hawk population. *Wilson Bulletin* 61:26-35.

———. 1953. A life history study of the yellow-throat. *Wilson Bulletin* 65:99-115.

Stocek, R. F. 1970. Observations on the breeding biology of the tree swallow. *Cassinia* 52:3-20.

Stokes, A. W. 1950. Behavior of the goldfinch. *Wilson Bulletin* 62:107-27.

Stout, G. D., ed. 1967. *The shorebirds of North America*. New York: Viking Press.

Summers-Smith, D. 1963. *The house sparrow*. London: Collins.

Sutherland, C. A. 1963. Notes on the behavior of common nighthawks in Florida. *Living Bird* 2:31-39.

Sutton, G. M. 1949. *Studies of the nesting birds of the Edwin S. George Reserve*. Part I. *The vireos*. University of Michigan Museum of Zoology Miscellaneous Publication no. 74.

Tanner, W. D., Jr., and Hendrickson, G. O. 1954. Ecology of the Virginia rail in Clay County, Iowa. *Iowa Bird Life* 24:65-70.

———. 1956. Ecology of the king rail in Clay County, Iowa. *Iowa Bird Life* 26:54-56.

Tate, D. J. 1973. Habitat usage by the chipping sparrow (*Spizella passerina*) in northern Lower Michigan. Ph.D. diss., University of Nebraska.

Tate, J., Jr. 1970. Nesting and development of the chestnut-sided warbler. *Jack-Pine Warbler* 48:57-65.

Taylor, W. K 1971. A breeding biology study of the verdin, *Auriparus flaviceps* (Sundevall) in Arizona. *American Midland Naturalist* 85:289-328.

Taylor, W. K., and Hanson, H. 1970. Observations on the breeding biology of the vermilion flycatcher in Arizona. *Wilson Bulletin* 82:315-19.

Terrill, L. 1943. Nesting habits of the yellow rail in Gaspe County, Quebec. *Auk* 60:171-80.

Thomas, R. H. 1946. A study of eastern bluebirds in Arkansas. *Wilson Bulletin* 58:143-83.

Tinbergen, N. 1959. Comparative studies of the behaviour of gulls (Laridae): A progress report. *Behaviour* 15:1-70.

Todd, R. L. 1977. Black rail. In *Management of migratory shore and upland game birds in North America,* pp. 71-83. Washington D. C.: International Association of Fish and Wildlife Agencies.

Tomlinson, D. N. S. 1976. Breeding behaviour of the great white egret. *Ostrich* 47:161-78.

Tompkins, I. R. 1959. Life history notes on the least tern. *Wilson Bulletin* 71: 313-22.

――――. 1965. The willets of Georgia and South Carolina. *Wilson Bulletin* 77: 151-67.

Tramontano, J. P. 1864. Comparative studies of the rock wren and the canyon wren. M.S. thesis, University of Arizona.

Trapp, J. 1967. Observations at a cerulean warbler nest during early incubation. *Jack-Pine Warbler* 45:42-49.

Trauger, D. L. 1971. Population ecology of lesser scaup (*Aythya affinis*) in subarctic taiga. Ph.D. diss., Iowa State University.

Trautman, M. B., and Clines, S. J. 1964. A nesting of the purple gallinule in Ohio. *Auk* 81:224-26.

Tuck, L. M. 1972. *The snipes: A study of the genus* Capella. Canadian Wildlife Service, Monograph Series no. 5. Ottawa.

Tutor, B. M. 1962. Nesting studies of the boat-tailed grackle. *Auk* 79:77-84.

van Camp, L. F., and Henny, C. J. 1975. *The screech owl: its life history and population ecology in northern Ohio.* U.S. fish and Wildlife Service, North American Fauna no. 71. Washington, D.C.

van Riper, C., III. 1976. Aspects of house finch breeding biology in Hawaii. *Condor* 78:224-29.

Verbeek, N. A. M. 1967. Breeding biology and ecology of the horned lark in alpine tundra. *Wilson Bulletin* 79:208-18.

Vermeer, K. 1970. *Breeding biology of California and ring-billed gulls: A study of ecological adaptation to the inland habitat.* Canadian Wildlife Service, Report Series no. 12. Ottawa.

Verner, J. 1965. Breeding biology of the long-billed marsh wren. *Condor* 67:6-30.

Walkinshaw, L. H. 1935. Studies of the short-billed marsh wren in Michigan. *Auk* 52:362-69.

――――. 1944. The eastern chipping sparrow in Michigan. *Wilson Bulletin* 56: 193-205.

――――. 1953. Life history of the prothonotary warbler. *Wilson Bulletin* 65:152-68.

――――. 1966. Summer biology of Traill's flycatcher. *Wilson Bulletin* 78:31-46.

Walley, W. J. 1973. A study of seven warblers in Riding Mountain National Park, Manitoba. *Blue Jay* 31:158-66.

Watson, A. T. 1977. *The hen harrier.* Berkhamstead, England: T. & A. D. Poyser.

Watts, C. R., and Stokes, A. W. 1971. The social order of turkeys. *Scientific American* 224(6): 112-8.

Weaver, R. L., and West, H. L. 1943. Notes on the breeding of the pine siskin. *Auk* 60:492-503.

Weller, M. W. 1958. Observations on the incubation behavior of a common nighthawk. *Auk* 75:48-59.

––––––. 1961. Breeding biology of the least bittern. *Wilson Bulletin* 73:11-35.

Welsh, D. A. 1971. Breeding and territoriality of the palm warbler in a Nova Scotia bog. *Canadian Field-Naturalist* 65:99-115.

––––––. 1975. Savannah sparrow breeding and territoriality on a Nova Scotia beach. *Auk* 92:235-51.

Welter, W. A. 1935. The natural history of the long-billed marsh wren. *Wilson Bulletin* 47:3-34.

Wemmer, C. 1969. Impaling behavior of the loggerhead shrike. *Zeitschrift für Tierpsychologie* 26:208-24.

West, D. A. 1962. Hybridization in grosbeaks (*Pheucticus*) of the Great Plains. *Auk* 79:399-424.

Weston, H. G., Jr. 1947. Breeding behavior of the black-headed grosbeak. *Condor* 49:54-73.

White, H. C. 1953. The eastern belted kingfisher in the Maritime Provinces. *Fisheries Research Board of Canada Bulletin* 97:1-44.

––––––. 1957. Food and natural history of mergansers on salmon waters in the Maritime Provinces of Canada. *Fisheries Research Board of Canada Bulletin* 116:1-63.

Whitson, M. 1975. Courtship behavior of the greater roadrunner. *Living Bird* 14:215-56.

Wiese, J. H. 1976. Courtship and pair formation in the great egret. *Auk* 93: 709-24.

Wilcox, L. R. 1959. A twenty year banding study of the piping plover. *Auk* 76:129-52.

Williams, L. 1952. Breeding behavior of the Brewer blackbird. *Condor* 54:3-47.

Willoughby, E. J., and Cade, T. J. 1964. Breeding behavior of the American kestrel (sparrow hawk). *Living Bird* 3:75-96.

Willson, M. F. 1964. Breeding ecology of the yellow-headed blackbird. *Ecological Monographs* 36:51-77.

Wolf, L. L. 1977. Species relationships in the avian genus *Aimophila*. *A.O.U. Monographs* 23:1-220.

Wood, N. A. 1974. The breeding behaviour and biology of the moorhen. *British Birds* 67:104-15, 137-58.

Woolfenden, G. E. 1956. Comparative breeding behavior of *Ammospiza caudacuta* and *A. maritima*. *University of Kansas Museum Publications* 10:45-75.

––––––. 1975. Florida scrub jay helpers at the nest. *Auk* 92:1-15.

Zimmerman, J. L. 1966. Polygyny in the dickcissel. *Auk* 83:534-46.

INDEX

Numbers in **bold** type refer to principal accounts of each species. The appendixes have not been indexed.